ROUTLEDGE LIBRARY EDITIONS:
GEOLOGY

Volume 20

THE HISTORY OF GEOMORPHOLOGY: FROM HUTTON TO HACK

THE HISTORY OF GEOMORPHOLOGY: FROM HUTTON TO HACK

Binghamton Geomorphology Symposium 19

Edited by
K.J. TINKLER

Routledge
Taylor & Francis Group
LONDON AND NEW YORK

First published in 1989 by Unwin Hyman, Inc.

This edition first published in 2020
by Routledge
2 Park Square, Milton Park, Abingdon, Oxon OX14 4RN

and by Routledge
605 Third Avenue, New York, NY 10017

Routledge is an imprint of the Taylor & Francis Group, an informa business

British Library Cataloguing in Publication Data
A catalogue record for this book is available from the British Library

ISBN: 978-0-367-18559-6 (Set)
ISBN: 978-0-429-19681-2 (Set) (ebk)
ISBN: 978-0-367-46449-3 (Volume 20) (hbk)
ISBN: 978-1-00-302889-5 (Volume 20) (ebk)

History of Geomorphology

From Hutton to Hack

Edited by
K. J. Tinkler

The Binghamton Symposia in Geomorphology: International Series, no. 19

Boston
UNWIN HYMAN
London Sydney Wellington

This book was typeset, proofed, and passed for press by the editor and contributors.

Unwin Hyman, Inc.,
8 Winchester Place, Winchester, Mass. 01890, USA

Published by the Academic Division of
Unwin Hyman Ltd
15/17 Broadwick Street, London W1V 1FP, UK

Allen & Unwin (Australia) Ltd,
8 Napier Street, North Sydney,
NSW 2060, Australia

Allen & Unwin (New Zealand) Ltd in association with the
Port Nicholson Press Ltd,
Compusales Building, 75 Ghuznee Street, Wellington 1, New Zealand

First published in 1989

Library of Congress Cataloging in Publication Data

The history of geomorphology: from Hutton to Hack / edited by
Keith J. Tinkler.
p. cm. – (The "Binghamton" symposia in geomorphology;
no. 19)
"Proceedings of the 19th Annual Binghamton Geomorphology
Symposium" – Pref.
Includes bibliographical references.
ISBN 0-04-551138-1
1. Geomorphology – History – Congresses. I. Tinkler, K. J.,
1942– . II. "Binghamton" Geomorphology Symposium (19th: 1988:
Binghamton, N.Y.) III. Series.
GB400.7.H57 1989
551.4'1'09–dc20 89-16594
CIP

British Library Cataloguing in Publication Data

The history of geomorphology. – (The "Binghamton" symposia.
International series; 19)
1. Geomorphology, history
I. Tinkler, Keith J. II. Series
551.4'09

ISBN 0-04-551138-1

Printed in Great Britain by The University Press, Cambridge.

Preface

TERRA INCOGNITA

it will be some time before .... historians and philosophers awaken from their dogmatic slumbers and see the history of geology as an area with broad implications for the study of science generally.

(M.T. Greene, 1985, p. 97.)

It is a great pleasure to present this book, the proceedings of the *Nineteenth Annual Binghamton Symposium*, to the geomorphological public to commemorate and honour both the publication of James Hutton's extended essay *Theory of the Earth* in the 1788 *Transactions of the Royal Society of Edinburgh* (their very first volume), and the twentieth century career of John Hack of the *United States Geological Survey,* whose particular contribution to geomorphology has been an explicit formulation of an equilibrium theory of landscape development (Hack 1960). Although Hack is approximately two centuries later than Hutton eighteenth century 'geologists,' especially in Britain, might not have been surprised to read a theory committed to viewing landscape as a state of equilibrium achieved within a system of competing forces, as I hope I demonstrate in my own essay (Chapter 3).

I am convinced that it constitutes a volume worthy of both individuals for it is the first set of essays whose determined focus is the history of geomorphology, although it is by no means the first book on the history of geomorphology. For good measure those few of us who have written books on the history of the discipline (and many of whom were at the Symposium) are frequently taken to task in this volume for having neglected substantial fields and significant areas of concern! We can all retort that nothing in our own essays can be found in our books, and that surely is very satisfactory!

At present geomorphology is showing signs of out-growing a lengthy adolescence and may soon adopt a formal international structure which will be both independent of, and competitive with, similar structures in geography and geology to which its parent bodies belong. This move is likely to be seen as one of great significance by those involved in its organisation and by many beyond it, and it may well signal structural changes within the discipline whose actual outcome and meaning will not be clear for several decades.

My perspective on such a move is cool, because the adoption of an historical perspective with respect to the practice of 'geomorphology' in times past brings the recognition that times present are in no sense different. That which seems appropriate, fundable, official and achievable is just that: the essays of Stoddart (Chapter 8) and Hewitt (Chapter 9) both neatly illustrate, in their different ways, the distinction between 'official' and 'scientific' approaches to the same problem. But institutional expansion, Vitek's paper not withstanding (Chapter 14), may not be why our geomorphology will be of interest to future generations. What that interest will be is a matter for speculation. Will it be contemporary interplanetary equivalents of the great explorations of the nineteenth centuries? Will it be our ability to collect vast quantities of real-time process data, and to manipulate it with mathematical models and GIS systems, on computers whose memory capacity is already outgrowing our imagination? Might it be that we have achieved disciplinary consciousness and have begun to write histories of ourselves? Or will it be our realisation that despite an almost total failure to communicate significant quantities of information between major language groups, we have nevertheless managed to muddle our way to a reasonably coherent picture of how the world's surface works? (see, for example, Starkel's essay, Chapter 12) We cannot know: if we did we could be prophets. We must keep an open mind, and we should read the opening essay by Gordon Herries Davies not just for its splendid rhetoric, but to realise that its logic is irrefutable.

I hope the reader will approach these essays with an equally open mind and try to share with the authors their sense of wonderment and empathetic understanding. Gerald Friedman (1988), in a recent editorial in *Earth Sciences History*, points out the tensions and difficulties that exist between geologists and historians when either sets out to write the history of geology, or any part of it: the chauvinistic views that the one is short on history, and the other is short on geology. The inevitable consequence is that either, or both, may miss something of significance that the other has to offer!

Yet this is particularly paradoxical since geologists have every reason to be sensitive to the difficulties of reconstructing a convincing and reasonable historical narrative - even if their means and materials are different. I should add as a corollary to this that Martin Rudwick's book *The Great Devonian Controversy* (Rudwick 1985) is of particular interest because here we have a geologist, long since turned historian of geology, who has used a narrative technique in a strict form for a specific historical purpose: to relate in strict chronological sequence the events which led to the uncovering of a major gap in the chronology of the geological column itself.

In one view, historians of science can be placed on a polarised scale with the externalists at one end, the internalists at the other. The former stress the external structural apparatus of society and nation that shape the actions of men, even men of science, whereas the internalists focus on the internal debates within particular sciences. In discussions of the nineteenth century, for example, the externalist would stress the national need to push westward in North America, with the consequent outcome of substantial exploration and the development of state and federal surveys. These, they would argue, gave rise incidentally to valuable geology. The internalist writing about the same century might describe the plethora of theories for mountain building (Greene 1982), or the debates about the reality and extent of Ice Ages.

It may be of interest that the funding which brought the speakers (herein the authors) together for the *Nineteenth Binghamton Symposium* was supplied by the *Social Science and Humanities Research Council of Canada*, and had to be obtained in competition from a panel of historians, not geologists. That a grant larger than the average was obtained attests to the fact that these issues were addressed satisfactorily in the application.

I should like to be able to claim that this volume has a coherence and drive entirely of my own design - but that would be to distort the truth considerably. I did attempt to get a broad coverage of topics and times, of spaces and places. In my initial choice for speakers I approached individuals whose published record was plain for all to see. But I was also approached by others who had a particular paper to offer. The outcome is a happy amalgam of these two components. Each paper was refereed by a historian of geology, and the list of referees at the start of the book speaks for itself. Because I was able to attract Mott T. Greene as the principal discussant for all the papers, and because he was able to see a copy of every paper before the conference, I had additional and valuable input from that source. Finally, all this would be to no avail if the authors had not been willing to make considerable modifications in the light of the comments received. They *were* willing, and they often went to considerable lengths, as many will testify.

All of this is part and parcel of the usual process of peer review, it may be said, but in the context of the strong ambivalence that exists about how the history of science should be written, and the intrinsic differences between historical and scientific writing, remarkable concessions were asked and granted. For this I thank both authors and referees and offer the opinion that a good deal of the middle ground was covered quite amicably once the referees asked the right questions, to which usually the authors could provide penetrating, and to them, obvious replies.

The volume has coherence and drive, but it comes from a serious scholarly effort on the part of everybody, and not because of any planned content that I foresaw. The common themes that emerge do so because they are latent in the material. They are not obfuscated by hero worship or a pre-judged position, and the freshness which characterises the collection stems, in large part, from the fact that these matters have not previously been aired. I shall not review the papers in detail, they are all too complex for that, and Mott Greene provides a perceptive afterword from his position as discussant. However, I will cite a two examples.

From at least four papers (Hewitt, Sack, Stoddart and Alexander) we see how pre-conceived views are brought to a problem by the scientists themselves. Ii may be an Anglo-American prejudice about how geology should be done, an imperial, frontier, approach to the delineation of territory and problems, or a technological bias in which every available technique is brought to bear on a problem, and must perforce provide a 'better' (equals more detailed?) result.

Nevertheless the landscape shines through - it forces new perspectives on those who try to pry loose its secrets.

As a second example consider the eighteenth century. It is almost a commonplace in the history of geology that the subject emerged in the period 1770-1820 - both in name and as an institutional discipline (Porter 1977, Laudan 1987). It is reassuring to find, then, in my own contribution (on British writers), from that of François Ellenberger (on the French), and from work in Italy described by David Alexander, that there is ample evidence, in sources not previously examined, of capable field workers developing a convincing regional and temporal perspective on landscape evolution from the evidence of their eyes. They used well established rules of reasoning, and the power of their own imaginations. From the same sources we see that the popular misconception of that century as one in which writings on 'geology' were entirely composed of cataclysmic or biblical theories is in dire need of revision Thus we can provide from geomorphology an independent thrust propelling the inception of geology, and one which shows that a golden age in geology did not arise from a few gifted minds, or by suddenly replacing misguided theories with rational insight.

Before I conclude I should like to say few words about our handling of the paper by François Ellenberger. We decided when doing the translation that all the original citations should be given in the original eighteenth century French, as well as in translation. And we extended this to individual words in order that nothing should be lost, for although the words are often identical after translation the reader may not otherwise know this. Many of the citations are from manuscripts not readily available except in Paris, and in view of the originality of the material we thought it best to make them as widely available as possible, at the expense of being pedantic. The orthography has been checked very carefully by Professor Ellenberger against the originals, and the reader is warned that accents were treated more freely in the eighteenth century than is now the case, and obviously there are differences in style, as well as spelling, from modern French. We should point out too, that in common with eighteenth century English writing on 'geological' matters, there is frequently a total lack of a technical vocabulary particular to the subject. Thus the translation, in preserving the character of the original, will also preserve its circumlocutions. In addition, it will be apparent that Professor Ellenberger is frequently paraphrasing closely some of his sources, without actually citing them *verbatim*. Where this is the case we have stayed close to the eighteenth century style in making the translation.

I should like to end with a few words about James Hutton (1726-1797), Hutton was born the year before Newton died, and he died the year that Lyell was born. When I chose the year 1988 for this Symposium I did so deliberately, with the knowledge that it was the bi-centennial of the appearance in the *Transactions of the Royal Society of Edinburgh* of Hutton's *Theory of the Earth* paper. It was partly to expand, explain and rebut contemporary criticisms of this paper that Hutton wrote and published his book *Theory of the Earth* in 1795, with the original paper as the first chapter.

It was never my intention to use the Symposium as a platform for a critical review of Hutton's work; to have done so adequately would have taken us far beyond geomorphology, and would have required a different cast of speakers and writers. Nevertheless, you will find that several authors make reference to

Hutton in their efforts to place him properly within the milieu of late-eighteenth century science, and one author (Dennis Dean, Chapter 4) considers specifically his subsequent influence on the development of our subject and traces his impact as far as William Morris Davis.

We should do no service to Hutton, or to ourselves, if we heaped uncritical adulation on his name. In consequence I ask you all to keep an open mind on what his achievements may really have been, and to re-read his work not just for the insight he had on the workings of the world, but also with an eye to the constraints that the eighteenth-century world placed upon him. Then ask yourself what constraints the present century places upon our perceptions of the world.

Keith Tinkler
Brock University

References

Friedman, G. F., 1988. Editorial in *Earth Science History*, Volume 7 (2), p. *i*.

Greene, M.T., 1982. *Nineteenth Century theories of Geology*, Cornell University Press.

Greene, M.T., 1985. History of Geology, *OSIRIS*, Second Series, Volume 1, 97-116.

Hack, J.T., 1960. Interpretation of erosional topography in humid temperate regions, *American Journal of Science*, 258A, 80-97.

Hutton, J., 1788. Theory of the Earth, or an Investigation of the Laws observable in the Composition, Dissolution and Restoration of Land upon the globe, *Transactions of the Royal Society of Edinburgh*, 1 (2), 209-304 (This forms Chapter 1, with minor changes, of his book *A Theory of the Earth*, 1795.(.

Laudan, R., 1987. *From Mineralogy to Geology*, University of Chicago Press.

Porter, R., 1977. *The Making of Geology*, Cambridge University Press.

Rudwick, M. J., 1985. *The Great Devonian Controversy*, University of Chicago Press.

JOHN TILTON HACK
UNITED STATES GEOLOGICAL SURVEY

Contents

Acknowledgements

For financial support I should like to thank the following:

Social Science and Humaities Research Council of Canada for a substantial grant in aid of the Conference which enabled me to support a variety of speakers, including several from Europe. Additional funds ,which enabled me to bring Mott T. Greene as the Conference Discussant, were provided by Dean Soroka, *Division of Social Sciences.* The *Department of Geography*, Brock University has supported my day by day running costs, and has provided essential computing facilities for the preparation of the manuscript.

For help in the preparation of the manuscripts I should particularly like to thank the following:

Margaret Tinkler translated Ellenberger's paper, and undertook extensive proof reading and logistic support. *Barbara J Bucknall* gave invaluable advice on the translation of Professor Ellenberger's paper. *Sophia Attick* proof-read an early page proof of Leszek Starkel's paper with an eye to Polish orthography. *E. Virgulti* helped with a tricky problem in Italian orthography. *Colleen Catling* provided secretarial help far beyond the call of duty. *Loris Gasparotto* was a mine of information about illustrations, and was responsible for their placement on the camera ready pages of the manuscript. He also drafted, or redrafted, several figures. The *Computer Users Centre* enabled the electronic transfer of files from a variety of non-Macintosh disk formats, and *David McCarthy* accessed some obsolete Jazz files. *David Hughes*, Chairman, *Department of Computer Science* arranged for the loan of a Mac Plus during the summer of 1988, and at Christmas 1988, to help ease the burden of editing.

Mott Greene, as Conference Discussant, very willingly undertook the job of reading and commenting on all the papers, and gave endless encouragement in the months before the Conference. In addition he provided the Afterword (Chapter 15) as a final perspective on the Conference. I must also thank *Mary Winsor* (Toronto) and *Rhoda Rappaport* (Vassar) for valuable advice relating to referees and the Conference program. I thank the *United Sates Geological Survey* who provided the Frontispiece of John Hack.

Referees

The following have acted as referees on the papers in this volume. Some refereed more than one paper. I am very grateful for their help.

A Secord
M. Rossiter
J. Menzies
D. R Currey
R. Laudan
K. L. Taylor
C. Albritton
M. T. Greene
G. Grinnell
W. O. Kupsch
S. G. Kohlstedt
A. V. Carozzi
C. G. J. Oviatt
E. Mills

Contributors

David Alexander
Department of Geography and Geology, University of Massachusetts, Amherst, Massachusetts 01003, USA.

Robert P. Beckinsale and R. D. Beckinsale
8 Park Road, Abingdon, Oxfordshire OX14 1DS, UK, and Oxford Polytechnic, Oxford, UK.

Brian T. Bunting
Department of Geography, McMaster University, Hamilton, Ontario L8S 4L9, Canada.

Frank F. Cunningham
958 Sinclair Street, West Vancouver, British Columbia V7V 3VP Canada.

Dennis R. Dean
Humanities Division, University of Wisconsin - Parkside, Box 2000, Kenosha, Wisconsin 53141 - 2000, USA.

François Ellenberger
Vent du Large, 7 rue du Font Garant, 91440 Bures-sur-Yvette, Paris, France.

Mott T. Greene
The Honors Program, The University of Puget Sound, Tacoma, Washington 98416, USA.

Gordon L. Herries Davies
Department of Geography, Trinity College, Dublin 2, Eire.

Kenneth Hewitt
Department of Geography, Wilfrid Laurier University, Waterloo, Ontario N2L 3C5, Canada.

Waite R. Osterkamp
United States Geological Survey, Box 25046, Denver, Federal Center, Denver, Colorado 80225, USA.

Dorothy Sack
Department of Geography, University of Wisconsin - Madison, Wisconsin 53706, USA.

Leszek Starkel
Institute of Geography, University of Kraków, Poland.

David R. Stoddart
Department of Geography, University of California, Berkeley, California 94720, USA.

Keith J. Tinkler
Department of Geography, Brock University, St. Catharines, Ontario L2S 1P4, Canada.

John D. Vitek
Graduate College, Oklahoma State University, Stillwater, Oklahoma 74078 - 0050, USA.

1

On the nature of geo-history, with reflections on the historiography of geomorphology

Gordon L. Herries Davies

Indeed I am, perhaps wrongly, inclined to look upon all geological theories as having their being in a mythical region, in which, with the progress of physics, the phantasms are modified century by century.

(Letter from Alexander von Humboldt to Louis Agassiz, Berlin, 2 March 1842, Quoted in Elizbeth Cary Agassiz, *Louis Agassiz : his life and correspondence*, London, 1885, volume 1, pp. 345-6).

Julius Cæsar's Gaul was divided into three parts; so too is my paper. In the first part I will discuss the significance which the year 1669 must hold for us. In the second part I will suggest that we have not thought sufficiently deeply about the methodology which earth-scientists have employed since 1669. In the the third part I will pose the question 'What is the import of these conclusions for the historiography of geomorphology?'

Part the first

I begin in a spectacular location - in the Basilica of San Marco in Venice. In the atrium of the Basilica one of the cupolas is adorned with a magnificent thirteenth-century mosaic dedicated to the *Genesis* story, and it is beneath that cupola that I wish you to join me. Above us we see beautifully depicted all the familiar events recounted in the three opening chapters of *Genesis* down to the expulsion from Eden. Nearby another series of mosaics carries the *Genesis* story forward in a glorious sequence illustrative of Noah, the Ark, and the events of the Flood. Those mosaics encapsulate the history of the earth as that history was understood within the Christian World for a period of more than a millennium and a half. It was an interpretation of geo-history which displayed, and which, through Creationist Science, continues to display, a remarkable quality of durability. That interpretation of geo-history was, and still is, grounded upon the Scriptures, and there we have the key to an understanding of its durability. It is an interpretation of earth-history which rests upon the secure foundation of a supposedly divinely inspired text itself possessed of eternal validity. It is an interpretation of geo-history which therefore holds for the faithful a satisfying immutable quality. But, as we all know, during the second half of the seventeenth century the Scripture-based interpretation of geo-history began to face a challenge - a challenge arising from the emergence of a novel idea which eventually was to revolutionise the writing of the story of our earth. To us that idea appears to be both obvious and simple, but in the seventeenth century the idea seemed to be startling in its originality. The idea was simply this: first, let it be assumed that the rocks, minerals, fossils, and landforms of the earth's surface bear trace of the historical vicissitudes through which they have passed, and second let it be assumed that human observers may learn to decipher those historical messages so as to compile for themselves synoptic accounts of geo-history. In short, the idea was to treat rocks, minerals, fossils, and landforms as lenses through which we might peer into the past.

This was the novel approach pioneered in Nicholaus Steno's *Prodromus* of 1669[1] and Steno's familiar series of six diagrams depicting the history of the Tuscan landscapes forms a fascinating contrast to the history of the earth as it is depicted in the mosaics of the Basilica of San Marco. In view of Steno's effective advocacy of this fresh approach to earth-history, I have elsewhere suggested that this revolutionary ascription of historical significance to rocks, minerals, fossils, and landforms should be the subject of eponymy and should be known as 'THE STENONIAN REVOLUTION.'[2] That is a term which I propose to employ here, and lest there be any doubt I will reiterate that THE STENONIAN REVOLUTION is to be defined as the emergence and general acceptance of the notion that from rocks, minerals, fossils, and landforms there may be read an account of past events in the history of the global surface.

It is through our adoption of the principles of the Stenonian Revolution that we have discovered a powerful alternative to the Scripture-based interpretation of geo-history. It is through our adoption of the principles of the Stenonian Revolution that we have been able to compile the wondrous tale which the modern historian of our earth has to tell - a wondrous tale of marine transgression and plate collision, of fiery vulcanism and icy glaciation, and of fearsome reptiles being catastrophically exterminated by the arrival of bodies from space. This we now hold to be the true pattern of geo-history. It is the kind of history which is taught in our classrooms and which is enshrined in our literature; it is the kind of history which is demonstrated at the interpretative centres of our national parks and which is presented upon our screens in the latest T.V. science spectacular; it is the kind of history which constitutes the very warp and weft of our prevailing paradigm within the earth-sciences. And this entire ornate geo-historical superstructure rests upon a principle enunciated

1 *De Solido Intra Solidum Naturaliter Contento Dissertationis Prodromus*. An English translation by Henry Oldenburg was published in London in 1671 under the title *The Prodromus to a Dissertation Concerning Solids Naturally Contained within Solids*. Another translation, by J.G.Winter, is in University of Michigan Studies: Humanistic Series, XI, part 2, 1916, pp. 165-283.

2 The term was first used in a lecture delivered at the William Andrews Clark Memorial Library in Los Angeles on 3 November 1984. The lecture will shortly be published by the library.

within the pages of a small volume published in Florence in 1669. It is for that reason that the year 1669 must hold for us a very special significance.

Part the second

Arriving at the second part of my paper, I wish to pose a series of simple and related questions. May we rest assured that through the adoption of the dialectical strategy basic to the Stenonian Revolution we will indeed be led to an understanding of the true nature of the earth's past? May we accept with confidence Steno's notion that rocks, minerals, fossils, and landforms do indeed encapsulate a decipherable record of geo-historical events? By embracing the principles of the Stenonian Revolution have we now arrived at an understanding of all the salient elements within the history of our earth? The thirteenth-century artist who created those mosaics in the Basilica of San Marco was satisfied that in his work on the Creation and the Flood he was depicting actual events from the earth's past. May we feel equally confident in our beliefs when we write of the closure of the Iapetus Ocean or the catastrophe of the Lake Missoula floods? In sum, does our attempt to extract geo-history out of rocks, minerals, fossils, and landforms really bring us into contact with a past as it actually existed or are our histories so devised really nothing more than elaborate fictions? Are we, living as we do in the aftermath of the Stenonian Revolution, really able to achieve geo-historical truths which were denied to the Venetian mosaic artists of the thirteenth century? To pose questions such as these will perhaps leave you aghast. Your hackles rise. With knowing glances one to the other you begin to doubt my sanity. You take pity. You recommend an appointment with a psychiatrist. You soothingly seek to cool my aberrations by reminding me that our modern visions of earth-history are at every point supported by painstaking field-observation and careful laboratory analysis. Of all this I am not unaware but in this context there are two topics upon which I would wish you to reflect.

First, I submit that the past can never be known to us. We have left it. Like our own youth, the past is gone beyond recall. Here I must allude to a publication which in recent years has been much in my mind. In 1819 Richard Whately, a future Archbishop of Dublin, published a pamphlet entitled *Historic Doubts Relative to Napoleon Buonaparte*[3] in which he sought to demonstrate the difficulty that we face in endeavouring to prove that there once existed a Corsican artillery officer who became Emperor of the French, and who, in July 1815, was to be seen in captivity pacing the decks of H.M.S. *Bellerophon* as she lay at anchor in Plymouth harbour. Whately was surely correct. There truly is no manner - no really conclusive manner - in which we may prove that the events with which we populate the past are events which really did take place. We may believe that an English major-general was killed by a French musket-ball outside Quebec on 13 September 1759 or that something unfortunate happened to an American President at Ford's Theatre on 14 April 1865 - I may believe that I was born on 18 January 1932 and that I flew across the Atlantic last Thursday but how would such beliefs withstand their being probed by an aggressive prosecutor in a court of law? I suggest that a skilled counsel would have little difficulty in

[3] The pamphlet was published anonymously and it passed through numerous editions. The eighth edition appeared in London in 1846.

making us look somewhat foolish for entertaining beliefs such as those just mentioned. The past is like a rainbow. We believe it to be there; we are satisfied that we see it in its every spectral detail. But when we reach out to confirm our beliefs through a closer encounter, then the entire phenomenon melts away and denies to us that satisfaction which we seek. I hold beliefs about what happened in 1865 and 1759 - about what happened during the Cretaceous and during the Cambrian - but there seems to be no way in which I am able to convert my beliefs about the past into indubitable, diamond-hard truth about the past. Indeed, did past time even exist? Is past time itself just a figment of the human imagination? I would remind you that Bertrand Russell once observed that there was no logical manner in which we could refute the view that the entire world might have been created just five minutes ago and that we ourselves might have been brought into existence imagining that we were able to recollect a personal past which in reality never existed.[4] And there you have the first of the two topics upon which I invite you to ponder. We may hold beliefs about the past and we may test those beliefs against the mettle of all the available evidence. That evidence may seem to offer complete confirmation of the validity of our beliefs, - may seem to confirm that the past was indeed exactly as we conceive it to have been - but I would urge that no matter how strong the confirmation, we are not really discovering the actuality of the past. The actuality of the past, if the past even existed all, is entirely lost beyond our recall. All that we may ever hope to possess is our reconstruction of what we *believe* to have happened in the past. When we enter into the realms of synoptic geo-history we enter into the realms of fallible human opinion and not into the realms of immutable natural truth.

That brings me to the second topic upon which I invite your reflection. I have just urged upon you the view that when we outline the course of events in geo-history we may or may not be outlining the course of *real* events. All that we can know and claim is that we *believe* the events we outline to be events which really did feature in the earth's past. This distinction between what we *believe* to be true and what really *is* true is a distinction of importance. It is a distinction which I now wish to explore further. If disciples of the Stenonian Revolution really were capable of reading pure geo-history out of the earth's rocks, minerals, fossils, and landforms, then surely, through time, there should be a general agreement as to the character of the story being discovered. If we really did read pure geo-history out of the earth's surface, just as we read the plot out of a novel, then clearly two students of the earth's past should draw more or less the same geo-historical conclusions out of a given set of geological phenomena. But, as you are only too well aware, the history of the earth-sciences reveals that since the Stenonian Revolution burst upon us there has been no agreement as to the nature of the geo-historical messages reputedly encapsulated within the component elements of the earth's surface. All that the last three hundred years in the history of the earth-sciences reveals to us is scientists acting in accordance with the truest principles of the Stenonian Revolution and studying the identical field evidence, but arriving at geo-historical conclusions which are both incompatible and enormously varied. The

4 *An Outline of Philosophy*, London, 1927, p. 7.

disciples of the Stenonian Revolution, like all fervent revolutionaries, have regularly proclaimed their ability to lead us to a new Elysium of truth. But where is their Elysium? Where is their truth? The Scripture-based geo-histories of yesteryear remained consistent in character for more than a millennium because they were founded upon a supposedly infallible source, but ever since we swept Moses aside - ever since we began trying to read geo-history out of the earth itself - all comfortable consistency has gone. In place of that former consistency a type of anarchy has prevailed as each new generation of earth-scientists has claimed to discover within the earth's crust yet newer and more significant geo-historical truths. Perhaps here there is an interesting if rather surprising analogy to be drawn between what has happened within the earth-sciences since 1669 and what has happened within the Christian Church since the Reformation. Within the earth-sciences the Pentateuch was dethroned as scientists sought to discover a new and purer truth inside the very materials of the earth's surface; within the Reformed Church the Pope was dethroned as the faithful sought to discover a new and purer truth inside the Scriptures. In both cases the result has not been the emergence of the expected fresh and lasting consensus. Instead there has come into existence a scene of turmoil characterised by dissension, by dispute, and by the regular substitution of what are hailed as ultimate truths in the place of what are dismissed as misguided fallacies. Within the earth-sciences you may see this process at work by looking at the shelves of our campus bookstores. How long does it take for any particular text within our own field of science to become outmoded? Five years?

This regular substitution of the new for the old over the last three hundred years has left the course followed by the earth-sciences strewn with the wreckage of innumerable rejected concepts. Among the jetsam we see de Luc's natural chronometers, Werner's Neptunism, Cuvier's catastrophism, James Hall's geosynclines, Lyell's marine erosion theory of topography, de Beaumont's theory of mountain building, the diluvial theory of drift formation, the Davisian Normal Cycle of Erosion, and Penck's fourfold division of the Pleistocene into the Günz, the Mindel, the Riss, and the Würm. Please let us be quite clear that none of these concepts was the child of a deranged mind. They were all of them concepts associated with geologists who intellectually were fully the equal of any modern earth-scientist. They were all of them concepts which, in their day, were believed to enshrine ultimate truths of geo-history, but above all else they are concepts which had been framed only after a painstaking examination of the appropriate field-phenomena. De Luc, Werner, Cuvier, Hall, Lyell, de Beaumont, Davis, and Penck all sincerely believed that they were actually deciphering a geo-history encapsulated within the component elements of the earth's surface. We would hold that they were doing no such thing. We would hold that they were merely allowing the field evidence to inspire their imaginations into flights of intellectual fancy. And there I arrive at the crux of this particular argument. I submit that all those geologists just mentioned, from de Luc to Penck, are holding aloft a mirror in which we may see ourselves reflected. We may believe that we have now attained at least some of the ultimate truths of geo-history but such a confidence in ourselves is no more justified than was the self-confidence of our predecessors. There is no reason to believe that we are any more successful at reading geo-history out of the earth's surface than were the geologists of yesteryear. And the reason for this continued want of success? I suggest that in all probability the task is impossible. I suggest that the promise held out by the Stenonian Revolution is really a chimera. Rocks, minerals, fossils, and landforms cannot be persuaded to reveal to us whatever geo-historical secrets they may hold. Rocks, minerals, fossils, and landforms are inanimate. They are mute. They exist in a state of stony silence. Geo-history is not a natural substance which seeps little by little out of the earth's surface, there to be harvested by an objective mankind. What we call geo-history is just as much a human creation as is the U.S.S. *Missouri* a statue of Queen Victoria. What we call geo-history is a substance of purely human origin which seeps generation by generation not out of the earth's rocks, minerals, fossils, and landforms, but out of the cerebrum of *Homo sapiens*. Geo-history is what results when a certain type of human being encounters the phenomena of the earth's skin. If we place a would-be Beethoven at a pianoforte we get musical composition. If we place an artist at an easel we get a painting. If we place a geologist in a quarry we get geo-history. Geo-history is a purely human construct. What we do is position ourselves before what we hold to be the evidence germane to geo-history and there we play the game of inventing geo-historical events which would seem to be capable of explaining the phenomena disposed before us. We can never hope to know whether or not the geo-historical tapestry of our weaving truly represents the events of a past which can itself never be anything more than a postulate. The rainbow will ever remain elusive. All that we can know is that our geo-historical inventions will bear ample trace of human fallibility in observation, of human weakness in reasoning, and of presuppositions acquired both during our scientific training and during our upbringing within the wider social communities whence we came. Erwin Schrödinger once asked the intriguing question 'Does the world cease with my bodily death?'[5] As I conclude this, the second section of my paper, you might like to muse upon the question 'Will the earth still possess a history once the last geo-historian has passed from the scene?'

Part the third

My paper would make poor drama because here in the final act I offer to you no thrilling *dénouement*. What I have to say now is neither startling nor particularly original. I have earlier sought to convince you that all geo-history is of human creation and that we can never know whether our geo-histories bear any relationship to what may or may not have happened in the past. I now want to explore the significance that such conclusions must hold for the historian of geomorphology, and in this context there is a trio of points that I wish to make.

The first of these points relates to the geomorphologist's involvement with geo-history itself. All students of landforms explain those landforms in terms of sets of natural processes operating within a variety of palæo-environments. A student of landforms therefore of necessity is forced into speculation about the environmental conditions which formerly prevailed upon the earth's surface. Geo-history thus becomes a component of geomorphology, and as we explore the history of geomorphology what we are really discovering

[5] *My View of the World*, Cambridge University Press, 1964, p. 12.

is the different types of palæo-environment which various scholars have invoked as the basis for their explanation of given suites of landforms. Burnet invoked the Noachian Deluge, Kirwan postulated a universal menstruum, Hutton and Lyell considered the past to have been much like the present, Agassiz invented *die Eiszeit*, Gilbert wrote of Utah as having been affected by 'secular cycles of climate,' Baulig was in love with the idea of falling Tertiary base-levels, and Lester King found it convenient to explain the morphology of Africa in terms of the continents having participated in a global cruise programme. It is impossible for the historian of geomorphology to avoid contact with the geo-historical tapestry which we have woven for ourselves and in which past students of landforms have been so intimately cocooned. This is a point which you will doubtless feel to be patently self-evident, but I introduce it because I wish to remind you that those palæo-environments which geomorphologists have invoked are palæo-environments created within human minds. Over the decades we have invoked widely differing palæo-environments in explication of precisely the same set of field-phenomena, and we can never hope to know whether any of our postulated palæo-environments bears a relationship to the actual events of the past. Today, for instance, we believe in the reality of Pleistocene glaciation not because it is a proven fact of geo-history; it is treated as a proven fact of geo-history because it is an event in which we happen to believe. I emphasise that point because it leads me forward to the two other observations which constitute the final points of my paper. These two points relate to two pitfalls which await anybody who seeks to study our changing styles in geo-history - two pitfalls with which I am only too familiar because into both of them I have myself tumbled.[6]

The first pitfall is that of the *Terminal Fallacy*. Since there can be no certainty about the veracity of our geo-historical conclusions, we must never regard ourselves as being the privileged owners of sets of ultimate geo-historical truths which were denied to our predecessors. For us to claim such a status would be an unpardonable conceit. In our histories of geomorphology there must be no implication that we have at last beaten Nature into a final submission or that she has now allowed us access to intimate secrets which she had defended against the scientific thrusts of our ancestors. I can see psychological, social, and even political reasons why earth-scientists might wish to pretend that they really are privy to Nature's ultimate geo-historical secrets, but we would be poor historians if we allowed ourselves to be deceived into thinking that today's geomorphic textbooks are the repository of the final geo-historical truths. We must remember that since proof positive must ever elude us in the area of geo-history, then the geo-historical truth of today will doubtless be dismissed as the geo-historical falsehood of tomorrow. The geo-historical truth of today is just

> ... a poor player,
> That struts and frets his hour upon the stage,
> And then is heard no more.

[6] G.L. [Herries] Davies, *The Earth in Decay : A History of British Geomorphology 1578-1878*, London, n.d. [1969].

There you have my first pitfall. To avoid its dangers we must ever be aware of our human fallibility, we must display modesty about the scope of our understanding, and we must recognise that, as the river of knowledge flows on beyond us, its waters will continue to reflect perpetually changing patterns of geo-historical belief.

My second pitfall is intimately linked with my first. My second pitfall is the pitfall of Whiggishness. Many years ago Sir Herbert Butterfield cautioned us about the dangers of studying the past through the eyes of the present.[7] He warned us of the perils of emulating the great English Whig historians of the last century and of writing history as though it had been a simple struggle between progessives and reactionaries - a conflict between 'goodies' eager to establish a world such as our own, and 'baddies' dedicated to retarding the march of progress. In his own words:

> The whig historian stands on the summit of the twentieth century, and organizes his scheme of history from the point of view of his own day.

Butterfield of course had in mind the field of political history but his exhortation must be taken to heart by all historians of science and it has a particular relevance for the historian of the earth-sciences. Hitherto we have viewed the history of the earth-sciences in a decidedly Whiggish manner. We have told the story in triumphalistic terms as a glorious advance from the 'ignorance' of Mediæval times down to the 'enlightenment' of our own age. We have accorded heroic status to those figures whom we regard as the pioneers of the ideas which we cherish as truths. In 1908 a statue was erected in Paris in the Jardin des Plantes in honour of Lamarck; it is inscribed 'Au fondateur de la doctrine de l'évolution.' In 1947 a plaque was added to Hutton's grave in Edinburgh; it hails him as 'The founder of modern geology.' Conversely, those supposed to have hindered the advance of our science we band as bibliolaters, Wernerians, catastrophists, diluvialists, or Davisians. There are many reasons why such a simplistic approach to the history of the earth-sciences will not suffice. From among those reasons I select just two for mention. First, by populating history with a cast consisting of either 'goodies' or 'baddies' we are tempted to dismiss the 'baddies' merely as impediments to progress. In reality the 'baddies' are deserving of far more sympathetic attention than they have ever received at the hands of our Whiggish historians. Within the field of the history of geomorphology, for example, there is a great deal to be learned about reasoning in science by studying the views of those who opposed fluvialism during the first half of the last century and of those who opposed the Glacial Theory during that century's middle decades. We need to be far less concerned with unearthing our own intellectual roots and far more concerned with studying what might be termed the intellectual ecology of those entire former communities of scholars which displayed interest in the genesis of landforms.

My second reason for concern at the persistence of Whiggishness brings me back to the issue of the *Terminal Fallacy.* I have repeatedly urged upon you the view that our geo-historical beliefs are no more than mere beliefs - beliefs which can never be tested against reality and beliefs which are likely to change

---

[7] *The Whig interpretation of History*, London, 1931.

generation by generation. Now if our geo-historical beliefs at any given moment are just intellectual ephemera, then is it not absurd of the historian to treat those beliefs as the ultimate nuggets of truth produced by a long process of scholarly refinement? Look, for instance at Plate Tectonic Theory. It is exactly twenty-five years since Vine and Matthews published their famous paper upon the subject in *Nature*,[8] and today the theory enjoys universal acceptance. Already many historians have given their attention to the theory and they have traced for us the route whereby we attained our possession of this new gem of truth. But is the theory really a gem of truth? Only this year a doyen of Australian geology - Professor S. Warren Carey - has published a book in which he dismisses Plate Tectonic Theory as a false dogma (he applies the adjective 'myth' to subduction) and in its place he invites us to subscribe to the notion that we inhabit a globe which is steadily expanding at an exponential rate.[9] Let us suppose that Carey is eventually admitted to have been correct. Will those who have written Whiggish histories of Plate Tectonic Theory not then be left looking rather foolish? Will they not be left in a position comparable to that of those who wrote learned discourses on the subject of the Piltdown skull during the decades before it was demonstrated to be a fake? Similarly, look at the story of our changing views on the age of the earth as that story has been told by many a recent historian. They have presented the story in a thoroughly Whiggish manner. They take us from Ussher to uranium; they take us from darkness into light; they take us from falsehood into truth. But again I have to ask you, since we have rejected the ages of the earth computed by every generation before our own, then why should we expect any future generation to accept our computation? What justification is there for the notion that our figure of 4,600 million years represents the final word in geo-chronology?[10] I know all about the precision now being achieved in our radio-metric laboratories, but in the fullness of time will future generations of scholars smile at our methods, at our conclusions, and at our gullibility just as we smile at what we perceive to have been the geo-chronological absurdities of the likes of Ussher, Buffon, Kelvin, or Joly? Will future generations of scholars have a far more sophisticated notion of time than that vouchsafed to us and will they wonder why, in an age of Relativity, we were so naive as to persist with a belief in time as a simple linear phenomenon? In science there is a tendency for the most fundamental of our beliefs to remain immune from question and it may be that one day there will have to be queried our belief in the Arrow of Time - a belief fundamental to the earth-sciences as they are at present known to us.

Be that as it may, my point for the moment is that we must beware of all history which smacks of Whiggishness. What we need is not histories which trace the development of what we hold to be geo-historical truths, but rather histories which adopt a contextual approach. We need to explore why, at different periods, varying geo-historical interpretations have been accorded the status of being ultimate truths. Why, for example, was the concept of evolution applied

---

[8] Vine, F.J. and Matthews, D.H., 1963. Magnetic anomalies over oceanic ridges. *Nature*, 199, p. 947-9.

[9] *Theories of the Earth and Universe: A History of Dogma in the Earth Sciences*, Stanford University Press, 1988.

[10] *Editor's Note:* A nice spice is added to this remark by a recent report in the *New Scientist*! Nigel Henbest, 1988, Universe sheds 10 billion years, *New Scientist*, 1624, p. 34-5.

in so many spheres after 1859? Why do we today believe in moving continents when the idea was viewed as risible fifty years ago? Why did events of extra-terrestrial origin feature in geo-historical reconstructions made during the seventeenth and eighteenth centuries? Why did such events disappear from geo-histories during the nineteenth century, and why have they been reinstated in our geo-histories during the last twenty years? Why was denudation chronology all the geomorphic rage during the 1930s and why has that type of study now gone to join the dodo? Why did the character of geomorphology change so dramatically in the aftermath of World War II? To explore such questions is far more difficult than is the writing of the traditional Whiggish 'let's trace the origins of the present' sort of history. The search for answers to such questions must take us far beyond the realms of our science and it may be that such contextual history demands a level of historical expertise which is difficult of attainment for those mainstream earth-scientists with whom the history of the earth-sciences is merely an interesting sideline.

I am clearly sailing my gondola into dangerous waters. I must conclude. We began within the Basilica of San Marco and there I propose to leave you, standing again beneath the atrium mosaics but now musing upon this question: if all geo-history is of human origin, then what distinction is to be made between the history of the earth as it was depicted by the thirteenth-century mosaicist and the history of the earth as we tell that story to our classes of first-year undergraduates?

2

Les Méconnus: eighteenth century French pioneers of geomorphology§

François Ellenberger

In eighteenth century France the organized science of geomorphology still did not exist, properly speaking. The present article discusses some authors who deserve to be counted amongst the pioneers of the discipline. First, however, it is appropriate to recall some general ideas relating to both the country and the period.

Weak biblical constraints

In France, Catholicism was the official religion, and indeed Protestants were persecuted. Consequently, unlike Great Britain, the Bible in France had authority only through the intermediary of Church teaching. Provided one was known to obey, in principle, the decrees and dogmatic truths, one could have in fact, a notable freedom where intellectual speculation was concerned. The movement of the Enlightenment gained ground as the century passed, and so it was of little concern to base *a priori*, theoretical, speculations on the imperatives of the limited biblical chronology and the occurrence of Noah's Flood. It was preferable not to talk about them, or alternatively to say in a purely formal manner that these are tenets of faith which one treats with due respect.

The differences between landscapes

I have treated this subject in a previous article (Ellenberger 1980) but I will review it briefly here. France offers three highly contrasted types of geomorphological landscape.

a) More than half the country (most of the plains of the north and west) bear very clear traces of periglacial conditions. Thus one cannot explain in a straightforward fashion either the geomorphological details or the gross morphology of the landscape by means of present processes (Figure 1, **A**).

§ Translated by Margaret Tinkler

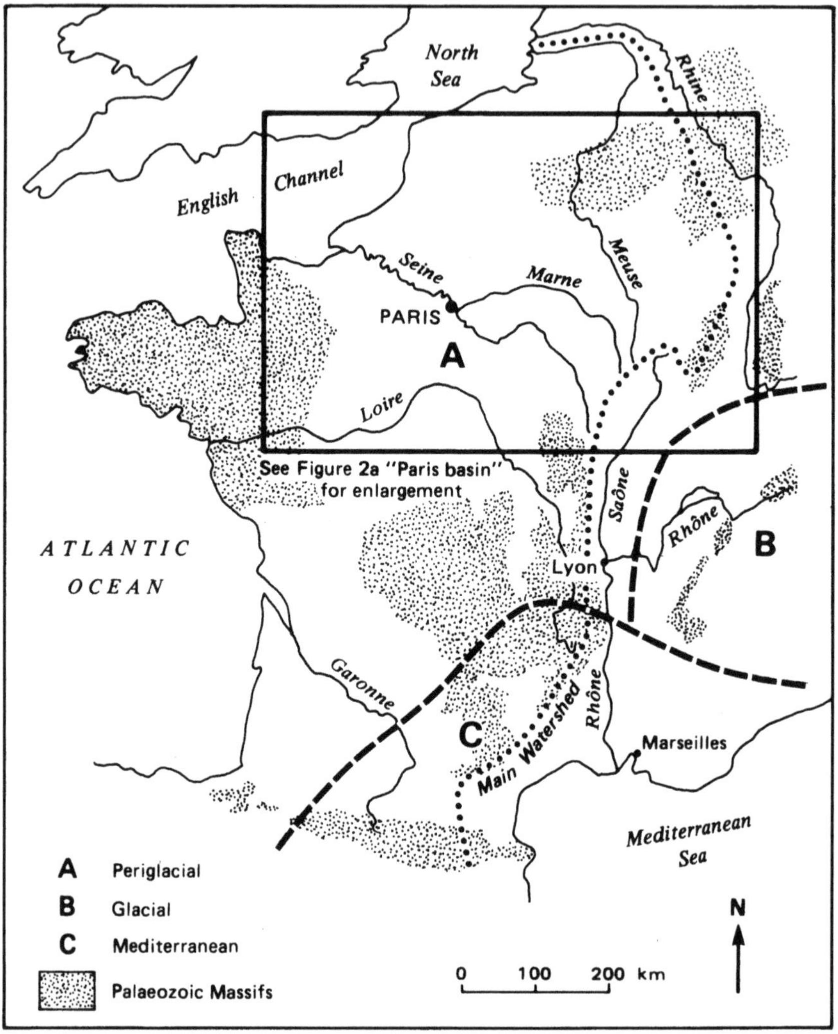

Figure 1

Three highly contrasted types of geomorphological landscape which have affected the ways in which the landscape has been interpreted. For details see the previous page, and this page, below.

---

b) The high mountain blocks which have undergone glaciation (the Alps, the Pyrenees, the Massif Central) are likewise inexplicable on the basis of present processes (Figure 1, **B**).

c) On the other hand the Mediterranean and neighbouring regions, where the soil is almost completely stripped and the meteorological conditions are severe, show landscapes which suggest quite persuasively that the present morphology has been fashioned by the same agents which are seen in action today (violent storms, etc.) (Figure 1, C).

The types of country in a) and b) did indeed suggest quite plainly the past intervention of very particular phenomena which alone are capable of explaining their diverse features such as dry valleys, substantial shingle terraces, large valleys disproportionate to the winding tranquil streams which flow in them today, alpine lakes and U-shaped valleys.

French cartographic endeavour

As early as the end of the seventeenth century a systematic program began which was to be pursued tenaciously throughout the eighteenth century with the aim of covering the French kingdom with a geodesic network as a basis for publishing a detailed cartographic cover at the scale of 1:86,400 (finally achieved in 1815). However, the representation of the relief was rudimentary and Cassini's map is of no interest for the history of geomorphology. In contrast, the Royal Administration had a body of well-trained Geographical Engineers *(corps des Ingénieurs-Géographes)* at its disposal from the beginning of the eighteenth century to carry out relief mapping with great precision using hachuring: but only for military use, their maps remained confidential.

There were two exceptions however. First, for the needs of the royal hunt an admirably precise and vivid map was published as early as 1727 of the Forest of Fontainebleau depicting its very strange alignments of parallel sandstone tors. It was subsequently re-issued, notably in 1778 and 1809 (Ellenberger 1983). Second, and more importantly, beginning in 1763, Nicolas Desmarest managed to obtain the cooperation of the Geographical Engineers to make a survey for an extremely detailed map of the Auvergne volcanic region. The hachured engraving, in black and white, is arranged to represent the topographical forms of the mountains and to bring out the nature of the volcanic materials: cones, craters and ancient lava flows. The complete map would not be published until 1823, but a partial map appeared by 1774 at a scale of 1:130,000, in the *Mémoires de l'Académie Royale des Sciences* for 1771. It is an authentic and noted masterpiece within the entire history of geology and geomorphology (Ellenberger 1983). We will see later the theoretical concepts of Desmarest on the interplay between erosion and volcanism.

An unrecognized pioneer: Henri GAUTIER (1660-1737)

This man, by profession Royal Engineer of Highways and Bridges, lived essentially in the Midi, and of professional necessity made a particular study of water courses from their sources to the sea between the Pyrenees and the Rhône. Normally he would have taken his place among the major figures of the history of Geology but his work remained ignored until recently (Ellenberger 1975, 1976, 1977). There are two reasons for this:

a) He only ever published his ideas on the Earth in a terse and little broadcast manner (1721, 1723), and
b) He illustrated them with a very naïve model of a hollow and empty earth and an already dated Cartesian physics.

To this we can add that at this time in France very few people were interested in such questions.

Henri Gautier was a resolute actualist. It was evident to him that valleys had been excavated by rivers. The products of fluviatile erosion are at first deposited in floodplains and then are re-deposited in the sea where they form new strata which gradually indurate. The beds of puddingstone within the mountains are ancient deposits of shingle, consolidated and tilted. Once the mountains are levelled by denudation and the seas are filled with sediment, sudden readjustment occurs: the seas become the new mountains, the previous land becomes the sea and these cycles follow on indefinitely. At the time of readjustment, pieces of the earth's thin crust break, float on the central fluid, and either disperse themselves, or pile up on one another. Gautier is then, in several ways, a definite forerunner to James Hutton.

The turbidity of the water courses pre-occupied him professionally (because of the filling in of canals, etc.). He measured the suspension load at midstream in the Rhône (his figure of 1 part in 1700 is still very satisfactory) and he demonstrated the possibility of calculating the time necessary to level a drainage basin by supposing the rate of solid discharge to be known and by putting forward a law for diminishing the rate of erosion with the progressive lowering of relief. But apparently he was afraid to publish the result of his calculations (certainly on the order of millions of years) and gave instead a figure that was obviously an under-estimate; 35,000 years. A Protestant converted to Catholicism, Gautier was by no means constrained by biblical chronologies; he ignored them! The Flood played no rôle whatever in the formation of rock layers. This event, which he only mentioned as a matter of form, he thought might correspond to an episode of crustal fracturing.

In 1722 a certain Tyssot de Patit published a *Discours* at the Hague where he interpreted in a very unorthodox fashion the story of Genesis which he unfolded over millions of years. He thought that inclement weather:

> denuded and broke up the mountains little by little, and in general all eminences.
>
> *émoussent, & écornent peu à peu les montagnes, & en général toutes les hauteurs*

very slowly but inevitably over time so that the ocean received the products of this fatal levelling of upraised lands. Eventually these would be covered by water and this will be the end, if it has not already been produced by the extinction of the sun. Here then there was no idea about the regeneration of mountains.

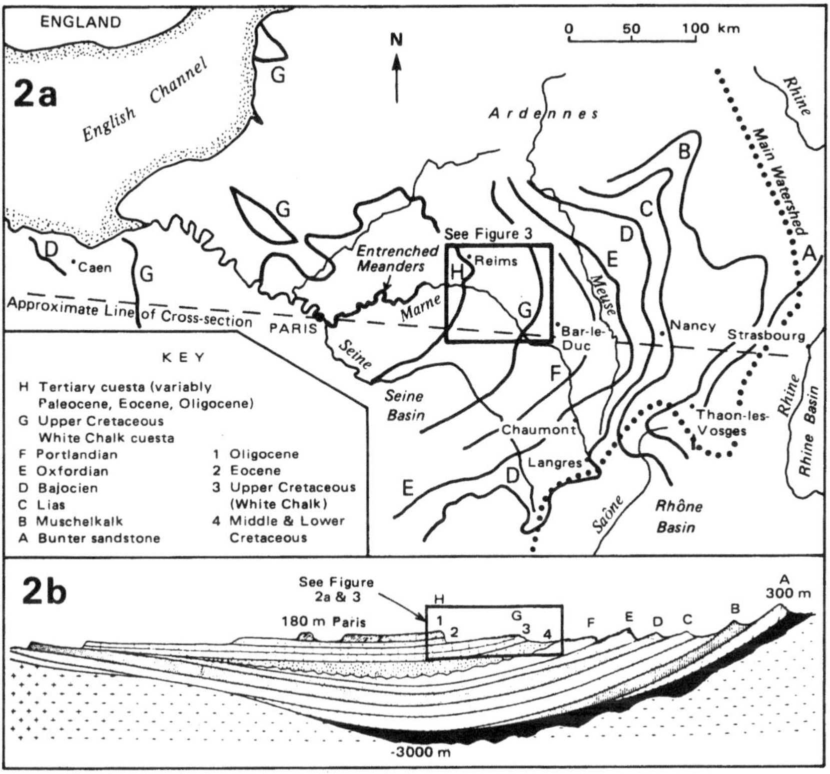

Figure 2

Geological plan and cross section of the Paris Basin (for inset, *see* Figure 3)

NICOLAS-ANTOINE BOULANGER (1722-1759): the greatest geomorphologist of the eighteenth century

The greatest, but the least known, for a major reason: the book where he gives us his remarkable views on the earth's surface (and also on the sub-surface) has never been published. This monumental work (614p) entitled "*Anecdotes de la Nature ...*" remains in manuscript (see Hampton 1955 for some extracts). But, in manuscript form, it appears to have been well circulated in France after the author's death, and in particular it greatly influenced Desmarest who owned a copy. Boulanger was attached to the division of Highways and Bridges and in his professional capacity he studied the Marne basin, from the headwaters around Paris up to the divide with the Saone-Rhône basin. His voluminous text embraced many problems simultaneously. Here is a brief summary of his theoretical, global, views of the Earth. Boulanger imagined that the Earth had undergone, at long intervals, a general catastrophe in which the collective mass of the continents collapsed and the ocean deeps emerged as dry land; thus the beds of fossiliferous marine sediments which have slowly accumulated are laid bare at the surface. The new continental masses are structured as shallow basins which drain progressively from one to another with the lowest ones being invaded by the sea, a scheme which accounts for the present geographical layout of the land. The latest catastrophe alternating land and sea corresponded, according to Boulanger, to that in the Book of Genesis called the Creation, and which was, therefore, only the re-establishment of peoples and civilizations.

As far as local and regional relief was concerned Boulanger did not think it could be explained by means of ordinary, natural, agents acting with their present power. He supposed that several repetitions of cataclysmic "*torrents*" had ravaged the earth, and these were the "*déluges*" which national traditions preserved as terrifying recollections of the events. These "*torrents*" were due to enormous aqueous eruptions gushing out of the earth from openings sited near the watershed. This vision was, in Boulanger's eyes, supported by various observations made in country which he had explored personally, namely the plateau of Langres. This is the part of the hydrological watershed between Dijon and the Vosges, formed from horizontal Jurassic limestones, with intercalations of marly formations; there the Seine and its tributaries the Aube and the Marne, and equally the Meuse, have their sources. Quite frequently the water courses there are born in a small topographic "*cirque,*" where one can also see karst openings presently dry: the supposed points of exit for the diluvial "*torrents.*" Further downstream several shaped landforms (in our opinion due to periglacial phenomena) justify in Boulanger's eyes his hypothesis of brief, ancient, flows of formidable power. This extreme concentration of time was no impediment to his having a remarkable perspicacity in perceiving features of the relief and in deciphering the successive episodes in their evolution.

The evolution takes its point of departure, therefore, from an ancient morphology of basins and a network of high ridges from which the present topography is inherited, strongly modified by the intense action of denudation. It presents everywhere "*Irrégularitiés*" (a key term) which are "*monuments*" of the history of ancient "*accidents.*" It is up to us to learn to read the "book of nature - *livre de la nature.*" According to Boulanger:

> the most natural method for studying the traces of the ancient operations of nature..., is to consider separately the surface of the earth, and portray it, and separately its interior, and to dissect its anatomy.
>
> *la méthode la plus naturelle pour rechercher les traces des anciennes opérations de la Nature..., c'est de considérer séparément la superficie de la terre et d'en faire le portrait, et séparément son intérieur, pour en faire comme l'anatomie* (p.10-11).

He insisted strongly on this duality and illustrated it with the image of a statue sculpted from a block of marble: the rock with its veins and other internal structures is evidently much older than the work of the sculptor and the subsequent damage inflicted on the statue. Again from the same source:

> to have a correct idea of mountains, one must consider them as parts of the mass left in relief, and the valleys as furrows excavated in the earth; a single cause producing the two effects.

> *pour avoir une idée juste des montagnes, il faut les considérer comme des partyes de la masse laissées en relief, et les vallées comme des sillons pris et creusés dans la masse; un seule cause à produit ces deux effets* (p.223-44).

The valley voids were, at earlier times, filled in by the same formations as those which exist on both sides, and which are composed throughout of regularly superposed layers.

There are, therefore, two histories completely separated in time; one of the relief, and the other of the formation of the mass of the earth. Reasoning as a lucid geologist Boulanger wished to be able to reconstitute what he termed the "subterranean geography - *géographie souterraine* " of the beds of the subsurface (p.317, for usage of the term *see* Ellenberger 1983b). He had the gift of seeing things in space, and grasped very well that in all the region of the Paris basin which he had explored, the suite of sedimentary rocks (that is, for us, from the Triassic to the Tertiary) had great lateral continuity; they were inclined gently but regularly downstream, more strongly than the watercourses, and therefore disappeared successively below the surface (p.325). A plan drawn on the surface, he explained, would give "only the plan of a section - *que le plan d'une section.* (p.318)." But what Boulanger wished was to know at a given moment:

> the former situation of the globe and the position of its axis with respect to different regions of the earth
> *l'ancienne situation de nôtre globe et de la position de son axe par raport aux differentes contrées de la terre,*

from the mineral composition of a given layer. In short, it is an inkling of what would later become palæogeography.

This strong distinction between the initial formation of the strata and their subsequent erosion by running water is directly opposed to the ideas Buffon had recently published in his "Theory of the Earth - *Théorie de la Terre* (1749)" namely that the valleys were contemporaneous with the strata of the neighbouring slopes, the result of marine action operating throughout the deposition of the sediments. Buffon had under his own eyes at Montbard, the Armançon valley, moulded as a shallow trough, and although an actualist, nevertheless he had reverted to this idea, only a little changed, of the radical diluvialist Louis Bourguet who was himself inspired by John Woodward. The "proof - *preuve*" of the shaping by marine action was for Bourguet, as for Buffon, the arrangement of the valleys as "salient and re-entrant angles - *angles saillants et rentrants,*" a highly successful idea in its time, although it astonishes us today!

One presumes that Boulanger had read Henri Gautier and gained some of his ideas from him, nevertheless he did have grounds for saying (p.172) that he (Boulanger) "gives the elements of a new science - *donne les éléments d'une science nouvelle.*" In our view this is incontestable as far as geomorphology is concerned. In what follows we are led to distinguish between what he says about valleys, and his daring speculations on the whole course of denudation.

The overall layout of relief is as follows: together with his geographical contemporaries, Buache for example, Boulanger accords great importance to what he called "the primary high divide - *sommet général*" which is to say the watershed for the principal rivers in France. But he did not think as Buache did that the watershed corresponded to the "global framework - *Charpente du Globe*" - which was so obvious on the geographical maps. From this "principal divide - *sommet de premier ordre*," "local divides - *sommets particuliers*" branched off, and they themselves ramified, the whole thing being like a tree trunk with its branches and roots. Similarly the valleys "which create all the divides by their spacing - *qui forment tous ces sommets par leurs intervalles*" can be distinguished as valleys of the first, second, third, and fourth orders; simple valleys forming the fifth order, and so on (p.177-179).

If a principal valley, such as the Marne's, is followed it is seen that its track is sinuous. It is the same in every region and Boulanger believes:

> probably that these bends have a behaviour all of their own, and that Geometry, so industrious these days, can determine it.
> *vraysemblable que ces courbes ont une nature qui leur est propre, et que la Géométrie si industrieuse aujourdhuy peut oser déterminer* (p.59).

He never employs the word "meander - *méandre*" but he describes very satisfactorily the nature of entrenched meanders. For him their formation is due to a raging torrent completely filling the valley bottom and which is thrown brutally from one bank to the other and by its force cuts steep slopes that are concave and bare:

> tossed from one bank to the other, in turn wearing away one side and increasing the other.
> *balotté d'une rive à l'autre, il a toûjours détruit alternativement un revers et augmenté l'autre.*

On the bank that was progressively abandoned by the *torrent* (in attacking the other bank) it deposited continually sediment and other matter brought down by the flood, deposits which prove all the more fertile as these banks:

> sloping very gently and normally with a circular base
> *en pente très douce et circulaire ordinairement par leur base* (p.43-4)

were more sheltered. An excellent observer Boulanger perceived that a single erosional event was insufficient to account for the formation of the landscape in all its details. He saw that often the top of the concave slopes displayed inflexions which are those of ancient valley cuts, now "truncated - *tronqués;*" ancient (alluvial) deposits which may well have been themselves reworked (p.122-24). He described to us very well how the "*torrent*" in following its course of demolition can open up a more direct way in leaving "in the middle of its courses, an isolated mass - *au milieu de son cours une masse isolée*" *:* after breaking through a meander neck, in modern terms. (At this point Boulanger refers his reader to plates which have been lost).

The inadequacy of present water courses to excavate their valleys was self-evident to Boulanger. The proofs are several: thus one sees the winding present river having lost all its energy, "encumbered by the silts of its grassy floodplain - *embarrassée dans les limons de sa prairie*," winding in every direction, independent of the general trend of the valley. Far from being able to excavate its bed, on the contrary it deposits its floodwater sediments. Except in mountain country our river beds:

> are as if suspended above the earth's rocky ground
> *sont comme suspendus au-dessus du sol* [rocheux] *des continents*

and are situated upon:

> superficial and superimposed deposits
> *déposts superficiels et postiches* (p.49).

Moreover there are numerous dry valleys today preserving many vestiges of their fluvial origin (p.49). Thus, Boulanger rejects the "false ideas - *fausses idées*" of too many people:

> who imagine that these deep excavations which we call valleys can be produced by the successive operation of our present rivers and streams with their present state of power and speed.
> *qui s'imaginent que ces profondes excavations que nous nommons vallées, peuvent être l'effet et l'ouvrage successif de nos fleuves et de nos rivières dans leur état présent de force et de vitesse* (p.46).

Equally he rejected collapse as a general cause of valley formation (p.52).

Now let us look at what is concerned in "grand denudations - *grandes denudations.*" Recall that the entire eastern part of the Paris Basin is a classic region of concentric escarpments, "*côtes,*" highly visible in the countryside; and the relative height of their steep eastern slope can be more than one hundred metres. They are a response to differential erosion on the more resistant members of the numerous Jurassic limestones, the white chalk of the upper Cretaceous, and the dominantly calcareous Eocene beds.

Boulanger understood this arrangement perfectly well; it is often compared to a piled up stack of plates of decreasing sizes. Already in this recognition he is well in the vanguard of French Geology. Furthermore he uses the enormous energy of his famous "*torrents*" in a most productive manner, an example of a false theory begetting remarkable results. He is especially interested in the Champagne chalk country, which in his time was to a great extent uncultivated on account of the poor soil. It forms an extensive north-south strip "more than forty leagues in length - *plus de quarante lieues de longueur*" and "ten or twelve in width - *dix ou douze de largeur.*" Now:

> the arrangement of the lands which surround the chalk ... give rise to the knowledge that all the area of the Champagne where this ground is seen was covered by other beds and by other deposits of a different nature and even with hard stones, which are very rare there at present.
> *les dispositions des contrées qui environnent la craye font ... connoitre que toute la partie de la Champagne où cette terre est découverte a été recouverte d'autres lits et d'autres bancs d'une autre nature et même de pierres dures, quoy qu'elles y soyent extremement rares de nos jours.*

To support this inference Boulanger provided three arguments, irrefutable in his own eyes, and for that matter to our eyes today!

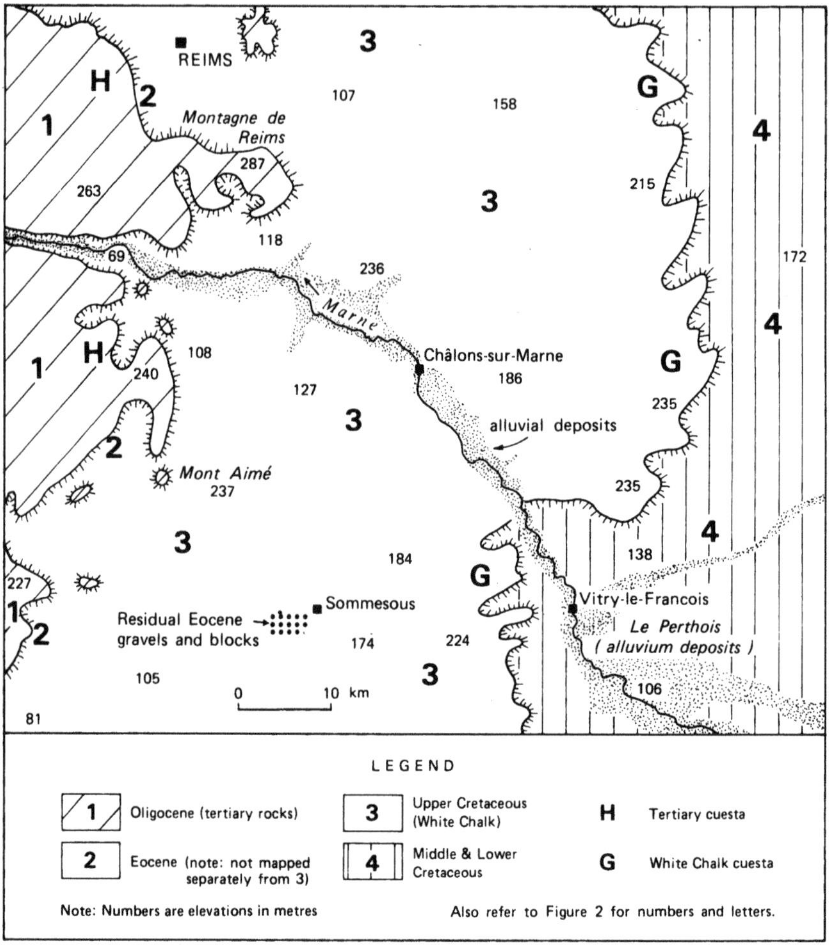

Figure 3

The region of the Champagne Chalk country, *see* text for description.

---

1. The Rheims Mountain as one can see on the map, is a large prominence jutting out above the country, and it is made of rocks quite different in nature from the chalk, and seems as if it used to extend much further than it does now towards Rheims and Chaalons and consequently covered formerly the surface which is now bare.

La Montagne de Rheims qui fait encore ainsi qu'on peut le voir sur la carte, une grande saillie en arrachement sur cette contrée, est construite de pierres d'une nature toute different de celle de la craye, et il paroit qu'elle devoit s'avancer bien plus qu'elle ne le fait actuellement vers Rheims et vers Chaalons, et recouvrir par conséquent autrefois la surface qui est à present découverte (p.322-23).

It is, in effect, a prominent outcrop of tertiary terrain overlying the chalk plain.

2. When the torrents have thus uncovered the lower deposits of the country, it is not without having left behind some remains of those layers directly above which have been destroyed. Mount Aymé which remains in the middle of the chalk country of Champagne is a striking monument to this, its rocks are not related to those surrounding it, but to mountains situated on the other side of the Bergeres and the Vertus. It was once connected to them ...

Quand les torrents ont ainsy découverts les bancs interieurs des contrées, ce n'est pas sans avoir laissé derrière eux des temoins des bancs superieurs qu'ils ont détruits, le Mont Aymé qui est resté isolé au milieu des crayes de la Champagne en est un monument frappant, la nature de ses roches n'a aucun raport avec le terrain qui l'environne, mais avec les montagnes scitüées de l'autre costé de Bergeres et des Vertus. Il en étoit une dépendance ...

that is, before it was separated from them by the torrents.

3. Moreover I have again found remains of the landmass to which Mt. Aymé was attached in places even further on in the chalk country ... in and around Sommesou some pebbles and broken stones in large quantities, remains of the upper beds ..

Bien plus, j'ay trouvé encore des vestiges du continent auquel apartenoit le Mont Aymé dans des lieux encore plus avancés sur la terre de craye .. Aux environs de Sommesou .. des cailloux et des pierres brisées en grande abondance, restes des derniers bancs .. (p.326-7).

I shall now give a few words of geological commentary. The entire recession of the edge of tertiary deposits is about forty kilometres according to Boulanger. The remaining outliers of indurated Eocene deposits left on the chalk near Sommesous are about twenty kilometres from the present escarpment. Mount Aimé is an outlying butte close to this escarpment, formed of chalk and capped with residual Tertiary deposits. All this constitutes an excellent geological interpretation by Boulanger.

For Boulanger, the limit of the chalk to the east has the same extensively reticulated relief; it:

must have extended over the fertile regions of Perthois, covering its plains

devoit s'étendre sur les fertiles contrées du Perthois et en recouvrir les plaines

now dominated in the west by "the sections and escarpments - *la coupe et les escarpements*" of the "upper chalk - *crayes supérieures.*" There the floodwaters have laid down deposits which form fertile ground. One more proof of the vast erosion that has affected the chalk is the abundance, whether resting on the surface, or whether within what might be termed "nature's warehouses - *magasins de la nature*" which are alluvial sands, of marcassites often broken, whereas in the chalk where they occur naturally they can be seen intact (p.328). That the hollowing out of the valleys must have happened long after the deposition of the strata is, *inter alia*, confirmed by the study of "the deposits of sand, gravel, and other detritus - *déposts de sable, de greves et autres débris*" in the valley of the Marne downstream of the Champagne country where one finds "pieces of chalk rounded like shingle - *des morceaux de craye arondis comme des galets*" as far as the outskirts of Paris (p.320).

The whole of this geological study of the Champagne region is excellent and we are fortunate that this part of Boulanger's manuscript (amongst others) has not been lost to science.

Having observed near Langres, along the watershed, some "isolated sugarloaves - *pains de sucre isolés*" .. "remains of terrains that no longer exist - *restes de terreins qui n'existent plus,*" Boulanger uses the small rounded buttes as the starting point to make a daring extrapolation. Starting from the "hog's back and platforms - *dos d'âne et des platteformes*" already isolated by the carving out, in the "main watersheds - *sommets généraux*" :, of early shallow valleys, still well apart

> the gullies were so deeply incised that each one of these uplands has formed nothing more than a single isolated peak.
> *les ravines se sont si profondément creusées, que chacune de ces éminences n'ont plus formé que des pics isolés.*

This gives rise to the peaks which stick up along the top of the present primary divides:

> The Alps, which have been thus so very much defaced, could originally have been country as continuous as the undulating plain of the Champagne region
> *Les Alpes qui en sont si considérablement défigurées, ont pû être originairement des contrées aussy unies que les plaines légerement ondées de la Champagne* (p.193).

The author held firmly to this explanation of the very large by the very small. Observe a small stream winding in the sand, it erodes progressively its banks which consist of cliffs, capes, promontories; where we direct its course into harder ground it narrows. We have there, on the scale of an ant, a working model of "*torrents*" with the narrowing of the valleys, and abrupt deviations due to powerful obstacles, the result of past events. For Boulanger there is a qualitative, though not a quantitative, unity about the process:

> Nature is always the same in her effects, but she has operated upon them more or less as the agents have been have more or less powerful.
> *la Nature a toûjours été la même dans ses effets, mais elle les a operé plus ou moins grands suivant que les agents ont été plus ou moins puissants* (p.189).

To conclude our appraisal of Boulanger I will say a few words about his evaluation of the geological timescale. Repeatedly he refers to the sixty four traditional centuries which have occurred since the last great catastrophe which interchanged land and sea. In this matter he is faithful to the biblical chronology, for it follows for him that this interchange is translated by the biblical author in terms of the Creation. The marine sedimentary beds are very slowly formed in the course of a former, very long, epoch and the materials for this deposit are supplied by the destruction of ancient uplifted lands, which resembled ours. Perhaps it was accumulating:

> little by little above more ancient ruins.
> *peu à peu par dessus de plus anciennes ruines* (p.536).

It is this vision of an indefinite succession of cycles to which Boulanger leads us. In fact, their duration is very much longer, in his real thinking, than the conventional 6.400 years. On two occasions he allows himself to speak of "thousands of centuries - *milliers de siècles*" (p.487, and if we decipher his writing correctly, on p.317.) There is no suggestion of an evolutionary direction in this endless repetition of worlds. The existence of living beings continues unchanged. Rocks from a previous cycle can be found redistributed in the sediments of another, just as one sees ancient building stones reused in the walls of the town of Langres:

> From a single monument one can deduce several epochs.
> *D'un seul monument on peut déduire plusieurs époques* (p.457).

Boulanger's cyclical interpretation of the earth is marked by a tragic pessimism, in contrast to the neutrality of Henri Gautier and the subsequent optimism of James Hutton.

The heritage of NICOLAS BOULANGER

As we see, the observations and ideas of Boulanger concerning geology and geomorphology make his book the "Anecdotes of Nature.... - *Anecdotes de la Nature,*" a document of considerable historical interest. But since his book remained unpublished; did he have an influence on the development of a science of the earth? One can reply in the affirmative. No less than two authors of the first importance have used it. The first is Buffon when he wrote his "Epochs of Nature - *Époques de la Nature*" (1779). He plagiarizes entire passages in his "Fourth epoch - *Quatrième époque*" (p.149-159) where he discusses the configuration of the Langres Plateau and nearby areas. But if he adopts the idea of a powerful scouring by water, what he had in mind was marine currents following the deposition of the layers, before the final retreat of the sea.

However we take it, Buffon had a vision of the evolution of the terrestrial globe which was short term and irreversible, radically different from that cyclic and potentially eternal view of Boulanger. And Buffon was not much interested in geomorphology: that is to say in the form of the surface.

The principal inheritor of Boulanger's ideas was Nicolas Demarest. We shall see later his major, personal, contribution to the geological and geomorphological history of the Auvergne around 1770. As early as this he acknowledged the existence of immense erosion as the work of running water. However, it is in l'"*Encyclopédie Méthodique - Géographie Physique*," an enormous compilation published late in his life, and left unfinished, that the heritage of Boulanger is clearly manifest. In the first volume (1794), the article dedicated to Boulanger is based on an early manuscript by the latter which is limited to a study of the Marne, and which Desmarest acknowledged having to hand during the lifetime of the author, with whom he had been able to converse. But it is in Volume 2, dated 1803, that the impact of Boulanger's ideas and text were fully revealed. A very long and detailed article entitled "Anecdotes from nature and from the history of the Earth - *Anecdotes de la nature et de l'histoire de la terre*" (p.523-586) is from beginning to end based on the work of Boulanger in 1753, although Desmarest doesn't cite his principal source. On almost every page he repeats passages, sometimes literally, sometimes correcting them by substituting rain and running water for the action of catastrophic diluvial currents. There were other articles which, at this crucial period at the beginning of the nineteenth century, revived a notable part of Boulanger's thinking. For example there is the article on CHALK (volume 3, 1809, p.524-536) where Desmarest (who appears to have revisited the field sites) takes over the excellent idea of a huge erosional retreat of the heads of upland masses, and of considerable correlated denudations.

It is interesting to note that in 1810-1811 Cuvier and Brongniart published their celebrated "Mineral Geography of the area around Paris - *Géographie minéralogique des environs de Paris*," a major step in the history of Stratigraphy, but an equally interesting and important work from the geomorphological point of view. They describe how the Tertiary formations presently at the surface are divided into plateaux, with each level separated by a steep slope. They acknowledge explicitly that this disposition implies enormous removals of debris with inequalities of removal between different districts and formations. Among the authors whom they acknowledge as having some small precedent over them, Desmarest is mentioned. Consequently one can suppose that in this case, as in others, the influence of Boulanger on the nineteenth century was not entirely lost.

NICOLAS DESMAREST and the Auvergne

In contrast to Boulanger Desmarest is a well known author. Only that part of his work relevant to geomorphology will be mentioned here. Essentially this aspect of his work is contained in his memoirs on the Auvergne volcanic region. The first is dated 1771 (but appeared in 1774) and is accompanied by the remarkable "Map of part of the Auvergne - *Carte d'une partie de l'Auvergne...*" In fact this is essentially of the Mont Dore massif which is, as is well known, an enormous volcanic complex of primarily Mio-Pliocene age. In this memoir Desmarest wishes to show that the columnar basalt, already well known in Saxony and elsewhere, for example Antrim, which exists there in the form of isolated masses, is a "volcanic product - *produit des volcans*": solidified lava. He demonstrates that four circumstances are always seen to be concomitant (p.738):

1) The complete or partial character of a lava flow.
2) Little or great alteration of the rock.
3) The preservation, or not, of the "mouth - *bouche*" of the ash or scoria cone.
4) The relative height.

The recent flows are completely preserved from the crater to their natural termination; their alteration is negligible or non-existent; they are restricted to the bottoms of valleys. On the contrary, the older flows are dismantled, and are often reduced to being the cap rocks of buttes or interfluves (to use modern terminology), stripped of superficial ash deposits, and accompanied by loose material resulting from degradation by rainwater or snow.

In 1775 Desmarest read to the returning public session of the Academy of Sciences a memoir which unfortunately was only published as a précis in 1779: "On the determination of several geological periods from the products of volcanoes... *Sur la détermination de quelques époques de la Nature par les produits des Volcans...*". This very important paper, despite its brevity, shows the method of analysis to follow in order to arrive at definite facts about the distinction between former epochs and the present state of affairs. The perfect overall agreement between the various "circumstances - *circonstances*" detailed above, "the states of volcanic matter - *états des matières volcaniques*" persuaded him that:

> Nature has been subjected to the same progress throughout more remote periods, as in recent times (Extrait, p.7).
> *la Nature a été assujettie à la même marche dans les siècles les plus reculés, comme dans les tems les plus modernes.*

This is the key phrase in the history of the uniformitarian doctrine because it is more than a simple act of faith or an attempt to beg the question. In fact, a profound and varied knowledge of landscape presented him with a diverse natural range of "a regular sequence of progressive alterations - *des suites régulières d'altérations.*" In taking as a frame of reference "the original forms of the latest products of fire - *les formes primitives des derniers produits de feu,*" which is to say the most recent lava flows, and comparing their subsequent alterations, Desmarest has been able to:

> delimit precisely the boundaries of each of the corresponding and parallel occurrences which developed or were altered in the same order.
> *circonscrire, dans des limites précises, chacune des circonstances correspondantes & parallèles qui croissoient ou qui s'altéroient dans le même ordre.*

His belief in this truth is explicit:

> It is only through a long succession of centuries that all these forms and occurences have changed.
> *Ce n'est que par une longue suite de siècles, que toutes ces formes & toutes ces circonstances ont changé* (p.12).

The gaps subdividing the ancient lavas:

> have very gradually become the first order valleys.
> *sont devenues insensiblement des vallons du premier ordre* (p.20).

The last of three periods which he distinguishes arbitrarily:

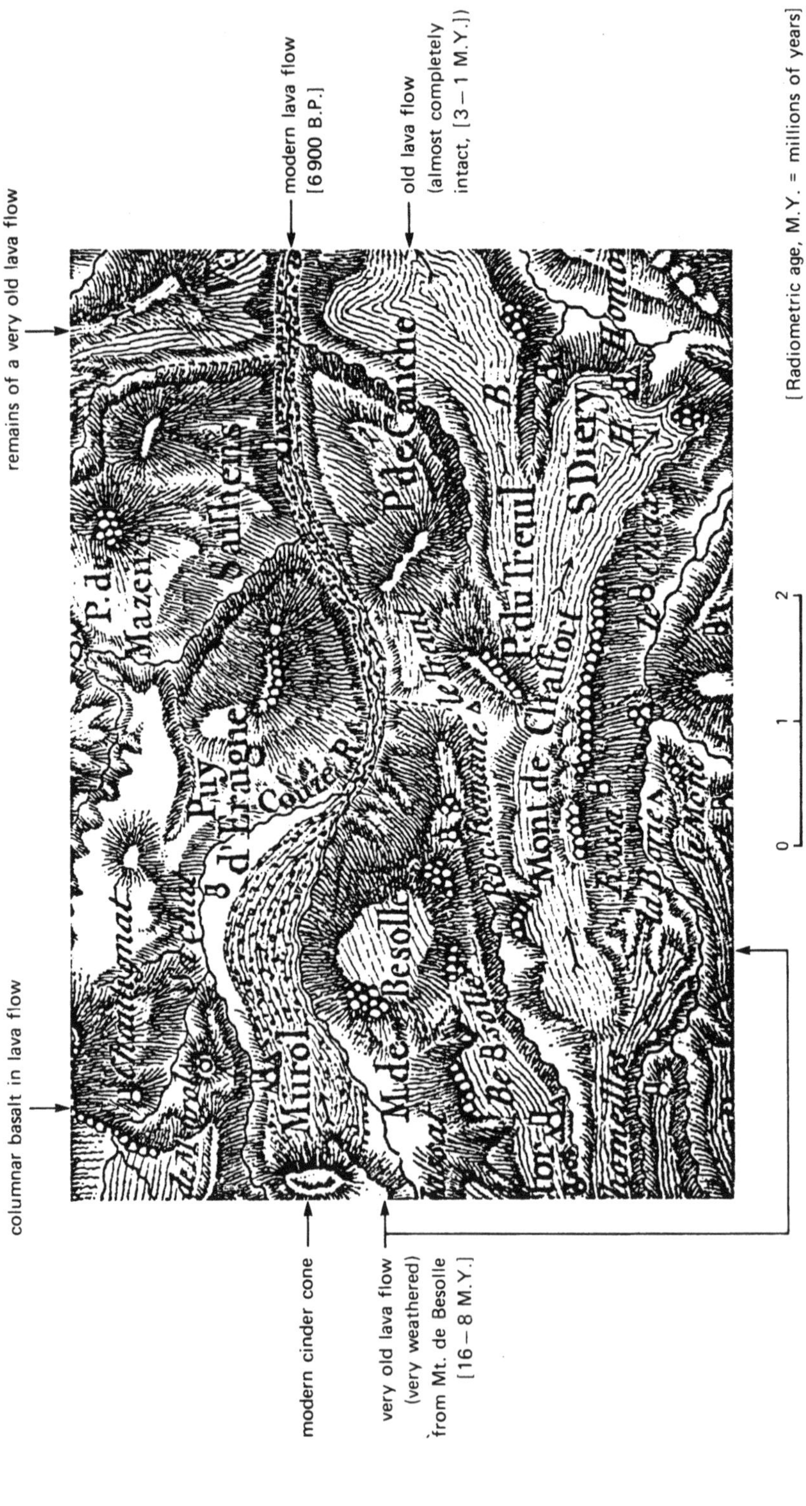
remains of a very old lava flow
modern lava flow [6 900 B.P.]
old lava flow (almost completely intact, [3 – 1 M.Y.])
columnar basalt in lava flow
modern cinder cone
very old lava flow (very weathered) from Mt. de Besolle [16 – 8 M.Y.]
0
1
2
kilometres
[Radiometric age, M.Y. = millions of years]

takes up all the time which rainwater needs to erode the valleys: it even shows the varying progress of this work by displaying the lava flow at all possible levels in the valley profile, and indicates to us that each point that acts as bed and site for the emplacement of a lava flow has been a valley bottom up to the time of the volcanic eruptions which produced these various flows.

occupe tout le tems qu'il faut accorder à l'eau pluviale pour creuser les vallons: elle nous montre même les différens progrès de ce travail, en nous offrant les courans à tous les niveaux possibles sur les croupes inclinées des vallons, & en nous indiquant, par là, que chaque point qui sert de base & d'emplacement aux courans, a été successivement un fond de vallon, lors des éruptions des Volcans qui ont produit ces divers courans.

The map drawn for Desmarest by the Geographical Engineers Pasumot and Dailley (published in 1774) already shows, in the clearest way, the stepping of the lava flows, with the oldest having now become the highest. In 1806, the *Mémoires de l'Institut* published an article in which Desmarest (1725-1815), already an old man, returned to and extended the theme expounded in his article of 1779, adding four maps which anticipated in the form of partial maps the large map published posthumously in 1823. Two of them illustrate the stepping of lava flows of various ages between the hills of the Puy range and the sedimentary lowland of Limagne. All that remained was to mark the absolute altitudes: this was to be the work of Ramond who published in 1818 a thorough lithological and geodesic description, the results of an enormous levelling project. Scrope had only to benefit by the fruit of these French researches in 1827 (although he was not without talent); and after that it was left to Lyell to play the part of the grand promoter of the uniformitarian doctrine. Now we shall see that in the last quarter of the eighteenth century in France, Desmarest was far from being alone in professing this doctrine. In finishing with Desmarest we will note that he established that since the first lava flows, the surface of the earth had been lowered by some 150 to 200 toises (390m); there had been "an immense destruction - *une immense destruction*" (1779, p.23-4). It is therefore in the knowledge of this fact that in the article "*Anecdotes...*" cited above from l'*Encyclopédie méthodique* (volume 2, p.571) he speaks of:

the need to reconstruct the often enormous erosion that has taken place in certain regions [if we wish] to know the original state of the uplands

la nécessité de rétablir les déblais souvent immenses qui ont eu lieu dans certaines contrées [si l'on veut] connoître l'état primitif des massifs.

He had no need in this case to base himself on the writings of Boulanger.

Figure 4 (<-- facing page)

Extract from the map of Desmarest, Pasumot and Pailley (1771) showing lavas of various ages in the Massif Central.

MONTLOSIER

In 1788 Count F. de Montlosier published an "Essay on the theory of the Auvergne volcanoes - *Essai sur la théorie des volcans d'Auvergne*" which does not lack insight. He describes the Mount Dore Massif as having formed from the outset "a solid and continuous mass which has only been divided by stream erosion - *une masse solide & continue qui n'a été divisée que par l'excavation des eaux.*" It is in fact a former basaltic shield of Hawaiian type which has built up a vast volcano, and:

> It is impossible not to see at once that Mont d'or was in the beginning nothing but an almost horizontal mass.
>
> *Il est impossible d'abord de ne pas voir que le Mont-d'or n'a été primitivement qu'un continent plein & presque horizontal.*

Flowing on this slight incline "running waters - *les eaux fluviatiles*" have bit by bit eroded deeper and deeper valleys, invaded by successive lava flows:

> In the Auvergne one can find medallions of all ages struck by nature to attest all the stages of her work and progress.
>
> *Dans l'Auvergne, on trouve à tous les âges des médailles frappées par la nature, pour attester toutes les gradations de ses travaux & de sa marche.*

These "medallions - *médailles*" are in this case the various plateaux formed by the old lava flows, such are the "markers - *témoins*" left in the excavations by the workmen. Montlosier, in comparison to Desmarest, supplies some details about the various rocks comprising Mount Dore, but professes the same uniformitarian fluvialism. He says that his ideas, "founded on long and mature reflections, ... have so to speak been my lifelong work - *fondées sur de longues et mûres réflexions ... ont fait pour ainsi dire l'étude de toute ma vie.*" It is evident, he writes of streams:

> that their erosive action finishes by attacking the mountains themselves ... In like manner nature progresses along this sure and steady track in the great task of levelling the landmass.
>
> *que leur action érosive finit par attaquer les montagnes elles-mêmes ... C'est ainsi que la nature marche par cette voie constante & sûre au grand ouvrage de l'abaissement du continent* (p.88).

He evokes the immensity of time, which poses no problem for him:

> All these alluvial and illuvial processes therefore change the face of the earth, and after an infinity of centuries must render it totally unrecognizable, and completely different from its original state.
>
> *Tout ce travail d'alluvion & d'éluvion change ainsi la face du Globe, & après une infinité de siècles, doit la rendre totalement méconnaissable, et la rendant tout-à-faire autre qu'elle n'était depuis son origine.*

Montlosier's book was republished in 1802. He died in 1838 and was well known in the Auvergne; in 1828 Lyell and Murchison paid him a visit (Wilson 1972, p.197). He had the benefit of other visitors too, and such visitors usually included another call in their itinerary: namely a visit to Bertrand de Doue of Puy en Velay, also a fervent fluvialist. Thus, despite the neo-catastrophism of Faujas de Saint Fond, Dolomieu, Cuvier and others, uniformitarianism continued to have influential supporters in France insofar as the process of erosion was concerned. Historiography does not have the right to suppress these truths.

MARIVETZ and GOUSSIER

In 1779 these two little known authors published a *"Discours préliminaire et prospectus"* of a general "Treatise on Physical Geography - *Traité général de Géographie-physique,"* which never actually appeared. They also professed the uniformitarian belief and exalted the time factor:

> The steady continuation of the same action necessarily produces some effects, which although neglected and uncalculated, are not incalculable.
> *La durée constante de la même action produit nécessairement des effets qui pour avoir été négligés & incalculés, ne sont pas incalculables* (p.5).

Water is "Nature's universal solvent - *le dissolvant général de la Nature."* It tends to:

> wear down the surface over which it runs. ..Nature's progress is uniform and constant, but slow [to produce] significant effects.
> *applanir la surface du globe qu'elle parcourt. ..La marche de la Nature est constante & uniforme, mais lente dans ses grands effets* (p.15).

And here is an interesting geomorphological proposition:

> We shall call primary forms those which have been produced below the ocean, and secondary forms, or furrowings, the modifications or alterations of these forms which we recognize to be the work of rainwater.
> *Nous appelleronts formes primitives celles qui auront été produites sous la Mer, & formes secondaires, ou sillonnemens, les modifications ou altérations de ces formes que nous reconnoîtrons pour être l'œuvre des eaux pluviales.*

In fact the work of these two authors was not very original and its failure to appear is only a slight loss, but their uniformitarian declarations illustrate well that this doctrine was prevalent and "in the air."

GIRAUD SOULAVIE (1752-1813)

Here, in contrast, is an author of major importance in many regards. He glimpsed transformism, laid down the principles of stratigraphic paleontology, and drew the first coloured French geological map (1780). He put forward a complete geological and geomorphological history of Vivarais (equivalent to the present Department of Ardèche) and neighbouring regions, describing in particular the volcanoes and lava flows of various ages and at the same time, relating them to the progress of valley erosion by running water, he attempted to calculate the time taken by the erosion. His chief work comprises the seven volumes of his "Natural History of Southern France - *Histoire naturelle de la France méridionale*," which appeared between 1780 and 1784.

It is necessary to add a few words about the nature of this region. It is on the southeastern edge of the Massif Central, and more precisely it is the high crystalline plateau of Velay (1000-1300 metres), against which are backed up the Mesozoic and Tertiary formations of the Rhône valley. The plateau edge has been vigorously attacked by Quaternary erosion; gorges eroded by tributaries of the Rhône have created an almost Alpine relief. The climate is one of great extremes, with summer thunderstorms that can yield more than 50cm of rain in a few hours. Powerful flashfloods in steep channels often cut into fresh rock. Under these conditions the idea spontaneously arises that V-shaped valleys, however wide and deep they may be, are the work of processes still working today. It is precisely this opinion which is so firmly defended by Giraud Soulavie (notably in Volumes 6 & 7) versus those who believed in sudden erosion by the waters of the Deluge. These believers included, amongst others, the Abbé Roux, whose letters the Abbé Giraud Soulavie loyally published, *in extenso*. It seems that this honest cleric, and fervent naturalist, was frightened of the enormous time periods overtly postulated by his resolutely fluvialist and uniformitarian colleague.

For example, Giraud Soulavie writes (Volume 7 (2), 1784, p. 127):

> All this destruction ... is explained by the gradual action of water: nature can well afford the time necessary for the carving out of our mountains and the deeply incised beds of limestone and lava. We see in the running waters of our rivers time's corroding chisel. We see before our eyes what has happened to the physical world in former times...
>
> *Toutes ces destructions ... s'expliquent par l'action lente des eaux: que coûtent à la nature les siècles nécessaires à la sculpture de nos montagnes, a la profonde coupure des couches calcaires, & des couches de laves ? Nous voyons dans les eaux courantes de nos rivières le ciseau rongeur du temps: nos yeux nous montrent ce qui s'est passé dans les anciennes périodes du monde physique...*

The Abbé Roux opposed this system with the observation that there has been very little river bed erosion since the building of the Roman bridges. Giraud Soulavie himself dared to extrapolate that it might need millions of years to excavate valleys (but at that time he kept his figures to himself).

As early as 1780, Giraud Soulavie understood, in three dimensions, the structural composition of Vivarais. There, the vulcanism has been less active than in the Auvergne. In modern views it is limited to an old basaltic shield that poured onto the plateau, which had its original slope towards the Rhône on account of a considerable neo-tectonic flexure. There are also a small number of volcanoes of late-Quaternary age with their ash cones and craters well preserved and their lava flows partly attacked by the erosion of torrents.

This situation had been well perceived by Giraud Soulavie. His reconstruction of the "periods - *époques*" forming the previous history of the area, in comparison to Desmarest, adds a history of marine sedimentation, dated by the principle of superimposition and by the faunas contained in the beds, to the twin history of vulcanism and erosion. After Füchsel and Arduino, Giraud Soulavie was the first author to compile a genuine and complete geological history of an area based entirely on a reading of the field evidence.

He misunderstood the rôle of global warping. For him, everything amounted to a progressive lowering of sea level, according to a model which paralleled that of Pallas, then of the Wernerian school. He writes that since their primordial formation:

> ocean currents have shaped the slopes of the continents, rainwater has subsequently worn into the surface and formed "departments" (= the main drainage basins, F.E.)
> *les courans des mers ont formé d'abord les pentes des continens, les eaux pluviales ont ensuite silloné la surface & formé les departemens* (= les grands bassins hydrographiques, F.E.) (Volume 1, 1780, p.90).

The outpouring of volcanic products may well have disturbed this initial geography, because:

> new slopes will result from these additions, which will derange the water courses.
> *il résultera de cette nouvelle apposition d'autres plans inclinés qui dérangeront les courans des eaux.*

Giraud Soulavie had discovered that the enormous lobe of ancient lavas preserved from the main eruptive centres on the high crystalline plateau, as far as the borders of the Rhône (called the Plateau des Coirons), concealed an ancient watercourse attested by cobble layers and freshwater fossils. Because of this he was able to describe with unusual precision the actual configuration of the original relief subsequently dissected by deep valley erosion. Instead of being an abstract notion, the initial surface which underwent an intense subsequent sculpture by fluvial erosion has already a clear history that can be read from the field evidence:

1) an ancient "physical geography - *Géographie physique,*"
2) its first "rilling - *sillonnement*" by running water, associated with "vast layers of rounded pebbles - *tables immenses de cailloux*" in the shallow river beds;
3) the modifications caused to this geography by volcanic eruptions.

We must reiterate the fact that Giraud Soulavie understood thoroughly this geological succession. Modern science fully confirms (*cf* Bout 1966, 1978) the *chronology* of Giraud Soulavie (a term which he explicated firmly). A Miocene age may be attributed to the peneplain on the crystalline plateau. The alluvial deposits preserved beneath the lava flows of the Coirons (more truly fluvio-lacustrine than solely fluvial) were deposited between the end of the Miocene and the Villafranchian, which is also the period of the vast outpourings of tableland basalts. The small volcanoes and lava flows at the bottom of the valleys are very recent (between 12000 and 15000 years ago where at least two of them are concerned). It is clear, from the little that has been said above, that the Abbé Giraud Soulavie must be counted amongst the major eighteenth-century pioneers of earth science. His principal work was widely circulated at the time it was written but unfortunately the author then turned to quite different interests. Cuvier's generation appears to have ignored it, although they cite De Saussure abundantly. His true influence remains to be determined: it seems as though he may have arrived ten years too early.

DARCET, PALASSOU, RAMOND and the Pyrenees

In the period 1775-1785 numerous authors in France were interested in the problem of valley cutting by running water. For example we could speak of the writings of Guettard on the Alps, De Gensanne on the Vivarais, etc.. We can say a few words about Darcet. In a small volume he discusses in particular the Pyrenees and asserts his uniformitarian view:

> universal and constant agents, ... are perpetually active everywhere, [and slight as they seem] they produce in the long term most significant effects.
>
> *des causes universelles & constantes, ... agissent sans interruption, dans tous les lieux [et pour faibles qu'elles paraissent] produisent à la longue les plus grands effets* (1776, p.30 and p.33).

In his view the Pyrenees perhaps have been reduced already by one half, and this overlooks the brief biblical chronology.

The Abbé Palassou is a naturalist of great merit. He explores the Pyrenean range and shows that on the large scale they present a ribbon-like structure in a longitudinal sense (an arrangement he tries to depict on maps in a rather rapidly executed manner). He believes it is possible to reduce them to an alternating series of parallel bands of marble and schist. At first, emerging from the sea, this range merely formed a continuous mass. Rainwater seamed it, the first streams cut their beds in the schist: resulting in the formation of [first] longitudinal valleys, and then lakes. Overflowing waters excavated the transverse valleys so characteristic of the present geography (1782, *passim*; Broc 1969, p. 159). In 1784, in a second book, he made so bold as to date the erosion (basing it on an estimation from De Genssane published in 1776, Volume 2, p.123, which used the annual sediment load of rivers): at a rate of ten inches a century; so that at the end of a million years the Pyrenees would be entirely levelled. In 1795, James Hutton copied this passage at length in the second volume of his fully developed *Theory of the Earth* (Volume 2, p.143-6).

In 1789 Ramond, a writer of pre-Romantic abilities and a brilliant observer, adopted Palassou's vision of temporary lakes, but he did not believe in final levelling. For him the process of erosion finishes by stopping spontaneously; the same idea was held by the Genevan, De Luc, who denied practically all erosion, and shaped all relief by subterranean movements alone.

JEAN-ANTOINE FABRE

It was with an engineer of the Department of Bridges and Highways, Henri Gautier, that we began this brief review of eighteenth-century French contributions to geomorphology. It is with another engineer of the same Department - Jean-Antoine Fabre - that we shall conclude. The intentions in his book are largely technical and utilitarian, as the title indicates: "*Essai sur la théorie des torrens et des rivières...,*" but he begins with an entire theoretical chapter on mountains. He too postulated that in the beginning each mountain range merely formed a single mass, more or less convex, bordering the sea (then higher than it is today) with a varying slope. For him too running water was responsible for cutting the valleys which dissect this single primordial mass (formed, so we are given to understand, by the Creator). So far this is not anything new. But he supplies two ideas which were ultimately developed to a very great extent. One is our modern concept of base level.

In verifying the existence of rounded pebbles at a height far above the river bed he writes:

> There existed then a time when the river beds were much higher than they are today.
>
> *Il a donc existé une époque où le lit de ces rivières étoit beaucoup plus élevé qu'il ne l'est aujourd'hui* (p.3-4).

Now:

> the height of the river mouth determines the height of all points along its course.
>
> *la hauteur de l'embouchure des rivières fixe celle de tous les points de son cours.*

It follows therefore:

> that formerly the sea level may have been much higher than its present level. We shall not undertake to examine the cause of the lowering of sea level.
>
> *que dans un tems les eaux de la mer s'élevoient beaucoup au-dessus de leur niveau actuel. Nous n'entreprendons pas d'examiner quelle est la cause de l'abaissement de la mer.*

The other idea is that of our concept of *gipfelflur*. In following downwards the skyline of a mountain range:

> It can be seen equally well on both sides that innumerable other mountains lower than the crest of the range are to be found, from which the summits heights are lowered progressively in proportion to the general progress towards the sea, from which it will be concluded that before these masses were broken down and dissected by the deep valleys that we have remarked on, the various planes, whose assemblage formed their original convexity, must have passed close to the summits of the different partial mountains resulting from all the degradation and all the incision which have happened there since the time of creation : planes which in consequence were all inclined towards the sea.

> *On verra que tant d'un côté que de l'autre il se trouve une infinité d'autres montagnes moins élevées que celles de la crête, dont le sommet ... s'abaisse progressivement à mesure qu'on s'avance de toutes parts vers la mer ...; d'où l'on concluera, qu'avant que ces masses fussent déchirées & sillonnées par les profondes vallées que nous y remarquons, les divers plans dont l'assemblage formoit leur convexité primitive, passoient à-peu-près par les sommets des différentes montagnes partielles résultantes de toutes les dégradations & de tous les déchiremens qui ont eu lieu depuis l'époque de la création : plans qui par conséquent étoient tous inclinés vers la mer* (p.6).

We note that Fabre says (on p.79) that the longitudinal profile of the river bed is "asymptotic - *assymptotique*," but he does not cite Du Buat. The latter is already well known, to quote in particular Chorley, Beckinsale and Dunn (1964, Volume 1, p.88-90). Let us simply recall the remarkable studies of Du Buat on river flows, their dynamics, their maturely graded river beds, the gradual cutting of valleys, the inevitable lowering of the relief, and the desolate perspective of the final levelling: and this final point reminds us of Tyssot de Patit.

CONCLUSION

Excepting the remarkable Boulanger, we have avoided talking about the other catastrophists, such as Faujas de Saint-Fond, Dolomieu, De Luc, because, in all, they contributed nothing to the comprehension of forms of relief. The reader of this paper may think that I tend to overvalue the contributions of actualist thought of the French naturalist school in the eighteenth century. It remains the case that this thinking was very strongly developed, and constitutes one of the least well known characteristics of the French geological and geomorphological school of the French eighteenth century. It is so true that James Hutton made abundant reference to it when, towards the end of his life, he found himself forced to defend his own doctrine, in a cogent fashion, against adversaries. To claim to trace back to Hutton the birth of the uniformitarian doctrine (even if he was great in other respects) is to demonstrate an overwhelming ignorance of the real history of geology.

It has been pointed out that most of our authors studied the dissection by fluvial erosion of an elevated global primitive mass, but only rarely do these same authors envisage a restoration of global relief after it has been levelled: even those who wished to follow erosion to its logical conclusion. The two principal writers who did are precisely the two authors who either published none of their ideas, or else very little: Henri Gautier and Nicolas Boulanger. It appears certain that they were unknown to James Hutton, who would preserve therefore the credit for being the sole promoter of repetitive orogenesis (and also of the related notions of plutonism and metamorphism which are rather remote from the subjects treated here); ideas which dominated the nineteenth century.

We have said nothing more here of those for whom the mountains are elemental and remained more or less unmodified after their original formation: they have not played such a large rôle in France as, for example, in Germany. Finally we left aside the francophone authors resident in Switzerland, such as Bourguet and De Saussure. The specific conditions of alpine glacial relief remove them too far from the positions adopted by the authors considered here, and would justify a separate study by a specialist in the history of alpine geomorphology.

References

Boulanger, Nicolas-Antoine, 1753. *Anecdotes de la nature Sur l'Origine des vallées, des montagnes et des autres irregularités exterieures et interieures du globe de la terre Avec Des Observations historiques et physiques sur toutes les vicissitudes qui paroissent lui etre arrivées.* Manuscript 869, Bibliothèque centrale du Museum d'Histoire Naturelle de Paris, 614p.

Bout, P., 1966. Histoire géologique et morphogenèse du système Velay SE - Boutières - Coiron. *Revue de Géographie physique et de Géologie Dynamique,* 2, Volume 8, fasc. 3, 225.

Bout, P., 1978. *Problèmes du volcanisme en Auvergne et Velay.*

Broc, N., 1969. *Les Montagnes vues par les géographes et les naturalistes de langue française au XVIIe siècle.* Paris, Bibliothèque Nationale, 298p.

Chorley, R.J., Dunn, A.J. & Beckinsale, R.P., 1964. *The History of the Study of Landforms, or the Development of Geomorphology,* Volume 1, London: Methuen.

Darcet (= D'Arcet), 1776. *Discours en forme de dissertation sur l'état actuel des Pyrénées, Et sur les causes de leur dégradation.* Paris, 60p.

De Genssane, A.F., 1776. *Histoire naturelle de la Province de Languedoc ....* 5 Volumes, 1776-79, Montpellier.

Desmarest, Nicolas, 1771 [appeared 1774]. Mémoire sur l'origine & la nature du Basalte à grandes colonnes polygones, déterminées par l'Histoire Naturelle de cette pierre, observée en Auvergne. *Mémoires de l'Académie Royale des Sciences,* 706-776. Accompanied (plate XV) by the *Carte d'une partie de l'Auvergne, où sont figurés les courants de laves, ...,* Levée par M.M. Pasumot et Dailley, Ingénieurs-Géographes du Roi, (Scale 1:130,000).

Desmarest, Nicolas, 1779. *Précis d'une mémoire sur la détermination de quelques époques de la nature dans l'étude des volcans, et sur l'usage de ces époques de la nature dans l'étude des volcans,* 24p. It appeared also in *Observations sur la physique, sur l'histoire naturelle et sur les arts,* Volume XIII, 1779, 115-126.

Desmarest, Nicolas, 1804 [appeared 1806]. Mémoire sur la détermination de trois époques de la nature par les produits des volcans..., *Mémoires de la classe des sciences mathématiques et physiques.* Volume VI, 1804 [appeared 1806], 219-289, plates VI-IX.

Du Buat, Louis Gabriel. 1779 [Second edition 1786]. *Principes d'Hydraulique, vérifiés par un grande nombre d'expériences faites par ordre du Gouvernement.* 2 Volumes, Paris.

Ellenberger, F., 1975. A l'aube de la géologie moderne: Henri Gautier (1660-1737). Premiere partie : Les antécédents historiques et la vie d'Henri Gautier. *Histoire et Nature,* 7, 3-58.

Ellenberger, F., 1976-77. A l'aube de la géologie moderne : Henri Gautier (1660-1737). Deuxième partie : la théorie de la Terre d'Henri Gautier (Documents sur la naissance de la science de la Terre de langue française). *Histoire et Nature,* 9-10, 3-149.

Ellenberger, F., 1980. De l'influence de l'environment sur les concepts : l'exemple des théories géodynamiques au XVIIIe siècle en France. *Revue d'Histoire des Sciences*. XXXIII (1), 33-68.

Ellenberger, F., 1983a. Recherches et réflexions sur la naissance de la cartographie géologique, en Europe et plus particulièrement en France. *Histoire et Nature*, 22-23 [appeared 1985], 3-54, 8 plates.

Ellenberger, F., 1983b. Documents pour une histoire du vocabulaire de la géologie : 1) le terme de géographie souterraine. *Documents pour l'histoire du vocabulaire scientifique*. Institut national de la langue française, Paris, 4, 35-42.

Fabre, Jean-Antoine, 1797. *Essai sur la théorie des torrens et des rivières ...*, Paris, XXXII + 481p.

[**Gautier, Henri**, 1721.] *Nouvelles conjectures sur le globe de la terre, où l'on fait voir de quelle maniere la terre se détruit journellement, pour pouvoir changer à l'avenir de figure : comment les pierres, les mineraux, les métaux & les montagnes ont été formez; les corps étranges, comme les carcasses des animaux, les coquillages, &C. qu'on y trouve, y ont été ensevelis* ... Paris, vi + 53p, 1 folding plate. {This text is reproduced, with other additions, in the work which follows.}

Gautier, Henri, 1723. *La Bibliothèque des Philosophes et des Sçavans, tant anciens que modernes, avec les Merveilles de la Nature, où l'on voit leurs Opinions sur toutes sortes de matières Physiques*... Paris, Volume 1, 716p & 2 plates; Volume 2, 678p & 4 folding plates; 1734. *La Bibliothèque*..., Paris, Volume 3, 563p. [A copy of *La Bibliothèque* (vol. I and II) existed as early as 1730 in the *Biblotheca Facultatis Juridica* of Edinburgh (and is now kept in the National Library of Scotland). Thus, in theory, the book was available to Hutton]

Giraud Soulavie, Jean-Louis, 1780-1784. *Histoire naturelle de la France méridionale*... Paris, 7 Volumes, Volumes 1 & 2 1780; Volume 3 1781; Volumes 4 & 6 1782; Volumes 5 & 7 1784. Numerous plates.

Hampton, John, 1955. *Nicolas-Antoine Boulanger et la science de son temps*. Genève, Lille 205p.

Marivetz, E.C. and Goussier, M., 1779. *Discours préliminaire et Prospectus d'un traité général de Géographie physique*.... Paris, 32p.

Montlosier, F. de, 1788 [second edition 1802]. *Essai sur le théorie des volcans d'Auvergne*. Riom, Paris, 155p.

Palassou, P.-B., 1784. *Essai sur la minéralogie des Pyrénées*. Paris 330p. with plates and maps.

Ramond de Carbonnieres, L.-F., 1789. *Observations faites dans les Pyrénées pour servir de suite à des observations sur les Alpes*... Paris.

Scrope, G. Poulett, 1827. *Memoir on the geology of central France*... London 198p with plates and maps. [second edition with modifications, 1858].

Tyssot de Patit, Simon, 1722. Discours de M. Simon Tyssot, Sr. Patot..., *Journal littéraire*, La Haye (The Hague), Volume 12, Article VI, 154-89.

Wilson, Leonard G., 1972. *Charles Lyell - The years to 1841 : The revolution in geology*. Newhaven and London: Yale University Press.

3

Worlds apart: eighteenth century writing on rivers, lakes, and the terraqueous globe

Keith Tinkler

"The eighteenth century was the century par excellence for theories of the earth" (J.C.Greene, 1984 in *American Science in the Age of Jefferson.*)

In this fashion even professional historians of science summarise eighteenth century 'geology,' and it may be added that most of those theories were seventeenth century theories spawned, at least in England, in the aftermath of the Newtonian revolution. Greene cites Woodward (1695) and Whiston (1696) as examples, and it has been argued by Davies (1969 p. 131) that the late eighteenth theories he also cites, such as Whitehurst's (1778), were merely delayed and enhanced echoes of those of the previous century (and which continued in print throughout the eighteenth century). It is argued too that even Hutton's theory is best seen as the culmination of this tradition, rather than as the initiation of a new one (Davies 1969 p. 196, 1985, Secord 1988, personal communication).

At the same time there is general agreement that during the eighteenth century geology 'came of age' and emerged in a form resembling the modern discipline by the early decades of the nineteenth century. Porter (1977) places this emergence in the years 1780 to 1830 and the more recent study by Laudan (1987) agrees with this judgement although she traces a different lineage: that of mineralogy rather than natural history.

It is not of my purpose to dispute these datings, but the two divergent perspectives, one backward (concerning theories with a seventeenth century cast), the other forward (concerning piecemeal field work and its evolution into modern geology), do suggest that some resolution is necessary. I hope to achieve this in pointing out that they may arise from considering two different, co-existing bodies of literature, whose public intentions are worlds apart, although as Neve and Porter (1977) have shown for the case of Alexander Catcott the private resolution of these problems could lead to serious intellectual tensions within individuals.

In the process I will examine the details of this emergence as they are revealed in some published writings not hitherto examined; to follow in these writings, as Laudan tried to, those "principles of reasoning" to which Lyell (casting back to Newton) alluded. In particular I will look at the rôle of 'geomorphology,' in helping to develop a regional approach to geological problems, an appreciation of what Gould (1987) terms "deep time," some insight into the accumulation of sedimentary units, and the inter-relationships among these three items.

Laudan's recent study is notable for its determined focus on the mineralogical lineage leading to modern geology. However, in her determination to avoid "received" accounts of the history of geology at least two themes were either neglected, or were not assimilated completely within the new account. One of these is the theme of time: the realisation that earth history might have to be measured in spans of time that are almost beyond human comprehension, that is to say within the context of recorded human history, and eighteenth century perceptions. The other is the importance of landforms studies, especially as it helped to reveal the immensity of time that an historical science would eventually discover was needed.

An eternal earth was standard with the Greeks, but it was in disrepute from the Renaissance onwards, and by the eighteenth century it was particularly associated with the Deists,[1] of whom Hutton (who thought the earth of indefinite age) is a famous example, and Toulmin, who believed in an eternal earth, was another (Porter 1978). Dean (1981) has shown that the age of the earth was an intensely debated topic in the eighteenth century, and I have illustrated the fervour with which it was discussed in the context of scientific and popular writing about the origin and operation of Niagara Falls in the period 1750 to 1845 (Tinkler 1987).

A Newtonian impulse

The Newtonian view of the universe inexorably required the conceptual expansion of physical space (over previous perceptions) in order to encompass the enormous distances necessary to keep astronomical bodies apart: lest they mutually and rapidly collapse under the operation of the universal force of gravity. The implications that this might have for the passage of time and the history of the globe were not all immediately apparent, but they were implied in passing at two points in *Principia*. In Prop. X, Theorem X, Newton concludes that:

> Jupiter revolving in a medium with the same density as the superior air would not lose ... a millionth part of its motion in 1000000 years.

with the result that:

> the planets and the comets will continue their motions ... for an immense tract of time.

Later, in the tail of a commentary on Prop. XLI, Theorem XXI (on parabolic orbits), Newton remarks that a:

---

[1] Dr. Johnson's *Dictionary* (1755): *deism* - the opinion of those that only acknowledge one God, without the reception of any revealed religion.

> globe of red-hot iron equal to our earth of 40,000,000 feet would scarcely cool in an equal number of days, or in above 50,000 years.

He says he:

> would be glad for the true ratio {of cooling rate to diameter}[2] to be investigated by experiments.

and it was this challenge that Buffon (a Newtonian *de rigeur*) apparently took up during the eighteenth century and reported in his *Histoire Naturelle* (1749), and which drew censorious attention from the French Academy.

The necessity to integrate all parts of the natural universe in time and space was clear to Hutton a century later, when he wrote:

> We cannot understand the system of the globe, without seeing that progress of things which is brought about by time; thus measuring the natural operations of the earth with those of the heavens (Hutton 1788, 1795 I-187).

These excerpts from *Principia* should not be read to mean that Newton believed in an earth that already was extremely old: but the implication was clear that it may persist for a very long time into the future, and the globe example hints strongly that *this* globe, if it cooled from a red hot ball (i.e. were torn off the sun) might be a good deal older than the theologians would allow, and Newton was not committed to a literal reading of days in the Biblical account of Creation (Stokes 1969).

All these extracts appear in that part of *Principia* that the average man of science in the eighteenth century actually may have read: the account of the *SYSTEM OF THE WORLD*, the third book of *Principia*, and a more popular account of the main findings, relatively light on mathematics (Andrew Motte's English translation of *Principia* appeared in 1729, and included a previously unpublished popular account of Newton's *System of the World).*

However, if the latent implications of time in *Principia* were easily missed because there was no easy way to measure or interpret them, there was one other implication that no one missed: the power of Universal Gravitation which explained in one fell swoop, the observable universe, the solar system, the falling of bodies, and last, but not least, the operation of tides: a very intimate part of the terraqueous globe to the maritime interests of western Europe, where an adequate theory of tides (and Galileo had tried before) guaranteed a general theory of physics of no mean power.

No theorist worth his salt could afford to neglect Universal Gravitation after 1687, and it may be said that none did, each in his different way. It is not without significance that the only seventeenth century theory in which universal gravitation did not figure so centrally was that of Robert Hooke, and its essential elements were conceived and lectured upon well before the appearance of *Principia,* although without prejudice to its scientific qualities (Stokes 1969, 1971, Oldroyd 1972).

[2] Throughout the paper remarks in curly brackets { ... } are my interpolations.

At the beginning of Book Three the same man of science would have encountered Newton's "rules of reasoning", which were by no means new, even then, but which have been echoed ever since by writers anxious to justify their methods and their results by reference to an impeccable source.

THE RULES OF REASONING IN PHILOSOPHY

(1) We are to admit no more causes of natural things than such as are both true and sufficient to explain their appearance.

To this purpose the philosophers say that Nature does nothing in vain, and more is in vain when less will serve; for Nature is pleased with simplicity, and affects not the pomp of superfluous causes.

(2) Therefore to the same natural effects we must, as far as possible, assign the same causes.

As to .... the descent of stones in *Europe* and in *America*; ....

(3) The qualities of bodies, which admit neither intensification nor remission of degrees, and which are found to belong to all bodies within the reach of our experiments, are to be esteemed the universal qualities of all bodies whatsoever.

For since the qualities of bodies are only known to us by experiments, we are to hold for universal all such as universally agree with experiments; and such as are not liable to diminution can never be taken away. We are certainly not to relinquish the evidence of experiments for the sake of vain dreams and fictions; nor are we to recede from the analogy of Nature, which is wont to be simple, and always consonant to itself.

(4) In experimental philosophy we are to look upon propositions inferred by general induction from phenomena as accurately or very nearly true, notwithstanding any contrary hypotheses that might be imagined, till such time as other phenomena occur, by which they may be either made more accurate, or liable to exceptions.

This rule we must follow, that the argument of induction may not be evaded by hypotheses.

(From Cajori's edition (1934) of Motte's English translation (1729))

To the rules themselves I have added some of Newton's textual commentaries which follow each rule. They can certainly be taken to be very generally applicable: i.e. to natural philosophy in all its guises. Rule (3) especially directs the reader to believe the experiential evidence he himself collects, while rule (2) (with its explicit geographical example) implies that these findings can, indeed should, be taken as universal. With whig-laden hindsight this is a very dangerous philosophical stance to take in the field of natural philosophy, but it would be the early nineteenth century before this danger became clear, and I shall show that there is good reason to believe that it was widely applied to geological arguments during the eighteenth century. Rule

(4) certainly tries to shift the onus away from the imagined hypothesis and onto the experimental data and the inferences made from it, but it does nothing to dictate which hypotheses shall be imagined in the first place, nor does the first rule's emphasis on simplicity help in this regard. Whatever their failings, or their interpretations by individuals, these rules underlay the workings of natural philosophers in the eighteenth century.[3]

However, as the excellent reviews by Stokes (1969, 1971) and Porter (1977) reveal, by the early eighteenth century there were clearly two traditions. One concerned global theories which engendered blazing polemics in public debates sufficient to have fuelled a minor publishing industry, in addition to warping several promising careers. The other related to small scale field investigations, which because they were more focussed, and discussed (with Newton's Rules presiding unseen) either within the confines of the Royal Society (Stokes 1969, 1971), or in private correspondence,[4] achieved some measure of mutual acceptance on topics of significance, such as on the organic origin of fossils which seems to have been accepted after 1699, if the lack of subsequent debate at the Royal Society is any guide. Likewise Halley showed from the observed fixity in the latitudes of places over several centuries, that Hooke's theory of axial tilt, which tried to account for the changes of climate necessary to explain the apparent evidence of fossils, was not an observable process, and likely not true on this account (Stokes 1969 p. 25). That these were two diverging traditions is demonstrated by Stokes (1969) who notes that there were few reviews, and very limited discussion of global theories in the *Philosophical Transactions*. This was in striking contrast to the lengthy debates on other matters, for example the long series of papers on the origin of columnar basalts exemplified by the Giant's Causeway (Stokes 1971).

But the second type of activity does not have the public notice that attends a book intending to explain the origin of the entire globe. Nor does it readily and quickly lead to theoretical syntheses. In consequence we should not be surprised to find that throughout the eighteenth century even short articles on geological topics necessarily referred to global theories as frames of reference or guiding hypotheses, while most global theories attempted to adumbrate and reconcile scriptural writings with the operation of natural terrestrial processes.

Two papers in the *Annual Register* 1760

Dodsley's Annual Register made a habit of printing or reprinting interesting papers in natural history,[5] usually from the Royal Society *Philosophical Transactions*, but in 1760 it reprinted from the *Scots Magazine* a letter dated December 1759 and signed J.M.. I have been unable to identify J.M., but he reports on limestone quarries near Kirkcaldy in Fife, opened up at a farm called Enderteel on land then belonging to General St. Clair. The location is about an

---

[3] Laudan (1973) gives an account of the importance of the Scottish philosopher Thomas Reid in propagating Newton"s *Rules*.

[4] For example, the John Ray / Edward Lhwyd correspondence.

[5] Note that the *Annual Register* has a separate pagination for the second part of each volume. My page references are to this second pagination.

English mile from the sea, and, he supposes, about 200 feet above it (an accurate figure judged by modern Ordnance Maps). He gives graphic details of the various fossil remains in "flakes of stone" which "lie horizontally dipping towards the sea": fish bones, sea-weed stalks (presumably crinoids), and vast quantities of shells (some "commonly found in our coasts", others "of very uncommon figure"), and he is impressed with the minute details preserved and the hardness of the induration.

Of these remains he makes two points: (i) they are all specifically heavier than water, even the sea-weed which has been stripped of its "broad leaves, that make it buoyant, before it has been lodged here," (ii) the shells are all found empty, and the rock appears to have been "gradually deserted by the sea."

The first point may make oblique reference to Woodward's theory (1695), which claimed that after the universal dissolution of the globe at the time of the Deluge, materials settled out in order of their specific gravities. The second point is greatly amplified in discussion for the author points out that the rock has been:

> for a long time, washed with the tide; for the upper surface is all eaten, and hollowed out in many places like a honey-comb, just as we observe in flat rocks exposed every tide to the access and recess of the waters.
>
> This rock proves beyond dispute the vegetation of stone, and a gradual retreat of the waters ever since the deluge. These two causes, over a long series of ages, alter the face of our globe entirely, or rather have reduced the earth into its present form, by creating rocks at the bottom of the sea, and then leaving them in dry land, where they turn into inland mountains. This seems to be the method which nature observes: for all along our coasts are lime-stone rocks, and some of them within low water mark, which have the very same inclination, and the same mixture of petrified sea-bodies, as in the quarry I have described; but since we see rocks of this kind arising out of the sea, we must, of necessity, ascribe the same origin to such as are more remote from the shore, and left up in the country (p. 98-9).

J.M. provides a lengthy footnote concerning the petrifying qualities of sea water on a blue clay; or the "vegetation of stone," as observed in Boulogne sur mer by a "Scotch gentleman" in the summer of 1750.[6] Whatever the actual process it suffices to convince J.M. that petrifaction is an ongoing process. He goes on to argue that:

> All rocks, therefore, where such extraneous bodies are found, seem to be formed from the common sediment of the sea, as sands of several kinds, with the bones of fishes, stalks of sea-weed, and empty shells, which are rolled into beds by the agitation of the waters. These different bodies, thus blended together, are, by the violence of the flux and reflux, banked up towards the shore: which is the cause of the inclination or dipping of the rock. No sooner is one stratum laid, than, by a continual accession of the same, a second is super-induced; and so on successively, till the mass has reached a certain height in water. These loose materials, as soon as the

6 From a modern point of view this was probably case hardening, rather than drying out of clay, with which people would have been familiar.

> vegetation commences, are fastened by a very strong cement, and, as at the sight of Medusa's head, begin to assume the consistency of stone. For the Petrific matter fills up all the interstices, pervades the pores of the solid bodies, and lodges every where the particles that enter its own composition: which seem to be a fixed salt, or else a very powerful astringent, together with a mixture of mineral juices, or metallic ores, which run in small veins, like wire, in several places of the rock.
>
> The shells, being of a close and compact texture, and therefore refusing admission to the grosser parts, seem to have received only the finer parts of the mixture, which has converted them into a transparent substance, something resembling crystal. The sea-weeds, of a more porous nature, have imbibed the whole lapidific matter; which has changed them into a fine white marble, capable of a very high polish. The like may be said of all the other bodies, as they are more rare, or dense in their texture, and fitted to receive more or less of the petrific matter (p. 99-100).

Thus J.M. erects a convincing argument by linking a sedimentary ensemble on the coast and the character of rocks inland.[7] over two hundred feet higher, by means of a petrifying process for which he provides considerable observational detail and anecdotal evidence authenticated not just by the inhabitants of Boulogne, but by a "Scotch gentleman." If Newton's Rule of universality is applied, a distant process observed may be used to explain a local conundrum. Even so:

> The only difficulty in this hypothesis, and what we must endeavour to surmount, is, that we must conceive the sea to be so high, as to cover the hills where such sea bodies are found. So in the present case, we must suppose it to have been above 200 feet higher than it is at present. Now, though neither history nor tradition could assist us in this enquiry, yet still the fact may be ascertained from indelible monuments, and more to be depended on than any human testimony whatever (p. 100).

At this point J.M. comes perilously close to a circular argument for he repeats the evidence of the "petrified sea-bodies ... turned into inland hills" as "undeniable proof" that "our earth hath risen arisen, inch by inch, from the sea." He reveals a uniformitarian outlook by remarking that:

> The age of man bears so small a proportion to the age of the world, that the insensible changes made on the face of nature, pass unobserved. We see so few alterations in our own times, that we conclude too hastily, that there are none at all; or, when the land makes any encroachment in one place, the sea, we imagine, takes her revenge by inundations in another, and in this manner their limits are pretty well secure. But this is a pretty lame account of the matter. For inundations seldom happen, and are but partial; where the recess of the waters is universal, and, like the other great laws of nature, acts incessantly at all times (p. 100).

---

7 The rocks in question are Carboniferous Limestone.

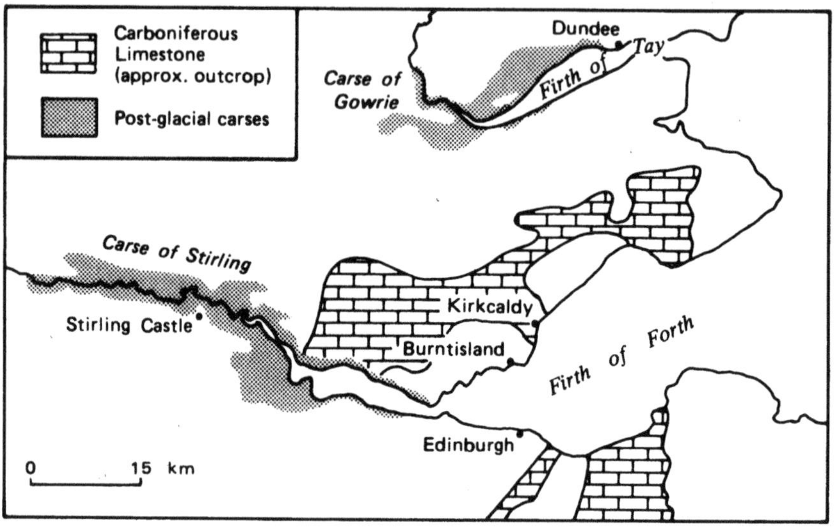

Figure 1

Location map of the area discussed by J.M., with postglacial carses and Carboniferous Limestone outcrops marked.

The recess of the waters is taken as a *universal* law, which rather contradicts the fact that we have been told it was a hypothesis! However, the author redeems himself by now providing evidence that the recess *is* universal by means of "two or three instances out of many, which, with equal facility, may be produced." Three well known examples from the classical world follow: Homer's Pharos, Alexander's Tyre, and Virgil's Lavinium, with Ostia on the Tiber, thrown in for good measure: all of these are examples of Classical ports now miles removed from the Mediterranean. He even speculates that London may have moved down river since Roman times as the sea recessed, but is on surer ground when he moves to local Scottish examples:

> Whoever views the Karse of Falkirk from Stirling castle, will think it extremely probable that champaign country, as the ancients believed of Lower Egypt, has been gained from the sea, by the vast quantity of of sand and mud brought down the river. To confirm this conjecture, whenever the the ground is digged in several places thereabout, they meet vast collections of shells, and other spoils of the sea.[8] A ship's anchor was found, some time since, buried under ground, at two miles distance from the Forth. These two circumstances put it out of all doubt, nor need we any farther proof of the matter (p. 101).

[8] The carses of eastern Scotland within the Midland Valley are all Late Glacial or postglacial raised beaches Sissons (1967).

J.M. goes on to call upon the place name evidence of Burntisland, now connected to the Fife coast by a neck of land, and the additional example provided by the "Karse of Gowrie" in the "frith of Tay," together with additional place name evidence: all as evidence of the sinking of sea level. He concludes that:

> These circumstances make it probable, that the land is continually usurping upon the sea, and it also reconciles us to what follows (p. 102).

What follows is a "tour of Great-Britain, and, by imagining the sea to be two or three hundred feet higher than it is at present" he concludes "that our island is larger, by a third at least, than it was at that time." He dates "that time" as at a time intermediate between the "deluge .. and our own times." The basis of the dating is:

> The bare rocks on our highest hills, {which} shew sufficiently, both the place where they have vegetated, and that for many ages they have borne the violence of that dreadful element: for it is impossible to conceive, that they could have come out of nature's hands in the miserable and ruinous condition in which they appear. Their ragged tops, shattered surfaces, and sifted sides, are the wounds they have received from an obstinate foe; who, tho' vanquished at last has made many furious attacks, and disputed every inch of ground, before he has retreated (p. 103).

Thus the heavily weathered uplands[9] are also attributed to marine erosion, and they are used to to suggest that "ever since the old world was drowned, the waters have fallen equally in equal times, and not faster at one time than another, as is commonly imagined."

Clearly J.M.'s account is fitted to a Mosaic theory of falling waters, but no specific age for the world is mentioned, nor is there any attempt to calculate it. In consequence there is little to hinder a strict uniformity being applied to natural processes operating in time, or an actualist approach being taken to the explanation of indurated limestone upon the basis of the present shore, and a petrifying process imported from an authenticated field site. Further evidence is then sought to support the recess of the waters. To our eyes the Scottish evidence may seem better than that from the Classical World, *because it is nearer*: but Newton's Rules (2) and (3) make no allowance for the possible latent complications of physical geography. Although we have no direct evidence that they were being followed, the remark that "we must, of necessity, ascribe the same origin to such as are more remote from the shore" betrays a powerful philosophical imperative at work.

From the point of view of the development of a geological way of thinking the example is interesting for several reasons. Local evidence is used to construct a chronology, linked locally by a small spatial shift, and then, perhaps *because* of Newton's Rules, these can be, and are, developed into a regional, even a continental perspective. *If* the Rules had explicitly *not* encouraged the *spatial* universality of the *local* conclusions about the 'universal qualities' of bodies, and the same causes for the same 'natural effects,' how might local site descriptions

[9] Now primarily attributed to periglacial conditions during or after glacial periods.

have fared? Who knows, but one can hazard a guess from the example of contemporary 'biology' in which the same idea applied to the species seems to have acted to inhibit ideas of spatial, and particularly temporal, shifts of species characteristics. In an age in which distant parts of the world were expected to provide curiosities of natural history, different from any known (and did), the global ubiquity of geological arguments comes as a remarkable surprise until its source in Newton's Rules is traced.

The paper is interesting too for its attention to detail, and for the way in which an imaginary geographical consequence with temporal implications (a *gedanken* experiment) is developed in such a graphically descriptive way that the compulsive picture itself serves as favourable evidence for the hypotheses under discussion, even though, like some of Newton's own 'experiments,' it may have had no basis in fact: a theme related to those discussed by Dear (1985).

The immediately preceding letter in the *Annual Register* for 1760 was from Robert Dixon, writing from Cockermouth in the Lake District. His letter is much more limited in scope, being, in modern terms, a graphic account of a débris flow resulting from a cloudburst attributed to waterspouts in nearby mountains. In the first instant an excellent geographical account sets the stage (so that it is easily located today on the Ordnance Maps: National Grid NY 155221). I forbear to quote extensively but next the spatial extent of the damage is described as:

> a pit of 2½ in. deep, and of 800 or 1000 yards in area. Several other pits, it is thought, were made, and afterwards filled up again with stones and sand, otherwise it is difficult to imagine how the vast quantities of stone, which composed the walls nigh the brook, should have disappeared (p. 96).

In addition one field was swept clear of débris down to bedrock, and another was entirely buried in a sand bank. Two meadows further down suffered a similar fate. The change in channel dimensions is described as a change :

> of the old channel, which did not exceed five or six feet in breadth, and one in depth, {to} a new one..at least 18 or 20 yards in breadth, and 1½ deep (p. 96).

The author:

> endeavoured, but in vain, to get *data* sufficient on which to build a calculation of the quantity of water which came down; for as it happened at midnight, neither the time of its continuance could be ascertained, nor could it be determined, whether it was constant and regular, or variable (p. 97).

In addition to noting various other measurements of flow depth he makes the perceptive observation that when the water fell into the River Cocker:

> Here was put an end to its fury: for though the channel of the river was far from being capacious enough to receive the whole of the water, yet, on account of the vast level plain on either side, its overflowings were innocent, as it could only deluge to be stagnant (p. 97).

In Dixon's account we see again the careful word picture building a convincing geographical account, sound inferences from the observable facts of destruction which are substantiated by precise measurements, and a frank confession of his failure to acquire some crucial *data* seen to be desirable for a complete description. As to the physical cause of the storm he is willing to leave it to "the adepts in natural history" being himself "distant from the seats of science," but he is not afraid to tell them that both the existing theories are inadequate in the light of local observations!

The account is also interesting because as a phenomena it can be classed as a British example of a "torrent" - whose behaviour is seen to be distinct from ordinary river flow. They were well known in the Alps where they frequently gave rise to life threatening débris flows, and they were one of the agencies later regarded by Hutton (1795) and Playfair (1802) as adequate for the transport of the very large boulders now termed *erratics* (Tinkler 1983). The common characteristics of torrents in mountain districts[10] were additional evidence in support of rules (2) and (3) acting in concert during the actualistic operation of the earth's surface.

In a rudimentary form it provided evidence of depositional sedimentary processes at work, and intimately connected with the operation of the land surface in an explicitly geographical context. It gains considerable significance from the fact that, as a contribution to knowledge, it is a very minor work whereas at the same time it demonstrates a close adherence to contemporary standards of scientific work by an 'ordinary' man of science.

John Gough of Kendal and the defacement of lakes

John Gough (1757-1825) was blind from early childhood but developed into a remarkable scientist - he taught both John Dalton and William Whewell. On June 16th 1790 he completed a paper on "REASONS *for supposing that* LAKES *have been more numerous than they are at present; with an Attempt to assign Causes whereby they have been defaced.*"

The paper (Gough 1793) has a closely woven argument well supported by accurate observations in natural history. He begins with the fact that is almost, so late in the century, a given: that most strata is formed on the sea bed, and is then uplifted to form land, in the process of which "if whole continents are torn up, and have their continuity every where broken" .... "Why, then are Lakes[11] so few in number?"

> This view of the subject certainly offers a formidable objection to the received theory of the formation of land, which ought to be attended to (p. 2).

Gough then attends to the business of repairing this conceptual damage to the "received theory," although which theory he has in mind he does not mention.

---

10 See below for Gough on torrents.

11 Davies (1969) has discussed the 'limnological objection' which the Huttonian theory faced. However, Huttonians had the problem of explaining how the lake basins came to exist in a theory committed to non-catastrophic processes. They would have agreed with Hough on the means by which the basins were eventually filled.

He notes that in the case of mountains "the resources of Nature" allow a "gradual progression from a rude to a more perfect state" by means of "slight but incessant causes" and he presumes that "that analogy of conduct and design, which pervades the whole visible system of things, at least authorizes the supposition" that the same is true for lakes. From an obviously detailed knowledge of the bio-geographical processes involved in the gradual infill of lakes and ponds Gough:

> *first,* enquire{s} what means are in the possession of Nature for producing such a revolution; (p. 3).

Then noting that "this method of converting a pond into land evidently points out a process that would diminish the inequalities of a disordered continent" he:

> *then,* endeavour{s} to discover, whether any proofs of such alterations having taken place are still extant. --- This method of proceeding being best calculated, either, to to remove the objection, or to establish it (p. 3).[12]

The theory (of pond infilling) which he has established by deduction and by applying the principle of "*Vegetation* .. a favorite process with nature" is then tested at great length against "what has passed in the world." The flat mårshes of north Europe provide support by the nature of their "peculiar soil whereby (they) are distinguished from the surrounding land" and he indulges in the *gedanken* experiment, that if all this soil could be removed from them "the cavity left after the operation would soon be converted into a lake." In addition, he notes that in "several marshes, very deep ponds are still to be met with", in some case "more or less concealed by a thin crust of aquatic plants."

Peat is discussed at length. The internal structure is "discovered to consist principally of flexible, branched fibres, variously interwoven, and twisted together", and of course peat is known to be combustible, leaving "a quantity of fixed Alkali; which is rarely, however, pure, or free from mineral salts." Gough states that impurities are due to the decomposition of leaves floating in the pond and the admixture of other vegetable débris floated in by streams so that:

> the extraneous matter of a swamp perfectly resembles the refuse of a river flowing through a woody country (p. 9).

From all this he deduces that one can work out what would now amount to the regional ecology by studying the remains in swamps, even to noting that frequently, embedded logs:

> all lie nearly in one direction, and are confined to a particular part of the marsh. {Implying that} the wind, which, in sheltered places, can only blow in certain directions {was responsible for the fact} that they have sunk in such a position, that the direction of the prevailing wind is commonly pointed out by the direction of their branches. The foregoing facts seem sufficient to shew how well the theory is supported by the evidence of nature (p. 10).

---

12 Note the echo of the last phrase in Rule 4.

There is no reason to disagree with Gough; he even suggests using the contents of swamps at different elevations on mountain sides so that an observing person:

> easily determine to what *height* the hills of this land have been anciently covered with *wood*. ... in some of these, which are more elevated, no tress are found; but I know a small one between two and three hundred yards above the level of the surrounding country, which abounds in *Birch*, and have been informed of another where *Fir* is plentiful (p. 11).

He is careful to note too, that sloping beds of peat on hillsides are different in structure to those of swamps. I quote these remarks so extensively to demonstrate the level of detail that Gough is able to provide in establishing the natural ecological chronology of lakes. Satisfied that the *process* of lake filling is now well tested against the deductive theory:

> it may be safely taken for granted, that the marshes of every country are similar to those of the north of England. Hence it follows that *Lakes* have once existed in every part of our Globe; and that they have been defaced by the same causes which have produced like effects in this part of the world. But nature is not confined to the process described above. She has other resources (p. 2).

Once more it is hard *not* to see the Newton's rules being very specifically applied. Rule (1) is subsumed by the combined operation of deductive theory and confirming evidence (which I have quoted at some length); then Rules (3) and (2) are virtually paraphrased, and the logical "Hence it follows" is flawed only from the geological viewpoint of the early nineteenth century and afterwards. Gough was a mathematician; he doubtless knew his Newton, and he would not fall prey to elementary logical flaws unwittingly. He *must* have truly believed in universality, and his authority is not likely to have been less than Newton. Finally rule (4) gives primacy to induction from the evidence.

We might well quarrel with Gough at this point. There is a strong suspicion that what he knew about lakes (which abound in the Lake District where he lived!) led him into the whole train of thinking. He induced a proposition about an essentially *temporal* principle (*vegetation*) from (*spatial*) evidence known to him, and then tested its temporal consequences in a primarily *spatial* way against evidence elsewhere. In essence he is practicing a form of empirical hypothetico-deductive reasoning which was in disrepute amongst philosophers at the time.[13] The flaw lay in the assumption that *all* landscape was initially all *equally* chaotic: but then he could invoke universality (a powerfully supported principle in view of the success of Universal Gravitation) to assert that this must have been the case, given that the broken strata is universally marine and disturbed.

Gough is not finished. He next shows that weathering and erosion in mountainous regions enables torrents to demolish:

---

13 See Laudan (1973).

> the hardest *strata* to a depth scarcely to be credited. Now it is evident, that, where the same causes have been applied to the bank of a lake, they could not fail in producing like effects as when exerted on the side of a hill. It may, therefore, be taken for granted, that the outlet of every lake has been more or less injured in the manner described above. This conclusion being admitted, the following consequences must immediately be assented to: because they depend on the simplest laws of Hydro-statics (p. 14).

Again we see the application of rules (2) and (3). Gough goes on to show that, as a consequence lakes which have filled up are cut through by streams which by erosion and deposition achieve:

> a gentle declivity, down which the current will glide, no longer capable of disturbing the impediments lying in its way: consequently the form and dimensions of its bed will become permanent (p. 15).

However, rules (2) and (3) now produce theoretical trouble, for:

> after these destructive operations have ceased of themselves, there is reason to apprehend, from what has been discovered by Philosophers, that the same process would be continued with equal certainty, though not with equal effect, for the constituent particles of water are sufficiently hard to abrade the surfaces of very compact bodies; it having been proved by experiment, that the cohesion of glass itself is not strong enough to resist their actions (p. 15).

Gough cannot ignore the experimental evidence, and the logic that:

> the bed of a river undergoes insensible changes from the friction of its own stream, after it ceases to be exposed to the more manifest ravages of a torrent[14] p. 16).

The solution lies in a "simple preventive" in the "œconomy of nature":

> These singular productions of the vegetable kingdom {which} are enabled by some peculiarities of their constitution to bear the friction of the stream without receiving the least injury (p. 16).

He names the species in question, *conserva rivularis* and different species of *tremella,* and notes how they cover over all the crevices, prevent the least grains from moving, and in short form "a sort of cement. Thus is permanence given to the course of every river by this provision."

Gough therefore establishes a state of landscape stasis by continual recourse to the rules of reasoning and a very detailed appeal to actualist processes.[15] He concludes by quoting evidence from the north of England which he hopes "is not contradictory to the experience of any one, who has had an

---

14 Thus Gough sees torrents as a chronologically prior state of rivers: I am not sure to what extent this was general thinking.

15 The power and clarity of Gough's argument establish him as a type of eighteenth century John Hack, a whiggish comment that I hope will be forgiven in the circumstances of this symposium!

opportunity to make similar observations." You will have take my word for it that this evidence is interpreted with characteristic astuteness, informed by a combined recourse to process and form.

I am convinced that a close reading of Gough leads to similar conclusions to those I reached over J.M.. The rules of reasoning are being followed assiduously, they are never far below the surface. Given that discussions of surface events have an intrinsic geographical component by virtue of physical continuity quite independent of any prevailing theory, then the rules of universality, and of same causes = same effects, induce an inevitable regional, indeed global, unity. It is of interest that matters of rocks, lakes, hillslopes, rivers and sea shores are being taken as bodies; as entities endowed with qualities taken to be universal once established. Why not? The Newtonian system of mechanics applied to the solar system, to the observable universe, where the bodies in question were planets, comets, and stars, and on earth, small falling bodies, mountains, and the oceans: not as we are liable to think today: quarks, atoms or molecules. It seems quite realistic then to similarly endow inert bodies on earth with experimentally established 'qualities,' and to follow identical modes of reasoning pertaining to them. When the same principles are applied to plants and animal the results are very different, for species are often clearly regionalised, contrary to universality, yet universality applied to time prevents species transformation.

James Rennell, R.H. Colebrooke and the Indo-Gangetic plains

In 1767 James Rennell (1742-1830),[16] who had risen from being a midshipman in the navy, with a developing talent for hydrographic surveying, to a Captain for the East India Company, became Surveyor-General for the Company. He had already completed several maps at varying scales, at least one as large as 3 miles to the inch, and when he finally left India in 1777 the excellence of his maps was already well known in India, and his reputation became international in 1779 with the publication of the Bengal Atlas (Rennell 1779). On his return to England he settled in London, joined the Royal Society, and in 1781 published his "*An Account of the* Ganges *and* Burrampooter *Rivers*" in the *Philosophical Transactions.*

Although this title sounds as though it might constitute an explorer's account, it is in reality an accurate account of the physical behaviour of the rivers. It is well to bear in mind that sixteen years *later* John Robison writing the article entitled *Rivers* in the 3rd edition of the *Encyclopædia Britannica* was moved to remark with respect to theoretical understanding of the motion of "water in a bed or conduit of any kind, that":

> Although the first geniuses of Europe have for this century past turned much of their attention to this subject, we are almost ignorant of the *general laws* which may be observed in their motions.... As to the uniform course of the streams which water the face of the earth, and the maxims which will

[16] Rennell's life is conveniently summarised in Markham (1895), and some additional evaluation of his work as a geographer is to be found in Baker (1963). A brief biobibliography is in Downes (1977).

> certainly regulate this according to our whims, we are in a manner totally ignorant....Who can say what form will cause, or will prevent, the undermining of banks, the forming of elbows, the pooling of the bed, of the deposition of sands? Yet these are the most important questions. (Robison 1797, p. 46).

Robison was setting up no straw target, hydrodynamics was in dire straits at a time when rational mechanics was in its hey-day (Bos 1980). This is perhaps exemplified by the fact that in 1776 the Mantua Royal Academy of Sciences and Letters set the problem "to indicate the best and the cheapest method of freeing navigable canals from the banks of sand and earth formed in their beds which render them too shallow" (quoted in Mann 1779 p. 598). In consequence Rennell had no theory upon which to call, and the only two scientific authorities whom he cites are Buffon and la Condamine. Despite, or perhaps because of these theoretical deficiencies, Rennell gives a convincing account of what we would call now the fluvial geomorphology of the Ganges valley. He quotes the gradient of the river in low stage, and in flood stage, the velocity of the stream, and the rate of the progress of the flood peak. He refers to:

> an experiment of my own on record in which my boat was carried 56 miles in eight hours; and that against so strong a wind, that the boat had evidently no progressive motion through the water (p. 94).

It might be thought that Rennell's recent entry to the Royal Society sensitised him to the need to adapt his language to that of formal science in making reference to his experiment. However, elsewhere in the paper he often speaks of deliberately made field observations[17] as "experiments," as for example when he notes that:

> few people have made experiments on the heights to which the periodical flood rises in different places ...{and that it is} a fact, confirmed by repeated experiments, that from the place where the tide commences, to the sea, the height of the periodical increases diminishes gradually, until it totally disappears at the point of confluence...(p. 109-110).

and when he remarks that:

> The quantity of water discharged by the Ganges, in one second of time, during the dry season, is 80,000 cubic feet: but in the place where the experiment was made, the river, when full, has thrice the volume of water in it, and its motion is accelerated in the ratio 5 to 3: so that the quantity discharged in a second at that season is 405,000 cubic feet. If we take a medium the whole year through, it will be nearly 180,000 cubic feet in a second (p. 110).

---

[17] Dr. Johnson's *Dictionary* (1755): ***experiment*** - Trial of anything; something done in order to discover an uncertain or unknown effect.

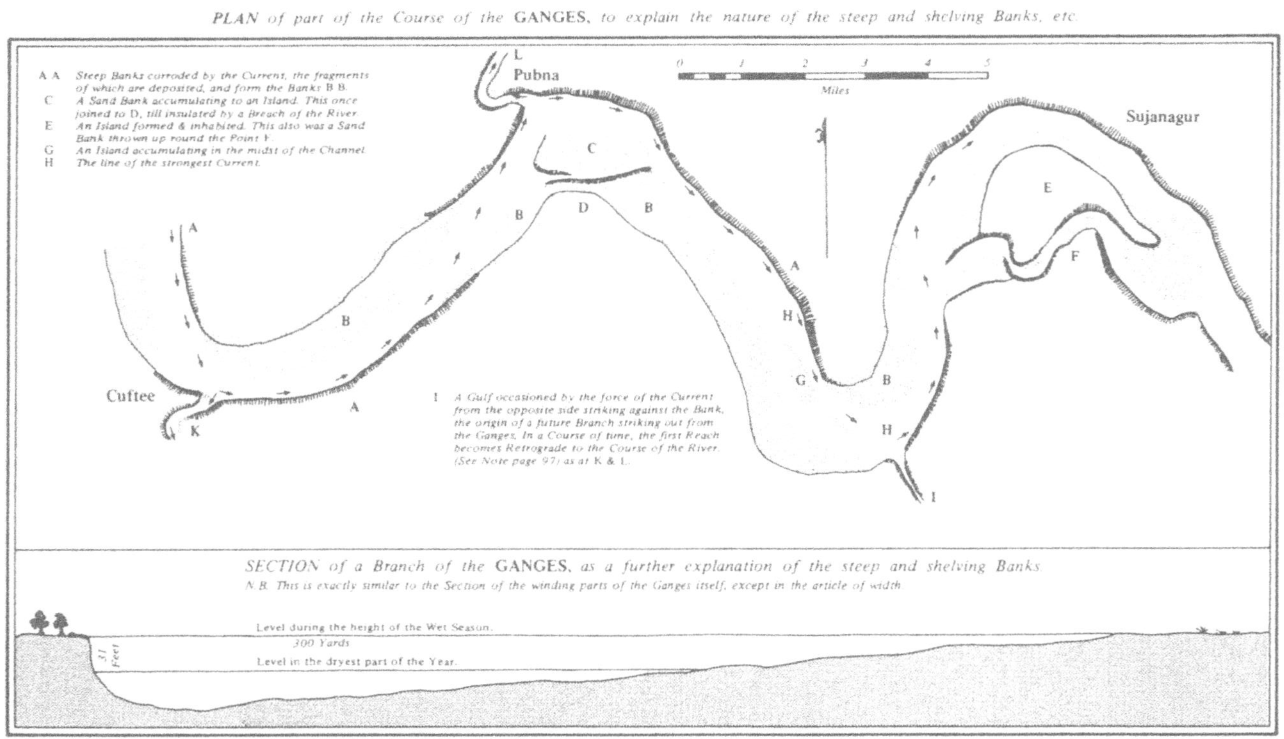

Figure 2

Rennell's map of the Ganges to illustrate meander behaviour (redrawn for clarity) . Note that the section, showing a vertical range of 31 feet in water level, is only for a branch of the Ganges, for the main channel see Figure 3. The scale is greatly reduced from the original.

There are several other examples too, which without doubt indicate a careful consideration of empirical findings, precisely and deliberately made. As an illustration he provides a survey of the Ganges in the area around Pubna "to explain the nature of the steep and shelving Banks, etc." The map, together with a cross section, is published in the *Philosophical Transactions* at scale of almost one inch to a mile, ample to illustrate a river whose width is 1.5 miles (Figure 2). He notes that declivity supplies little to velocity since velocity changes so slightly with the stage, and declivity remains almost constant. Instead he attributes the increase to the deepening of the water profile and the impetus supplied by adventitious waters pouring in to the main channel.

He describes the form and process of meandering with great precision, and from a sure knowledge of the changing pattern of the river which he had surveyed so accurately, he could determine that:

> During the eleven years of my residence in Bengal, the outlet or head of the Jellinghy River was gradually removed three quarters of a mile farther down: and by two surveys of a part of the adjacent bank of the Ganges, taken about the distance of nine years from each other, it appeared that the breadth of an English mile and a half had been taken away. This is, however, the most rapid change that I have noticed; a mile in ten or twelve years being the usual rate of incroachment, in places where the current strikes with the greatest force, namely, where two adjacent reaches approach nearest to a right angle (p. 96-97).

Rennell speaks specifically of "a meander"[18] and he attributes meandering in general to two causes (i) irregularity of the original surface and (ii) "the looseness of the soil, which yields to the friction of the border of the stream." For the former case the meanders "are, of course, as digressive and irregular as the surface" but in the latter case he notes that:

> they are so reducible to rule, that rivers of unequal bulk will, under similar circumstances, take a circuit to wind in, whose extent is in proportion to their respective widths: for I have noticed that when a branch of the Ganges is fallen so low as to occupy only a part of its bed, it no longer continues in the line of the old course; but works itself a new channel, which winds from side to side across the former one...(p. 97-98).
>
> The windings of the Ganges in the plains are, doubtless, owing to the looseness of the soil: and (I think) the proof of it is, that they are perpetually changing; which those, originally induced by an inequality of the surface, can seldom, or never do (p. 98).

Without question this is sound inductive reasoning,[19] and in other passages he describes the conservation of the channel form and dimensions as it shifts in space "by the mere operation of the stream," and the way in which sediment eroded from the outer bank goes to form the shelving bank immediately downstream at the next bend, and on the same side. Each meander tends to

---

[18] 'Meander' as a noun is noted in Dr. Johnson's *Dictionary*, 1755.

[19] Playfair compliments "the acuteness and accuracy of that excellent geographer" in discussing Rennell's remarks on the Ganges Delta, and Kirwan's mis-interpretation of them. I think, with universality and justice, we might apply the compliment to the entire paper (Playfair 1802, p. 429).

deviate further from its course until "till either the opposite bays meet, or the stream breaks through the narrow isthmus, and restores a temporary straightness to the channel." Because islands in the Ganges form and are destroyed well within the lifetime of man "accordingly, the laws respecting alluvion are ascertained with great precision" (his footnote, p. 101). In consequence he does not hesitate to go a step further:

> I can easily suppose, that if the Ganges was turned into a straight canal, cut through the ground it now traverses in the most winding parts of its course, its straightness would be short duration. Some yielding part of the bank, or that which happened to be the most strongly acted on, would first be corroded, or dissolved: thus a bay or cavity would be formed in the side of the bank. This begets an inflection of the current, which, falling obliquely on the side of the bay, corrodes it incessantly. When the current has passed the innermost part of the bay, it receives a new direction, and is thrown obliquely towards the opposite side of the canal, depositing in its way the matter excavated from the bay, and which begins to form a shallow or bank contiguous to the border of the canal. Here then is the origin of such windings as owe their existence to the nature of the soil. The bay, so corroded, in time becomes large enough to give a new direction to the body of the canal: and the matter excavated from the bay is so disposed as to assist in throwing the current against the opposite bank, where a process, similar to that I have been describing, will be begun (p. 98).

I quote this at length for two reasons. Primarily it is another example of a *gedanken* experiment.[20] It calls upon the geographical imagination: the temporal development of a spatial pattern behaving according to known laws, and by implication (and from common experience) it has universal application within the purview of the Ganges. Secondarily, it is remarkable, from a modern perspective, that neither this passage, nor indeed any of those concerned with the details of meandering (including the map), are cited by subsequent work, even those that cite the paper. The *gedanken* experiment is also used in a *reductio ad absurdum* situation when Rennell wishes to justify his views on the reduction of the vertical range in the periodical flood (and the exposed banks of the Ganges) as it approaches the ocean:

> Could we suppose, for a moment, that the increased column of water, of 31 feet perpendicular {which is the vertical range of the flood away from tidal influences}, was continued all the way to the sea, by some preternatural agency: whenever that agency was removed, the head of the column would diffuse itself over the Ocean, and the remaining parts would follow, from as far back as the influence of the Ocean extended; forming a slope, whose perpendicular height would be 31 feet. This is the precise state in which we find it. At the point of junction with the sea, the height is the same in both seasons at equal times of the tide (p. 108).

In a footnote to this passage Rennell chides M. de Condamine for failing to note the state of the river "at the time of the experiment" when he

[20] A version of this 'experiment' is now commonplace in the laboratory, and has also been reported in the field, see Leopold *et al.* 1964.

reported tides in the Amazon at an elevation of 90 feet, 600 miles inland. For his own part he is willing to extend his logic:

> Similar circumstances take in the Jellinghy, Hoogly, and Burrampooter Rivers; and, I suppose, in all others subject either to periodical or occasional swellings (p. 109).

This is universality, but it is less confidently assertive than that of J.M. and Gough, although he obviously harboured no doubts about the "laws respecting alluvion."

Rennell provides equivalent detail about the Ganges Delta:

> as a strong presumptive proof of the wandering of the Ganges from one side of the Delta to the other, I must observe, that there is no appearance of *virgin* earth between the Tiperah Hills on the east, and the province of Burdwan on the west; nor on the the north till we arrive at Dacca and Bauleah. In all the sections of the numerous creeks and rivers in the Delta, nothing appears but sand and black mould in regular strata, till we arrive at the clay that forms the lowest part of their beds. There is no substance so coarse as gravel either in the Delta or nearer the sea than 400 miles (at Oudanulla).... but out of the vicinity of the great rivers the soil is either red, yellow, or of a deep brown (p. 102).

Once more we have the careful geographical description, which can be coupled with earlier remarks that the area has a uniform slope with a scarcely detectable declivity of about:

> nine inches descent in each mile, reckoning in a straight line, and allowance made for the curvature of the earth. But the windings of the river were so great, as to reduce the declivity on which the water ran, to less than four inches *per* mile: and by a comparison of the velocity of the stream at the place of the experiment with other places, I have no reason to suppose, that its general descent exceeds it (p. 93).

The quantitative data is carefully related to the description and the whole generalised.[21] Rennell also describes the progress of the annual flood, and the way in which the particles in suspension in the flood waters give rise to:

> several strata of sand and earths, lying above one another in the order in which they decrease in gravity.....I have counted seven distinct strata in a section of one of these islands. Indeed, not only the islands, but most of the river banks wear the same appearance (p. 102).

It is as a footnote to this passage that he gives the detail that "a glass of water taken out of the Ganges, when at its height, yields about one part in four of mud." This remark, and indeed all the Delta passages, were cited frequently in the following century (Bakewell 1813, and see Baker 1963), and that, taken with the passage quoted about the regular strata, and the implications of its thickness

---

21 Especially when consulted in the context of Rennell's *Atlas of Bengal* (1781) to which the paper was later added as an appendix.

(from the vertical range of the flow) and its geographical extent, excited the interest of Hutton, and later, Lyell. Hutton uses all this information as follows:

> The description now given is from the rivers of this country, where it is not unfrequent {sic} to see relicts of three or four different haughs[22] which had occupied the same spot of ground upon different levels, consequently which had been formed and destroyed at different periods of time, but the same operation is transacted every where; it is seen upon the plains of Indostan, as in the haughs of Scotland; the Ganges operates upon its banks, and is employed changing its beds continually as well as the Tweed {at this point Hutton cites Rennell's paper}. The great city of Babylon was built upon the haugh of a river. What is become of that city? nothing remains;-- even the place, in which it stood, is not known (Hutton 1795 II - 210-11).

The Gangetic case was a perfect example of the type Hutton needed. It implied huge amounts of regular strata uniformly accumulated, and deposited as a consequence of the existence of huge mountains from which the sediment obviously came. Playfair abstracts the same passage more briefly in the *Illustrations* (Playfair 1802, p. 103, §100), and the reference to Rennell is omitted. Hutton certainly applied the universal rule to this observation, and the existence, in an undoubted fluvial environment, of such vast spreads of sediment was of interest to both Hutton and Playfair as a helpful analogy in explaining similar spreads of sand and gravel to be found in Great Britain; many of which contained erratics derived from distant mountain sources.[23] In a uniformitarian, and universal, interpretation such deposits could be attributed to river deposition at a higher level (when the mountain sources were more elevated), which have since been cut through as the rivers have lowered themselves into the landscape by incessant, if slow, erosion.

Twenty years later, Major R.H.Colebrooke (1762-1808) published a paper entitled "*On the Course of the* GANGES, *through* BENGAL" (Colebrooke, 1801) since he supposes that even after Rennell's paper "additional remarks .. on so curious a subject, will not be thought superfluous nor uninteresting." Colebrooke re-iterates many of Rennell's themes, and adds some historical depth based on an additional twenty years of precise surveying, especially his re-survey of the Ganges in 1796 (which could be compared in parts to Rennell's map of 1764), and his own considerable knowledge. In particular, he details the enormous rapidity with which new islands develop, and others are consumed, and he adds more detail about the sedimentary wedge first described by Rennell (Figure 2). Colebrooke finds in one instance that, allowing for the depth of the river and the height of its banks (Figure 3) "that a column of 114 feet of earth had here been removed by the stream" (p. 12). He describes the growth of an island (the island of Sundeepa, *see* Figure 4) in the river from nothing, to eight miles long and two miles wide within twenty years, and estimates:

> The quantity of land which has been here destroyed by the river, in the course of a few years, will amount, upon the most moderate calculation, to

---

22 *haugh*: Scots and north of England - flat alluvial land in a river valley; floodplain, and river terrace.

23 Playfair (1802 NOTE XVII §105, p. 371 et seq.) discusses the various origins of the "*Remains of Decomposed Rocks*" and in the succeeding note he discusses how they may be transported (NOTE XVIII, § 112, p. 381)

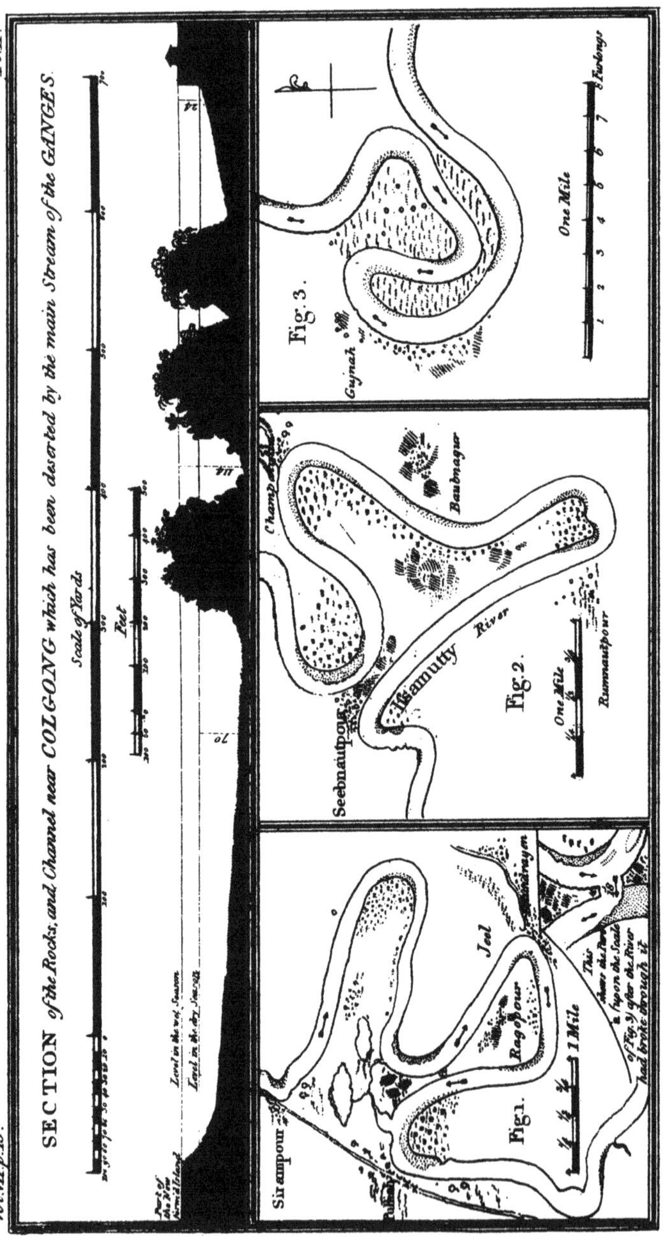

Figure 3

Extract from Plate II in Colebrooke illustrating extremely contorted meander bends (very different in character to meandering on the Ganges itself), an actual breakthrough (lower left detail), and a deserted branch of the Ganges (upper section) indicating the vertical wedge (114 feet) of sediment it was building by migrating. (Reproduced at the original scale)

> 40 square miles, or 25,600 acres; but this is counterbalanced, in a great measure, by the alluvion which has taken place on the opposite shore, and by the new island of Sundeepa, which last alone contains upwards of 10 square miles (p. 14).

In essence Colebrooke confirms Rennell's picture of the Delta, but his word picture is even more explicit:

> The GANGES, in its course through Bengal, may be said to have under its dominion a considerable portion of the flat country; for not only the channel which, at any given time, contains the principal body of its waters, but also as much of the land, on each side, as is comprehended within its collateral branches, is liable to inundation or to be destroyed by encroachments of the stream, may be considered as belonging to the river. We must of course, include any track, or old channel, through which it had formerly run, and into which there is any probability of its ever returning again; (p. 21)....
>
> If corresponding sections of the bed of the river and neighbouring ground were represented, it would probably appear, that all the land is disposed in regular strata; whence we might with certainty conclude, that the whole had been at some former period deposited by the stream.
>
> The strata in general consist of clay, sand, and vegetable earth the latter of which is always uppermost, except when in some extraordinary high flood, a new layer of land is again deposited over it, by which means the ground becomes barren, or is at least materially injured.

In these summary paragraphs Colebrooke conjures up the regional picture well supported by the evidence he has cited in great detail, and by means of a *gedanken* experiment, adds to it a significant vertical component: the horizontal strata vertically stacked and with 'obvious' regional continuity. It was this aspect of both Rennell's and Colebrooke's work which subsequent geologists seized upon. It was an example of the contemporary construction of strata on a sufficiently large scale to have spatial significance for the explanation of similar strata elsewhere. Colebrooke, in addition to excellent maps (Figures 3 and 4), had more remarks to make upon the meandering behaviour of the smaller streams, compared to the main channel of the Ganges, but as with Rennell's remarks on the mechanics of meanders, little, if any use was made of them by subsequent writers.

Lyell (1830), in the first volume of his *Principles of Geology*, based a considerable part of chapter XIV, "*Oceanic Deltas*," on the papers by Rennell and Colebrooke. He put great emphasis on the quantity of sediment in the river water, even to the extent of using Rennell's estimate, quoted above, as a basis for establishing a common measure of denudation (or indeed renovation) using the Great Pyramid of Egypt (Lyell 1830, p. 248). By this means he quantified Rennell's remark that "Next to earthquakes, perhaps the floods of tropical rivers produce the quickest alterations in the face of our globe." He also laid considerable stress upon the "*Grouping of Strata in Deltas*'" (p. 249), again basing himself on the Ganges example, amongst others, because "the same phenomena is exhibited in older strata of all ages" (Lyell 1830, p. 255). He returned to the Ganges in Volume II (1832, p. 204) when he remarked that:

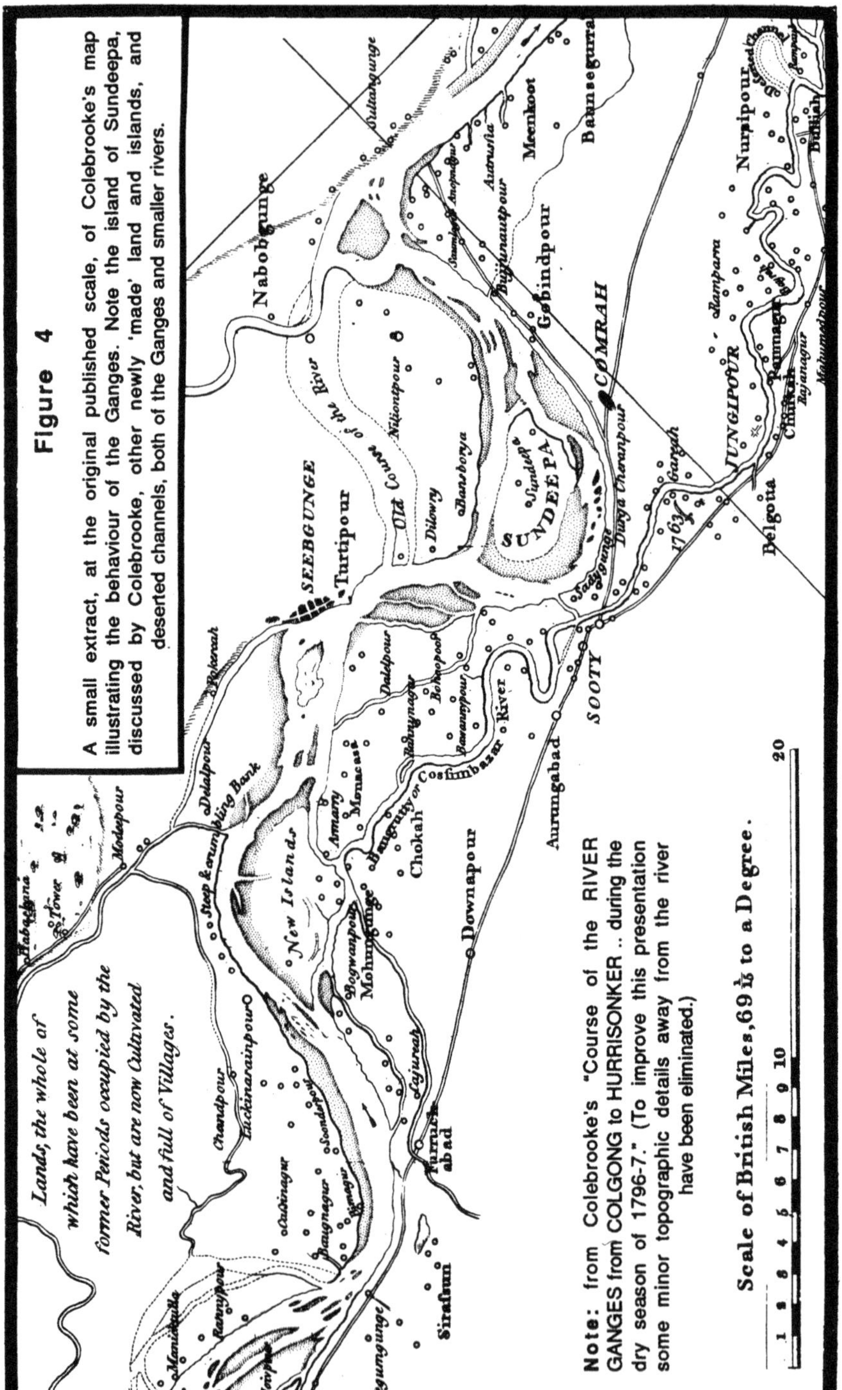

Figure 4

A small extract, at the original published scale, of Colebrooke's map illustrating the behaviour of the Ganges. Note the island of Sundeepa, discussed by Colebrooke, other newly 'made' land and islands, and deserted channels, both of the Ganges and smaller rivers.

Note: from Colebrooke's "Course of the RIVER GANGES from COLGONG to HURRISONKER .. during the dry season of 1796-7." (To improve this presentation some minor topographic details away from the river have been eliminated.)

> To the geologist, the Gangetic islands, and their migratory colonies, may represent an epitome of the globe as tenanted by man.

This introduces a final element common to Rennell's and Colebrooke's account: the theme of teleology, which permeates most eighteenth century writing. Of the two of them, Colebrooke lays more stress upon it than Rennell (writing for the Royal Society) who merely noted with reference to the mud supplied to the floodplain that "it is indeed a part of the œconomy which nature observes in fertilizing the lands in general" (p. 101). Colebrooke uses the "bounty of Providence" to explain the differences between the Ganges and the smaller streams:

> for had the great rivers been decreed to wander like the smaller, they would have encroached too much on the land; while the current being considerably retarded, would have rendered them more liable to overflow their banks, and less able to drain the smaller streams, and low grounds, of the superabundance of waters in high floods (p. 24).

However, this did not prevent Colebrooke from explaining the diurnal and annual workings of riverine processes in considerable detail in subsequent pages and we may conclude that the design hypothesis, while it provided intellectual comfort, did not interfere with the practical interests of man.

Description, data and the imagination: universality and time

The five texts that I have examined range from 1760 to 1801, and they obviously vary in character, yet they have some common threads that serve to characterise eighteenth century writings in 'geology.' I encountered them all accidentally, and in the first instance I found them intrinsically interesting in their own right: for what they had to say about 'geology.' In the course of analysing their methodological logic it seemed that despite their disparate scientific origins they shared a common attitude towards reasoning about the world.

All of them (apart from Gough) contain a thorough regional geographical description of the area in question, although usually I have not quoted it at length. Gough for his part provides an imaginary description of lake basins, and the processes which lead to their being filled, although this is undoubtedly based on his extensive knowledge of all types of lakes in the English Lake District. In addition all of the writers use *gedanken*[24] situations in a geographical context, and to a remarkable extent, although in Dixon's letter this is restricted to inferences about the depth of the flood, and the degree to which he presumed that débris had been eroded and re-deposited.

In all the papers the geographical component is eventually tied to a vertical component (over and beyond the variation in height of the land): a

---

[24] The tradition of *gedanken* situations in 'geological' thinking certainly goes as far back as Herodotus' famous speculation on the twenty thousand years it might take to fill up the Red Sea if the Nile were to be turned into it.

sedimentary unit[25] whose origin is carefully linked to actualistic processes observed by the author. In Dixon's paper this is merely a débris fan (in modern parlance) from an overnight cloudburst; in Colebrooke's (amplifying Rennell) it is a wedge of sediment up to 100 feet thick extending (in the imagination by virtue of the *gedanken* description) over the entire range of the Ganges and Burrampooter Delta. The authors are all meticulous in their description of the processes connecting the sedimentary unit and its origin in an actualistic process observed. Gough (in a passage I have not quoted, p. 18-19) is careful to show that the sands and gravels observed by him within existing infilled lake basins contain no shells at all, even in the fine sand beds, which is consistent with his reasoning that they have been filled by subærial, not marine, processes.

There is a common concern with precision in measurement: J.M. notes that some 'flakes of stone are all reflected from the perpendicular, and make a very obtuse angle with the horizontal, not under 120 degrees as nearly as I can guess.' I have quoted at length the other authors on this matter and although Gough is lightest on merely numerical information, it *is* used to make precise points, and he handles essentially quantitative information with great verbal skill. As is inevitable in a biogeographical subject, there is a wealth of detail on the character of the vegetation.

In the cases of Rennell and Colebrooke quantitative measurement reveals itself in a new way: in the precise survey, the accurate map, and in both their cases by re-survey after a period of years (extending over thirty years in all). Given the rapidity with which the Ganges moved (about two hundred yards a year) it is inevitable that an observant man would have discerned the main elements of what Rennell described. However, there can be no doubt that the quality of his maps added authority to his findings. With the possible exception of hydrographic charts, town plans and private estate maps there was nothing comparable to Rennell's maps in Britain until well into the nineteenth century. In any case, only on the coast were physical changes fast enough to be noticed on large scale maps in Britain. Playfair, whose library contained copies of Rennell's maps of India, must have had Rennell and Colebrooke in mind (as well as the newly formed Ordnance Survey) when he wrote in the *Illustrations* (1802, #134) that from the:

> advancement ... of other sciences, less directly connected with the natural history of the earth, much information may be received. The accurate geographical maps and surveys which are now in the making; the soundings; the observation of currents; barometric measurements, may all combine to ascertain the reality and to fix the quantity of those changes which terrestrial bodies continually undergo. Every new improvement in science affects the means of delineating more accurately the face of nature which now exists, and of transmitting, to future ages, an account, which may be compared with the face of nature as it shall then exist (p. 140).

Playfair placed this in the concluding paragraph of his précis of Hutton's theory, acutely aware of Hutton's inability to find or use much information of this type to measure rates of decay, or to estimate the age of this

25 My use of *unit* is roughly that of Krumbein and Sloss (1963) p. 332.

world (i.e the current cycle). Playfair himself was no armchair critic: he surveyed Schihallien (1811a, although the work was done much earlier) and Glen Tilt (reported in Webb-Seymour 1815), and there is extant a letter in which he regrets being unable to meet an instrument maker in Edinburgh due to a prior appointment, but sends two substitutes (Playfair 1813).

In all the papers (except Dixon, who can be forgiven) there is the mention of experiment, or the calling to mind of experiences. J.M. cites the evidence from Boulogne, Gough cites the wearing of glass by water (from an unnamed source), Dixon wished to make measurements, and Rennell and Colebrooke made repeated 'experiments' by making surveys and measuring the state of the river (and recall that Rennell's work was done in the period 1763-1777). In addition of course any deliberate collection of field data at field sites might also be regarded as an 'experiment,' and this seems to be the sense in which Rennell uses the word.

Finally, to hold all these components together there is the application of methodological rules: Newton's Rules. I believe I have shown in the case of each paper that the Rules were being followed assiduously, and their impact becomes apparent when they are applied to problems which have an intrinsic geographical component. Of these the one with the most powerful consequences is the principle of universality (Rule 2 and 3 acting together). Universality converts the verbal regional image into a solid three-dimensional reality: the image, and the reality, are not yet deep (vertically speaking), but they are there. It is no accident. The steps which have characterised the construction of each of these papers fit quite neatly into the framework developed in the early years of the Royal Society and analysed by Dear (1985), given that it had to be done 'in the field.' The careful establishment of the convincing setting; the calling upon experiential evidence (appropriately 'authenticated' in Gough's case), the description of experiments done, of observations made, the quantification of data, and finally, the use of agreed 'Rules.'

It is not surprising in itself that existing rules and procedures where followed, what is interesting is the effect that it had on knowledge of the terraqueous globe gained through piecemeal field work. Because the character of a geographical problem necessarily called for a regional description of the natural laboratory within which features would be examined, the net result when universality was applied to induction from observations made, was spatial generalisation at the scale originally described, and then beyond to the entire globe. To the extent that temporal elements where introduced these too were integrated into the picture: but there were, at that time, far fewer ways to add any precision to the temporal dimension, although that did not deter efforts to determine the age of the world, for example, by means of induction on the rate of recession at Niagara Falls (Tinkler 1987).

As no one knew how 'geology' should be done, what it was going to be like, or how to 'do' science outdoors, the first steps were achieved by mimicking existing methods, and in the circumstances this led to spatial generalisation; too generalised to be sure for later standards, but recognisably a fortuitous step in the right direction from the point of view of the future.

-o - o - o - o - o - o - o - o - o - o - o-

To what extent were these characteristics common to other contemporary writers? There is good evidence to suggest that others were doing geology according to the same rules: the arch opponents Hutton and Kirwan were at least agreed upon that, more or less, although it would do nothing to resolve their differences over facts, or their interpretations.

Kirwan began his *Geological Essays* with the following:

> In the investigation of past facts dependent upon natural causes, certain laws of reasoning should inviolably be adhered to. The first is that no effect shall be attributed to a cause whose *known* powers are inadequate to its production. The second is, that no cause should be adduced whose testimony is not proved either by actual experience or approved testimony ...
>
> The third is that no powers should be ascribed to an alledged cause but that it is known by actual observation to possess in appropriated circumstances.
>
> To these laws I shall strictly conform ...(1799, p. 1-2).

Although rather garbled in their statement, these amount to the essence of Rule 1 with Rule 2 perhaps implied. However, Rule 3 (universality) and Rule (4) (primacy of induction from phenomena) are not mentioned, and the reliance on "approved testimony" is a potential loophole, such as was called upon by J.M. in "authenticating" the Boulogne evidence. One might well argue that by the latter half of the eighteenth century universality was assumed as an undisputed *a priori*. Hutton, as one might expect, is strongly Baconian and Newtonian (the two often being conflated,[26] although they were by no means identical:

> Not only are no powers to be employed that are not natural to the globe, no action to be admitted except those of which we know the principle, and no extraordinary events to be alledged in order to explain a common appearance {Rule 1}.
>
> ... nor are we to proceed in feigning causes, when those seem insufficient which occur in our experience. {Rules 2 and perhaps 4} (1795-II-547).
>
> ... we must not allow ourselves ever to reason without proper data, or to fabricate a system of apparent wisdom in the folly of a hypothetical delusion {Rule 3 and 4} (1795-II-564).
>
> ... {it must} universally appear that causes have been concluded, upon scientific principles, for those effects {Rule 2 and 3} (1795-II-566).

These amount (as I have bracketed them) to all of Newton's Rules, and although they are by no means as clearly distinguished as the originals it will be seen on re-examining Newton's amplifying comments that Hutton appears to include parts of the comments within the rules themselves.

I do not find that Whitehurst (1778, 1786[27]) provides quite such clear guidance to his reasoning but he is indisputably in favour of "Lord Bacon" and the "immortal Newton," and he speaks of the fact that the "laws of nature are

---

26 e.g. by Reid, see Laudan 1973.

27 I quote from the second edition.

unalterable," and of "truths universally known." His belief in universality is implicit in his remarks that as the:

> book of Nature is open to all men, written in characters equally intelligible to all nations, and perhaps in no part of the world more so than in Derbyshire, ... for amidst all the apparent confusion and disorder of strata, in that mountainous country, there is nevertheless one constant invariable order in their arrangement ... (Preface).

He notes that he will avail himself as such parts of other systems as suit his purposes, viz. "to trace appearances in nature from causes truly existent; and to inquire after those laws by which the Creator chose to form the world, not those he might have employed, had he so pleased."

Thus we see that authors followed a variable path with some common threads when they were pursuing their goals with what they took to be common sense philosophy. I cannot over-emphasise the fact of regional description, the more important at a time when topographic maps were still in their infancy as tools of a geographical science; although Whitehurst made interesting use of a map when he remarked that:

> whoever, attentively views and considers the present state and condition of the terraqueous globe; its craggy rocks and mountains, its steep, angular and impending shores, subterraneous caverns, &c, will almost be persuaded, without any farther inquiry, that these romantic appearances are not the effects of a regular law, but of some tremendous convulsion, which have thus burst the *strata*, and thrown their fragments into all this confusion and disorder: nay, the very representation of sea and land, upon a geographical chart, seems alone sufficient to establish the truth of such a conjecture (1786 p. 61).

Here we see elements of the *gedanken* experiment again: the visual landscape image, reinforced by the fractured image of a coastline on an eighteenth century map, is used to induce belief in one hypothesis over another. It makes an interesting contrast to J.M.'s imaginary tour of the British Isles (which I have cited above) and which he concludes by appealing to reality (with a mere touch of *gedanken* experience) in support of his uniformitarian perspective:

> In reality, when one views the country around, from any of the high hills I have mentioned {the Cheviots, the Lammermuirs and the Lomonds} and observes it sloping gradually from the inland parts down to the sea, and on each hand towards the beds of the rivers, one can hardly forbear thinking it was once in such a situation, and that it still looks like the shore of a great sea, which has now, after a long succession of ages, almost disappeared. For, not only the earth slopes, as I said before, towards the sea, or towards the rivers; but the very rocks, contrary to their natures, conform themselves to this inclination. (J.M. 1760, p. 103).

As a final appeal to the *gedanken* idea, and the powers of its imagery, let me remind you of Hutton, who had himself been led round the Alps by De Saussure's *Voyages dans les Alpes*. In Chapter IX of Book II (p. 287) "*The theory illustrated, with a View of the Summits of the Alps.*" Hutton takes us to the top of the Alps in order to follow the products of decay and denudation as one descended from the mountaintop, via rivulets and rivers to the ocean. Perhaps Hutton's most powerful example was his imagining the Mediterranean drying up to a salt lake should the Straits of Gibraltar be closed by some natural event (1788, and 1795-I-75).[28]

The *gedanken* idea is a powerful one, perhaps more essential, yet philosophically more dangerous,[29] in a geographical and temporal context, than in the timelessness of the laboratory where repetition can assure consensus; where the actual falsity of the fabricated experiment, caricatured as a recipe, is less evident and perhaps in some sense less false because it may be replicated. Perhaps though, the very scale of 'geological' problems forced writers to adopt *gedanken* methods to achieve their objects, when no other recourse was available.

- o - o - o - o - o - o - o - o - o - o - o -

The full extent of 'geological' time, slowly expanding in the eighteenth century imagination, was quite indeterminate, and indeterminable, by available methods in the eighteenth century: J.M.'s reference to the "age of the world" is loose, but not apparently constrained, and the same can be said for Gough. Most serious writers seemed to feel that the conventional 6000 years was an undesirably low minimum. The words "thousands" and even "millions" were sometimes used in connection with the age of the earth, attached to centuries and years respectively. However, without any regular process to which the stepping of geological time might be synchronised these were, in essence, empty rhetorical phrases. Hutton, who would seem to have most to gain by supplying actual values, made no specific estimates, even of the duration of 'this world' (note: *not* earth): almost certainly because he could see no secure way to do it.

His concern with the matter is signalled by the fact that the final twenty pages of his 1788 paper (Chapter I in the 1795 book) are devoted to the problem, and in his book, so is the whole of Chapter VIII, Part II (p. 238-286) He notes that one can only estimate "a certain *minimum* of this quantity"; that is to say the time needed to construct the sedimentary beds which will form a new continent or world. He could do this either by noting the rate at which erosional processes operated, or by looking to see what remained of the land, relative to what was (he imagined) there to begin with. Like Whitehurst he resorts to maps and the *gedanken* idea again:

> Take the map of any country, provided it be sufficiently particular, and you will see the breaking of continents or islands, first, into promontories or peninsulas ... (1795-II p. 267).

---

28 A *gedanken* experiment subsequently confirmed two centuries later, on some interpretations, see my letter in *Scientific American*, November 1984.

29 In a paper I have just encountered, Hassler (1986 p. 35) suggests that "for speculative scientists, especially during the formative years for modern science in the 18-century Enlightenment, the voices of observational experience provided images that were so powerful, so sublime, that in spite of themselves those scientists toyed with speculative theories and implications that went far beyond what they would acknowledge, even to themselves, as plausible. In other words, images led them beyond belief - beyond what they were willing to state as true."

On page 263 he inserts a lengthy footnote noting the work of M. de Lamblardie, "ingénieur des ponts et chaussées" in France who:

> has made an estimate, seemingly upon good grounds, with regard to the wasting of a part of the coast of France, between the Seine and the Somme.

The wasting is estimated at one foot per year,[30] and Hutton remarks that:

> We may thus, perhaps form some idea of the time since the coast of France and that of England had been united, as one continued mass of those strata which are the same on both those coasts (p. 263).

With the charts then available Hutton could have made the actual estimate as readily as we can now: the figure in question is about 170,000 years, but he did not. As I have indicated this is likely because, although it is a figure which would "astonish us" - if we were an eighteenth century audience - it does not, even so, do more than supply, as Hutton noted "a certain *minimum.*" As well "astonish," and draw contemporary fire, it might only have confused the issue, rather than clarified it.

All the efforts he did make to secure and estimate such a minimum (and to which Playfair's remarks quoted above allude) were *all* based on denudation or erosional rates. These indicated, even with the evidence of very rapid depositional rates in the Ganges, that to build a bed of sediment only a few tens of feet thick (and concomitantly to erode commensurate amounts of a mountain) required æons of time beyond 6000 years. To this extent, therefore, surface studies *did* contribute very significantly to the stretching of time in the imaginations of men, and to tying that extension specifically to depositional processes.

A century of *universality*?

The nineteenth century has often been tagged as one in which there was a slow ascendancy of the concept of *uniformity*. By a similar token, I believe we might tag the eighteenth century, in Britain, as one characterised by a period of stasis for the Newtonian concept of *universality*, followed by its rapid demise as a legitimate spatial argument at the turn of the nineteenth century. The extent of the idea's applicability beyond Britain remains to be ascertained, although Cuvier (1817), for example, did not hestitate to endow his series of "catastrophes" deduced from strata in the Paris basin with world-wide validity, even in the early nineteenth century.[31]

I am not sure that this particular characterisation has been made before with respect to knowledge about the terraqueous globe, and its transformation into the 'science of geology' (Hutton 1795 - II - 259). However, an interpretation in terms of universality helps to explain Hutton's position over the igneous origin of granite. He has frequently been chided for making up his mind

30 A similar value is quoted these days for the corresponding cliffs between Brighton and Eastbourne.

31 Is it significant that Cuvier asked "why should not natural history also have one day its Newton?" (1817 p. 4).

before seeing the *in situ* field evidence for intrusion. But under the principle of universality this was not necessary:[32]

> We shall therefore only now consider one particular species of granite; and if this shall appear to have been in a fluid state fusion, *we shall be allowed to extend this property to all of its kind* (my italics, 1788, 1795 - I - 104).

Reaction to the sweeping spatial generalisations generated by the principle also helps to explain the sudden onset of atheorism that accompanied the establishment of the Geological Society in 1807 (a philosophical stance against which Playfair inveighed in his *Illustrations* (p. 527-528). Even so it is by no means clear that the disease had been clearly diagnosed, even though the symptoms were evident.

The persistence of universality well into the nineteenth century may be indicated by Murchison's eighteen-month delay in visiting North Devon in the mid-1830s, when the Devonian Controversy was in full swing (Rudwick 1985). From his knowledge of the Welsh Borders, and Pembrokeshire, (and possibly from a standpoint of Newtonian *universality* rather than Lyellian *uniformity*) Murchison was convinced that the north Devon exposures could not possibly be interpreted the way that de la Beche was insisting, therefore he had no need to waste valuable time in the field when he might be writing *The Silurian System*.[33] In contrast, Lyell, almost perversely it seems from a twentieth-century perspective, never extended particular findings to general rules, for example, he did not *generally* (i.e. *universally*) attribute the erosion of river valleys to the streams now to be found in them. He regarded each as a case to be considered individually (Tinkler 1987).

In the early nineteenth century, in the vacuum created by discarding or discrediting the spatial consequences of *universality*, and with no general agreement about the adoption of any alternative, be it *uniformity* or *catastrophism* (in any of their many guises), it is possible to understand why even Playfair's rather lame alternative, a "practical research program" (Porter 1977, p. 207), was better than nothing:

> Indeed, if the face of the earth were divided into districts, and accurately described we have no doubt that, from the comparison of these descriptions, the true theory of the earth would spontaneously emerge without any effort of genius or invention (Playfair 1811b, p. 209).

The two worlds were no longer apart; they were split asunder and one was discarded. No longer could a single piece of field work illuminate the cosmological history of the entire globe: but such a collection of works could show that there could be no such cosmology. The idea was still in the future, that *all* that different investigations might have in common is actualistic

---

32 On page 98 of Volume I Hutton intends to show firstly that strata has 'been consolidated by simple fusion; and *2dly*, That this operation is universal, in relation to the strata of the earth...' On page 162 he states that 'we have sufficient reason to believe, that, in knowing the construction of the land in Europe, we know the constitution of the land in every part of the globe.' On page 409 he states: 'I wish again to generalise these facts, by finding them universal in relation to the globe'. Thus his use of universality is quite common.

33 I am indebted to Mott T. Greene for this suggestion.

physical principles operating over vast tracts of time, not universal chronologies or 'qualities' readily generalised to lands still unknown. Hutton's *Theory* straddled the two centuries and tried to marry a cyclic *cosmology*, *universality* and *uniformity*. If the *spatial* consequences of *universality* were to be discarded, was it Lyell's achievement to transform *universality* into *uniformity* ?

References

Baker, J.N.L., 1963. Major James Rennell (1742-1830) and his place in the history of geography, *in:* **Baker, J.N.L.**, 1963. *The History of Geography (Collected Essays)*. Oxford University Press. 130-156.

Bakewell, R., 1813 (later editions 1815, 1828, 1833, 1838, the last three also form the three American editions). *Introduction to Geology*. London.

Bos, H.J.M. 1980. Mathematics and Rational Mechanics, *in:* (editors), **R.Porter** and **G.S.Rousseau**, 1980. *'The Ferment of Knowledge: studies in the historiography of eighteenth century knowledge'*. 327-356, Cambridge University Press.

Buffon, G.H.L Count de., 1749. *Histoire Naturelle*. Paris.

Cajori, F., 1934. see **Newton, I.**, 1729.

Colebrooke, R.H., 1801. On the course of the GANGES, through BENGAL. *Transactions of the Asiatick Society* (Calcutta), 7, 1-31.

Cuvier, G., 1817. *Essay on the Theory of the Earth*, English translation, 3rd Edition, with notes by R.Jameson. Edinburgh.

Davies, G.L.H., 1969. *The Earth in Decay*. MacDonald.

Dean, D., 1981. The age of the earth controversy: origins to Hutton. *Annals of Science*, 38, 435-456.

Dear, P., 1985. *Totius in verba*: Rhetoric and Authority and in the Early Royal Society. *Isis*, 76 (282), 145-161.

Dixon, Robert, 1760. *Extract of a letter from* Cockermouth *in* Cumberland, *dated* October 15th, 1760, *giving an account of the havoc made by a water-spout which happened in the village of* Brackenthwaite, *about six miles from that place, on the 9th of the preceding* September. In *Dodsley's Annual Register*, 1760, 95-8 (of the second pagination in that volume).

Downes, A., 1977.James Rennell, *in* Freeman, T.W., Oughton, M. and Pinchemel, P. (eds.) 1977. *Geographers: Biobibliographical studies*. Volume 1, 83-88.

Gough, J., 1793. REASONS for supposing that LAKES have been more numerous than they are at present, with an Attempt to assign Causes whereby they have been defaced. *Memoirs of the Literary and Philosophical Society (of Manchester)* Volume 4, 1-19.

Gould, S.J., 1987. *Time's arrow and time's cycle*. Harvard University Press.

Greene, J.C., 1984. *American Science in the Age of Jefferson*. Iowa State University Press.

Hassler, Donald M., 1986. The Scottish Reasoning of James Hutton: Poet in Spite of Himself. *Studies in Scottish Literature*, 21, 35-42.

Hutton, J., 1788. A Theory of the Earth, or an Investigation of the Laws Observable in the Composition, Dissolution and Restoration of the Land upon the Globe. *Royal Society of Edinburgh, Transactions*, 1(2). 209-304 (with minor amendments and some footnotes this forms Chapter 1 of the next item).

Hutton, J., 1795. *A theory of the Earth* (2 volumes). William Creech: Edinburgh.

Kirwan, R., 1799. *Geological Essays*. London.
Krumbein, W.C. and Sloss, L.L., 1963 (2nd Edition). *Stratigraphy and Sedimentation*, W.H.Freeman: San Francisco.
Laudan, L., 1973. Thomas Reid and the Newtonian turn of British methodological thought. in **Butts, R.E.** and **Davis, J.W.** (editors) *The methodological heritage of Newton*, University of Toronto Press, 101-131.
Laudan, R., 1987. *From mineralogy to geology*, Chicago University Press.
Leopold, L.B., Wolman, M.G. and Miller, J.P., 1964. *Fluvial Processes in Geomorphology*. W.H.Freeman: San Francisco.
Lyell, C., 1830-3. *The Principles of Geology*. J. Murray: London (v1: 1830, v2: 1832, v3: 1833).
M., J. (not identified further). 1759 or 1760. A letter dated Kirkaldy, Dec. 1759. Observations upon our lime-stone quarries. *Scots Magazine*, reprinted in *Dodsley's Annual Register*, 1760, 98-103 (of the second pagination in that volume).
Mann, A., 1779. A Treatise on Rivers and Canals. *Philosophical Transactions of the Royal Society of London*, 555-656.
Markham, C., 1895. *Major James Rennell and the rise of modern English Geography*. Cassell and Co.: London.
Neve, M. and Porter, R., 1977. Alexander Catcott: glory and geology. *British Journal for the History of Science*, 10, 37-60.
Newton, I., 1687. Philosophiæ Naturalis Principia Mathematica. London.
Newton, I., 1729. *Sir Isaac Newton's Mathematical Principles of Natural Philosophy and his System of the World*, translated into English by **Andrew Motte** in 1729, the translations revised and supplied with an historical and explanatory index by **Florian Cajori**. 1934. University of California Press, Berkeley.
Oldroyd, D.R., 1972. Robert Hooke's methodology of science as exemplified in his 'Discourse on earthquakes.' *British Journal for the History of Science*, 6, No22, 109-130.
Playfair, J., 1802. *Illustrations of the Huttonian Theory of the Earth*. William Creech, Edinburgh.
Playfair, J., 1811a. Account of a lithological survey of Schehallien, made in order to determine the specific gravity of the rocks which compose that mountain. *Philosophical Transactions of the Royal Society of London*, 101, 347-77.
[Playfair, J.], 1811b. Review of Transactions of the Geological Society of London. *Edinburgh Review*, 19, 207-29.
Playfair, J., 1813. Playfair to unknown, Edinburgh, 20th July 1813. Liverpool City Libraries, *Hornby Library*, Autograph letter collection, Box 10 #37.
Porter, R., 1977. *The making of Geology*. Cambridge University Press.
Porter, R., 1978. George Hoggart Toulmin's theory of man and the earth in the light of the development of British geology, *Annals of Science*, 35, 339-52
Rennell, J., 1781a. *A Bengal Atlas: containing maps of .... that side of Hindoostan, etc.*. London.
Rennell, J., 1781b. An account of the Ganges and Burrampooter Rivers. *Philosophical Transactions of the Royal Society of London*, 87-114.
Rennell, J., 1783. *Memoirs of a map of Hindoostan; or, the Moguls Empire, etc.*. London. (A second edition, 1788, included Rennell's *An account of ...Burrampooter*, see British Museum Catalogue of printed books, and **Baker**, 1963, above).
Robison, J. 1797. Article on *'Rivers'* in *Encyclopædia Brittanica* (3rd Edition), Edinburgh.

Rudwick, M.J.S., 1985. *The Great Devonian Controversy*. Chicago University Press.

Sissons, J.B., 1967. *The evolution of Scotland's scenery*. Oliver and Boyd: Edinburgh.

Stokes, E., 1969. The six days and the Deluge: some ideas on earth history in the Royal Society of London 1660-1775. *Earth Science Journal* (Waikato, New Zealand), Volume 3, (1), 46-70.

Stokes, E., 1971. Volcanic studies by members of the Royal Society of London 1665-1780. *Earth Science Journal* (Waikato, New Zealand), Volume 5, (2), 13-39.

Tinkler, K.J., 1983. John Playfair on an erratic in the Isle of Arran. *Scottish Journal of Geology*, 19, 129-34.

Tinkler, K.J., 1984. letter. *Scientific American*, November.

Tinkler, K.J., 1987. Niagara Falls: the idea of a history and the history of an idea. *Geomorphology*, 1 (1), 69-85.

Webb-Seymour, Lord, 1815. An account of observations, made by Lord Webb-Seymour and Professor Playfair, upon some geological appearances in Glen Tilt and adjacent country drawn up by Lord Webb-Seymour, *Royal Society of Edinburgh, Transactions*, 7, 303-376.

Whiston, J., 1696. *A new Theory of the Earth*. London.

Whitehurst, J., 1778 and 1786. *An inquiry into the original state and formation of the earth*. London.

Woodward, J., 1695. *Essay towards a natural history of the earth*. London.

4

James Hutton's rôle in the history of geomorphology

Dennis R. Dean

It has been two hundred years since James Hutton of Edinburgh published in the transactions of his Royal Society the geological theory now commonly regarded as basic to all subsequent progress in geomorphology. Not unlike the case with other major scientific theories, however, Hutton's success was far from immediate. My essay, therefore, begins with a brief review of the theory itself and then traces the several distinct stages involved in its eventual acceptance. That acceptance, always a partial one at best, was undoubtedly delayed by the relative indifference of certain British investigators toward the origins of landforms. Even so, we have good reason to associate the effective founding of geomorphology with Hutton's name.

Born in 1726, Hutton attended Edinburgh University, eventually becoming a medical student but remaining primarily interested in chemistry. After stays in France and Holland he received his doctorate in medicine from Leyden in 1749. Rather than practice, however, he derived an adequate income from paternal legacies (including two farms) and a successful business partnership. His interest in geology grew out of his interest in agriculture, the ability of the Earth to maintain its fertility being central to his concerns. Hutton's earliest geological investigations were not in Scotland but at Yarmouth (where he had gone to learn the newest agricultural techniques) and throughout southern England. After returning to Scotland in 1754, he then explored the geology of that country assiduously. By December 1767 he had moved to Edinburgh and was soon part of a distinguished intellectual circle.

Hutton's geological *theorizing* began at Yarmouth with the realizations that many present-day rocks are made from the destruction of former ones and that all of the present ones are themselves disintegrating and being transported into the sea. He assumed, therefore, that new continents destined to replace the old were being formed from the old within the world's oceans. Related travels from Yarmouth to the Low Countries and the south coast of England called his attention specifically to the formation of chalk and flint, the consolidation of which he could not explain. Upon moving to Scotland, however, Hutton learned of the recent experiments on lime by Dr. Joseph Black (who quickly became Hutton's closest friend). In accord with Black's discoveries, and probably because of them, Hutton soon attributed the consolidation of rock masses to heat. His chief geological concern then became the origin of basalt and granite, both of which he eventually regarded as unerupted subterranean lavas.

In March and April 1785 Black and Hutton read to the Royal Society of Edinburgh a lengthy paper by Hutton "concerning the System of the Earth, its duration and stability." Having been retitled by a friend, it was then published in 1788 as *Theory of the Earth*, though author's separates had been available some months earlier. In the first of his paper's four parts, Hutton emphasized that the Earth is a unique creation, specifically designed by its infinitely wise Creator to serve as a habitable world for life in all its forms. Soil, necessary to the growth of plants, is created only by the destruction of land. Erosion, therefore, is a necessary function of the Earth and, paradoxically, the removal of fertile soil from the continents is actually part of a great cycle by which the continuing fertility of the earth is assured. That this process is extremely slow, involving lengths of time beyond human experience and comprehension, should not disguise the benevolence underlying it. In part two of his theory, Hutton first analyzed both landforms and individual rock types to establish the reality of consolidation through fusion. Having shown how strata deposited at the bottom of the sea are lithified, he next went on in part three to propose that the same force, of subterranean heat, also elevates these consolidated strata into continents. The fourth and final part then emphasized the on-going cycles of decay and renovation that Hutton believed fundamental to his divinely contrived Earth. His famous and often attacked conclusion - "The result, therefore, of our present enquiry is that we find no vestige of a beginning, no prospect of an end" - did not of itself affirm an eternal Earth but merely pointed out instead that comprehensive cyclicism would necessarily obliterate any evidence of a beginning and that its regularity would necessarily exclude any warnings of the end.

Though complacent historians have sometimes assumed the contrary, this 1788 paper of Hutton's - for many, the beginning of modern geology - is not an adequate synopsis of his geological thinking as a whole. Indeed, by the time of its official appearance in 1788, Hutton had already gone on to further insights. His most significant fieldwork, for example, was done *after* the presentation of his paper. Between 1785 and about 1790, Hutton enlarged upon his theoretical views in a series of ten essays, only one of which (his paper on granite) was published separately. Six others remained unpublished till 1899 and the remaining three were incorporated into the longer version of his theory that appeared in 1795.

In the meantime, reviews and other criticisms of his original 1788 paper had begun to appear, with major and prolonged opposition from Richard Kirwan and Jean Andre Deluc. Responding primarily to the former, Hutton revised and augmented his theory's exposition until it became the two-volume *Theory of the Earth* (1795) that so many scholars have cited and so few have read. The work as a whole is not simply a miscellaneous collection of geological writings but rather a well constructed (if sometimes badly written) argument on behalf of the theory (reprinted, with additions and changes) that forms its first chapter. The second replies specifically to Kirwan; chapter three discusses geological theorizing as such; and several more then criticize specific theoretical errors, including the supposition of either landforms or rocks having endured since the Creation. Other topics dealt with are petrifaction (a more specialized form of consolidation) and the formation of coal.

Volume two, primarily devoted to additional facts and observations on behalf of Hutton's own theory, is surprisingly concerned with topography. To an extent not commonly recognized, Hutton's continuing mental growth had augmented his earlier conceptualization of geological theory with this new awareness of landforms and the processes which shape them. Read by itself, this second volume may well be seen as having founded a modern science in its own right-the science of geomorphology.

According to its three-page "*Introduction*," Hutton's theory had represented Earth's present continents as having been formed originally as horizontal strata at the bottom of the ocean and then elevated into their present positions. Upon emergence, they must have looked very different from today's. (Rivers, for example, had yet to excavate their own channels.) All in all, he thought, three causes have determined the form of our present land: the regular stratification of materials; operations of the mineral region; and operations proper to the surface of the globe, including sun and atmosphere, wind and water, and rivers and tides. The effects of these three interacting causes are then examined in a series of chapters. Before discussing them, however, we must remember that Hutton wrote as a bedridden, sick, and dying man who, having no further opportunities for fieldwork, necessarily (and rather ingeniously) relied instead upon the meticulously objective descriptions of other authors, H. B. de Saussure in particular.

Thus, chapters one and two utilize lengthy quotations from Saussure and other authors to establish that strata originally horizontal have been bent and broken by the same force that elevated them. Chapters three and four then confirm the extent to which elevated strata are attacked by a variety of erosive forces, including both rivers and glaciers, over a vast period of time. Chapter five extends this analysis to the formation of gravels and soils, recalling Hutton's concern with agricultural fertility. Six and seven, defending the wise economy of nature, contain two of Hutton's most astonishing geomorphological analyses, as he chronicles the progressive changes made in a fertile plain by an encroaching river that eventually destroys it and imagines stop-camera changes in an Alpine landscape through time. Hutton then extends this grandeur of vision even further in chapter eight, where he speculates on the present configuration of the Earth's landmasses and envisions a distant past in which France was joined with England, Ireland with Britain, the Orkneys with Scotland, and the Shetlands with Norway, all of them being sundered eventually through marine erosion. Chapters nine and ten next consider the mountains and valleys of the Alps, both of which also attest to the slow but inexorable power of erosion. In chapter eleven, he surveys still other mountains and valleys, including those formed in other ways. Twelve and thirteen adduce further evidence on behalf of the theory worldwide and fourteen, the last, sums up his theory as a whole. Despite his illness, then, Hutton succeeded in writing a logically coherent book.

As I have shown elsewhere, it is not true that Hutton's theory failed to attract critical attention. Initially, however, much of the early reaction concerned itself with extremely broad issues, such as the intrinsic feasibility of geological theorizing and the supposed compatibility (or lack of it) between Hutton's theory and Scriptural assertions regarding the ages of man and the world. When geological arguments opposing Hutton appeared, they were usually subsidiary to theological positions, though the reality of Hutton's postulated central fire or heat and freely invoked immensities of time was also frequently questioned.

The most incisive of Hutton's early critics proved to be Richard Kirwan, who published his important "Examination of the Supposed Igneous Origin of Stony Substances" in 1793. Clearly uncomfortable with the seemingly mechanistic implications of Huttonian theory, Kirwan (a distinguished Irish chemist) opposed Hutton point for point in every major aspect of his conjectures. Thus, Hutton envisioned originally solid continents from which soils and gravels had arisen. Kirwan agreed that decaying rock augments soils but denied that all soils originate from rock. Hutton saw continental soils being washed away by waters flowing from the mountains to the sea. Kirwan denied that such erosion was constant and that all such water necessarily reached the sea, much being lost to evaporation. Deposition, moreover, takes place along the banks of rivers and at their mouths; much of whatever silt actually reaches the sea is then deposited along the coast, marine *erosion* being far less important than Hutton and others had supposed. Hutton stated that the solid parts of the globe are primarily composed of sand, gravel, limestone, and similar deposits. Below the surface, however (Kirwan replied), the basic rock is granite, not only in Scotland but throughout the world. The consolidating power, which Hutton imagined to be subterranean heat, finally, is more likely to be water. Kirwan cited the supposedly aqueous origin of granite in particular. In addition to these cogent scientific objections, Kirwan also echoed already familiar nonscientific ones: "Why," he asked, "should we suppose this habitable earth to arise from the ruins of another anterior to it, contrary to reason and the tenor of the Mosaic history? What do we gain by that supposition?" (63). Hutton then replied specifically to Kirwan in chapter two of his 1795 *Theory* and more generally in later ones.

As Kirwan persisted in attacking Hutton's theory, even after Hutton's death in 1797, he also became the primary opponent for Hutton's two chief defenders, John Playfair and Sir James Hall. Playfair's masterful *Illustrations of the Huttonian Theory* (1802) was primarily a more readable explication of Hutton's ideas than Hutton himself had been able to provide. In his "*Notes and Additions,*" however - actually the larger part of the book - Playfair went beyond Hutton in several respects, being particularly concerned to relate the theory to astronomical and paleontological contexts. Aside from his considerable attention to rivers, Playfair was noticeably less concerned with the origins of landforms than Hutton (in 1795, II) had been. For his part, Hall persuasively established the plausibility of Huttonian compression and fusion under heat - but was not yet concerned with landforms. Immediately ensuing episodes of Huttonian controversy, therefore, would only occasionally involve geomorphological issues.

From around 1800 to around 1815 the prevalent geological theory in both Europe and the United States was that of Abraham Gottlob Werner, which asserted the aqueous origins of basalt, granite, pumice, obsidian, and other rocks; advocated the aqueous origin of mineral veins; and postulated an invariable, sometimes universal series of strata deposited worldwide from an originally ubiquitous but receding ocean. Between 1815 or so and Werner's death in 1819, however, a series of well argued papers and observations demonstrated the inadequacy of Werner's conceptions. While a number of Huttonian positions were either confirmed or strengthened en route, the overall result was not Huttonian victory but pervasive skepticism regarding all geological theorizing. Throughout the 1820's, consequently, opinion favored the relatively undogmatic outlook of Georges Cuvier, who accepted parts of Wernerian theory but discarded the German's universal ocean in favor of recurring catastrophes, the cause or causes of which he did not attempt to determine. Meanwhile, Hutton's chief defenders had either, like Playfair, died or lapsed into exhausted silence, like Hall.

Though Hutton himself had been concerned with geomorphological issues to a remarkable extent, and though we can trace diluted reflections of that interest in the writings of such later combatants as Playfair and Robert Jameson, among others, it is still true that attention to the problem of landforms and their origins intensified considerably under the ægis of Cuvier, though he himself had done almost nothing with it. There were, no doubt, several reasons for this enhanced attention. One is simply that geology had become increasingly popular and better organized, so that the cumulative effects of collective endeavor were greater. A second is that the end of the Napoleonic wars freed the British (in particular) from both their insularity and their pent-up aggression. Following 1815, everyone in England and Scotland who could vacationed on the Continent, with Chamonix and Naples among the most favored destinations. Being no longer confined to their own island, British theorists were free to regard those highly impressive geomorphological questions that exposure to the Alps, volcanoes, and other classic phenomena of Europe would naturally engender. Finally, with the frustrations of war now over, and the two previously contending theories both weakened, geology became less of a confrontational and more of a co- operative endeavor.

Though the disinterested accumulation of geological facts now became fashionable, well defined specific controversies necessarily underlay reports from afield. His theory as a whole having been discredited, Werner's posthumous influence upon these activities was soon limited almost entirely to the still unresolved problems of the filling of veins and the origin of basalt. There remained two further legacies of importance - namely, stratigraphical and petrological investigations - but endeavors in both served almost entirely to further discredit Werner rather than defend him. Because his theoretical remarks had been so general, Cuvier was valuable primarily in eliciting opinions with regard to the duration and kind of geological causation. While this was no small matter, to be sure, such ambiguity did not yet invite comprehensive refutation. Huttonian theory, on the other hand, was *quite* specific in its major assertions and therefore became the touchstone of a number of geological investigations.

By the end of the 1820s, then, several often long-standing disputes had either been significantly refined or outrightly resolved. Among these frequently interlocking themes one may cite: the origins of valleys, the reality of the Deluge (or any other universal flood), the origins of basalt and granite, the identity of basalt with lava (which Hutton had denied), the geological significance of volcanoes, and the (increasingly dubious) reality of central fire within the Earth. Other disputes emphasized internal heat and the consolidation of strata, the topographical importance of uplift, the efficacy of erosion (coastal, diluvial, and subærial), and the nature and extent of geological forces and time. While the Huttonian and geomorphological nature of many of these issues will be apparent, we should also remember that all of them were being pursued within a milieu in which the primacy of fieldwork was now assumed.

Since we cannot investigate each of these controversies in detail, let us take for our example the new emphasis on uplift as a fashionable explanation for the origin of landforms. Several distinct kinds came to be so interpreted. Thus, in 1825, William Buckland proposed before the Geological Society of London that some valleys, at least, could not be accounted for by denudation alone. The Weald of Kent and Sussex, for instance, almost certainly owed its fundamental structure to "a force acting from below, and elevating the strata along their central line of fracture" (1829, 123). G. P. Scrope then not only agreed with Buckland that certain valleys had been formed by uplift (together, for him, with subsidence) but asserted two years later that most of the Earth's surface irregularities, including the basins of its seas, lakes, and rivers, were probably attributable to subterranean expansion rather than erosion. During the same years Leopold Von Buch, a former student of Werner's (though now confessedly Huttonian), proposed that even the craters of volcanoes had been pushed upwards from beneath (like domes) rather than constructed from accumulating ejecta. And Leoncé Élie de Beaumont, a French theorist, recklessly attributed the origins of whole mountain ranges worldwide to catastrophic uplift along predetermined compass lines. While the popularity of uplift theories naturally enhanced Hutton's reputation (as did several other of the controversies), the extravagance of some of them also made his gradualistic assumptions increasingly appealing.

For Charles Lyell, the 1820's were formative years, and by their end he had emerged as a firm disciple of John Playfair's, whose modified Huttonianism he took to be Hutton's own. Lyell then defended Playfairian positions at length in his famous *Principles of Geology* (I, 1830), which accepted aqueous and igneous causes as equal partners in geological explanation, endorsed both the mechanical and solvent powers of water, and advocated fluvialism. Two chapters on deltas emphasized the constructive aspect of rivers but others fully accepted (and indeed borrowed) Hutton's awareness of the efficacy of erosion. Lyell then dealt at length with both volcanoes and earthquakes, two significant topics about which Hutton had said very little. In discussing both, Lyell stressed the importance of uplift and other seismic movements, for, like Hutton, he believed the most powerful of all geological forces to lie within the Earth.

Like Hutton too, Lyell soon found himself surrounded by opponents, most of whom rejected his assumed duration of geological time, his equation of past geological causes with present ones, his plutonism, and in particular his fluvialism. Adam Sedgwick and others, therefore, spoke out against many of Lyell's broad generalizations and were answered in part in Lyell's second and third volumes (of 1832 and 1833, respectively). Volume three, especially, opposed both Werner (by now a strawman) and Cuvier while emphasizing landforms and fossils associated with the Tertiary period. Lyell's descriptions of Sicilian valleys and similar features, however, denigrated the role of running water in favor of ongoing subterranean uplift and some previous marine erosion. For him, then, the Weald of Kent and Sussex became a gradually uplifted landform excavated primarily by marine currents. Because of their catastrophic assumptions, he also opposed the uplift theories of Von Buch and Élie de Beaumont. Overall, Lyell emphasized the identity and uniformity of past and present geological causes rather than the balanced cycle of destruction and creation postulated by Hutton.

Whereas all the leading geologists of his time accepted the persistence of natural law - a position we would call *Actualistic* - few agreed with Lyell that the intensity of geological forces had remained unchanged throughout time or that geological history (especially in its biological aspects) had remained so resolutely unprogressive as he had argued. Certain other issues had by this time been resolved. Distorted strata, for instance, were now almost universally interpreted as having been deposited in an originally horizontal position and then subsequently reshaped by underlying or lateral pressures. The three major families of rocks - igneous, sedimentary, and metamorphic - were by this time well established, in part because of Lyell. Though specific estimates continued to be rare, and disagreements significant, the immensity of geological time (as opposed to any Scripturally-derived conception) was taken for granted. Similarly, almost everyone recognized that incessant geological changes (both destructive and creative) had taken place throughout the Earth's history, whatever its duration. While the shaping forces of the past may have been considerably greater (as some maintained), Lyell's demonstration of the efficacy of present-day ones achieved general agreement.

Despite his considerable success in *that* instance, however, Lyell failed to attain a degree of insight equivalent to Hutton's with regard to the evolution of landscapes. Though he studied both earthquakes and volcanoes assiduously, for example, Lyell never achieved a convincing theory of either. Similarly, he preferred to explain denudation through a combination of continental vacillation and marine erosion rather than through the fluvial and subærial agencies so often stressed by Hutton. Paradoxically, Lyell often attempted to expound traditionally Huttonian phenomena by postulating an undulatory sequence of slow-motion catastrophes in which the relative fixity of continental masses and ocean basins was assumed. The continents, for him, bobbed up and down, but they did not (as in Hutton's conception) regularly change places with the oceans.

In their responses to it, many of Lyell's critics characterized his theory as a restatement of Hutton's. To some considerable extent it was, but there were differences also and it should now be clear that major portions of Lyell's thought are less akin with our own than Hutton's had been. I am not, of course, assuming the finality of our own assumptions but merely describing one blip in the stock market of scientific opinion - which, for me, at least, is on the whole progressive but forever doomed to remain in search. Whatever the validity of our current understanding, it was arrived at in part by rejecting Lyell and reclaiming Hutton.

The transition of which I speak took place gradually, over more than half a century, and cannot be attributed to the influence of any one advocate or even of any one school. A very significant stimulus, however, was the renewed attention to glaciers and their associated geological phenomena. Without having ever seen one himself, Hutton had come to appreciate the power of glaciers (by reading of them in Saussure and other accounts); after 1815, Playfair and others inspected them at firsthand. Following a series of papers, and increasingly bolder conjectures, by several researchers, Louis Agassiz and others became convinced that extended glaciation at some time in the not too distant past had been responsible for many of the geological phenomena formerly attributed to catastrophic floods. Despite some initial enthusiasm for Agassiz's Ice Age, Lyell remained ambivalent toward the concept until 1858, when he more fully indicated his acceptance. Though glaciation successfully explained a number of formerly puzzling geological phenomena, it also challenged Lyell's uneventful view of the distant past and discredited both uniformitarianism and catastrophism as then understood.

A second counterinitiative gradually undermined Lyell's disparagement of fluvial erosion. Following some generally ignored precursors, G. P. Scrope and other geological writers of the 1850's began to speak out on behalf of rain and rivers, whose efficacy in shaping the land had been widely noticed in the latter eighteenth century, emphasized by Hutton and Playfair, and regularly denied for half a century thereafter. It is a measure of Lyell's residual influence that almost all of the most telling evidence originated outside England - in America, Ireland, and Scotland, primarily. Scottish geologists like Andrew C. Ramsay and Archibald Geikie, moreover, not only supported comprehensive fluvial explanations for geological phenomena but explicitly attributed the origin of such explanations to Hutton. Being historians of geology as well as practicing geologists, both Ramsay and Geikie - the latter more memorably - exhorted their fellow fieldworkers to read Hutton and Playfair in the original, thereby to correct Lyellian distortions and to pay just homage to a man both regarded as the effective founder of their science.

In 1862 J.B.Jukes read a paper analyzing river valleys in southern Ireland. Firmly restricting the Lyellian agency of marine erosion, he insisted once again that rivers had created their own valleys. Atmospheric denudation, moreover, was constantly at work degrading land masses everywhere and seemed easily capable of removing sediment layers even hundreds of feet thick. Upon reading Jukes' paper, both Ramsay and Geikie were immediately compelled to accept his theoretically significant conclusions. A series of publications by each then applied both fluvial and glacial analyses to landscapes.

If any doubts remained as to the explanatory power of Huttonian erosion, they were soon overcome by a glut of fresh, scarcely contestable evidence from the United States - in particular, John Wesley Powell's dramatic exploration of the Grand Canyon and his subsequent theorizing upon the formation of valleys (1875). With the incredibly efficacious potentialities of riverine wear and transport now established beyond question, both Powell and G. K. Gilbert then presented further timely defenses of subærial erosion, of which the American West contains so many fine examples. While Gilbert's *Geology of the Henry Mountains* (1877) is better known for its theory of laccoliths (also somewhat Huttonian), the same work was among the first on its side of the Atlantic to integrate and apply such concepts as land sculpture, drainage systems, planation, and terracing - though Hutton himself had recognized them all more than eighty years before. With the systematic approach to landforms undertaken by W. M. Davis in the 1880's and thereafter, it is generally agreed, a science called geomorphology had unquestionably come into being.

It was not Davis's primary intention to write a history of his own field, but in such essays as "*Plains of Marine and Subærial Denudation*" (1896) and "*The Physical Geography of the Lands*" (1900) he alluded to many of the most significant landscape theorists active during the last quarter of the nineteenth century. Here and throughout his work, he emphasized without malice that the founders of geomorphology had been Hutton and Playfair, and that their link to later workers had come not through Lyell but through Scrope and the Americans. Though Davis, it seems to me, unfairly slights the roles of such other significant contributors as Ramsay, Jukes, and Geikie, he nonetheless reminds us of a legendary ancestry that by all indications is also true.

To be sure, Davis's conception of the history of his own field-which he called "geomorphogeny" for a time, "geomorphology" being an American term - is not entirely that which such more recent scholars as Chorley, Davies, and Tinkler have urged us to accept. In his view, a fundamental step had been taken early in the nineteenth century by the German geographer Carl Ritter (1779-1859), who transformed geography from "the description of the earth and its inhabitants" to "the study of the earth in relation to its inhabitants." Throughout his voluminous works, Ritter did so within a context of uniformitarian geology. Despite such spectacular disasters as volcanic eruptions, earthquakes, and tidal waves, he believed, the more important geological changes - particularly from a human point of view - are those which have taken place gradually, almost imperceptibly. "The duration of the Earth," he affirmed in 1852, "outruns all measurement," and its history is that of "ceaseless transformation." Yet underlying this unending change, he insisted, are holistic, consistent laws and a teleology emphasizing the human condition.

In his philosophical outlook, Ritter was highly akin to Hutton and surely indebted to him. Not surprisingly, then, Davis rather explicitly equated the triumph of Huttonian gradualism with the emergence of his own discipline. "Theological preconceptions as to the age of the earth and the associated geological doctrine of catastrophism," he wrote in 1900, "although attacked by the rising school of uniformitarianism, were then dominant. They gave to the geographer a ready-made earth, on which the existing processes of change were unimportant ... In the victory of the uniformitarians over the catastrophists began the fortunate alliance of geography with geology." For Davis, the "emancipation of geology from the doctrine of catastrophism," a "necessary step" in the development of geography as a science, took place chiefly on two fronts: one of these involved the overthrow of Élie de Beaumont's theory of the sudden and violent upheaval of mountain chains; the other, the re-establishment of fluvialism by Greenwood, Scrope, and others. By 1858, then, Davis observed, "victory may be said to have been declared for the principles long before announced by Hutton and Playfair, which since then have obtained general acceptance and application" (1954, 71, 72, 77, 78).

Insofar as those two esteemed progenitors differed in outlook, however, Davis clearly preferred Playfair to Hutton. He specifically rejected teleology, for example, "which maintained the predetermined fitness of the earth for its inhabitants, and of its inhabitants for their life work." All this, he believed, "had to be outgrown before geographers could understand the slow development of land forms and the progressive adaptation of all living beings to their environments" (p. 72). What Davis wanted from his own field was not the God-centered assurances of a Hutton or a Ritter but rather the quasi-mathematical certainty of Playfair's Law, as he called it, regarding the accordant junctions of lateral and main streams and valleys. The "*Geographical Cycle*" announced by Davis in 1899, and for which he is so famous (or perhaps notorious), was in large part an attempt to restate broadly conceived Huttonian insights in formulaic Playfairian terms.

Though the importance of Hutton and Playfair to Davis's highly formative contributions is now clear, we are nonetheless left with a series of further questions, regarding which our historians of geomorphology have not yet responded adequately. Even if we set aside the as yet inadequately addressed problem of Ritter and his influence, for example, there remains the like problem of Alexander Von Humboldt (1769 - 1859), Europe's other great pioneering geographer along with Ritter. Of the two, Humboldt was also the more geological, achieving a stature within geology itself that Ritter (basically a closet philosopher) could never hope to equal. Though Humboldt eventually adopted Huttonian views in part, he began his career as a Wernerian; the broadly-based teachings of Werner, which often related the history of mankind with that of the physical world, had much to do with Humboldt's later expertise as a geographer. Where is the historian who can show us in outline the important Werner-Humboldt tradition? It probably culminated in the work of Charles Darwin, another surprisingly neglected figure (so far as geomorphology is concerned), though more important for his philosophy of nature - as we have seen in the case of Davis - than for specific landscape analyses.

Finally, there is the question of Davis himself. How sound do his views now appear and what relationship has he with the present science of geomorphology? We are now far enough away from Davis to perceive at least some of the limitations of his outlook. Yet it is not clear that a new, more adequate synthesis of equivalent scope and force has yet arisen. Thus, the position of Davis today is rather like that of Lyell within geology around 1860, when a certain obsolescence became apparent but a replacement was not. Presumably, we can eventually look forward to a more satisfactory science of geomorphology that will derive even more specifically than our present one does from our advancing knowledge of plate tectonics and the inner structure of the Earth. If so, I do not doubt, its ultimate derivation from the work of James Hutton and his *Theory of the Earth* (1795) will remain apparent.

Authors cited (or suggested)

Agassiz, Louis, 1840. *Études sur les Glaciers.* Neuchâtel.

Albritton, Claude C., 1980. *The Abyss of Time: Changing Conceptions of the Earth's Antiquity after the Sixteenth Century.* San Francisco.

Buckland, William, 1829. On the Formation of the Valley of Kingsclere and Other Valleys by the Elevation of the Strata That Enclose Them. *Transactions of the Geological Society of London,* NS 2, 119-130.

Chorley, R. J., A. J. Dunn, and R. P. Beckinsale, 1964. *The History of the Study of Landforms, or the Development of Geomorphology, I: Geomorphology Before Davis.* Frome and London. 1973. Ibid., *II, The Life and Work of W. M. Davis.* Frome and London.

Cuvier, Georges, 1813. *Essay on the Theory of the Earth.* Translation by Robert Kerr, with notes by Robert Jameson. Edinburgh. Further editions to 1827.

Davies, G. L. H., 1969. *The Earth in Decay: A History of British Geomorphology, 1578 to 1878.* London.

Davies, G. L. H., 1985. James Hutton and the Study of Landforms. *Progress in Physical Geography,* 9, 382-389.

Davis, William Morris, 1909. *Geographical Essays.* ed. D. W. Johnson. Boston. (Reprinted 1954, NY.)

Dean, Dennis R., 1973. James Hutton and His Public, 1785-1802. *Annals of Science,* 30, 89-105.

Dean, Dennis R., 1980. Graham Island, Charles Lyell, and the Craters of Elevation Controversy. *Isis,* 71, 571-588.

Dean, Dennis R., 1981. The Age of the Earth Controversy, Beginnings to Hutton. *Annals of Science,* 38, 436-456.

Dean, Dennis R., 1985. The Rise and Fall of the Deluge. *Journal of Geological Education,* 33, 84-93.

Dean, Dennis R., Forthcoming. *James Hutton and the History of Geology.*

Dott, Robert H., 1969. James Hutton and the Concept of a Dynamic Earth. Pp. 122-141 *in* Cecil J. Schneer, ed., *Toward a History of Geology.* Cambridge, Mass., and London.

Ellenberger, François, 1973. La Thèse de Doctorat de James Hutton et la Rénovation Perpétuelle du Monde. *Annales Guebhard,* 49, 497-533.

Geikie, Archibald, 1868. On Modern Denudation. *Transactions of the Geological Society of Glasgow,* 3, 153-190.

Geikie, Archibald, 1905. *The Founders of Geology,* London, second edition.

Gilbert, Grove Karl, 1877. *Report of the Geology of the Henry Mountains.* Washington.

Greene, Mott T., 1982. *Geology in the Nineteenth Century: Changing Views of a Changing World.* Ithaca and London.

Greenwood, George, 1857. *Rain and Rivers, or, Hutton and Playfair Against All Comers.* London.

Humboldt, Alexander von, 1849-1852. *Cosmos: A Sketch of a Physical Description of the Universe.* Translated by E. C. Otte. 4 vols. London.

Humboldt, Alexander von, 1850. *Views of Nature.* Translated by E. C. Otte and Henry G. Bohn. London.

Hutton, James, 1788. Theory of the Earth; or an Investigation of the Laws Observable in the Composition, Dissolution, and Restoration of the Land upon the Globe. *Transactions of the Royal Society of Edinburgh,* 1 (ii), 209-304.

Hutton, James, 1794. Observations on Granite. *Transactions of the Royal Society of Edinburgh,* 3 (ii), 77-85.

Hutton, James, 1795. *Theory of the Earth, with Proofs and Illustrations.* 2 vols. Edinburgh. Volume three, 1899, ed. Archibald Geikie (London).

Jukes, J. Beete, 1862. On the Mode of Formation of Some of the River Valleys in the South of Ireland. *Quarterly Journal of the Geological Society of London,* 18, 378-403.

Kirwan, Richard, 1793. Examination of the Supposed Igneous Origin of Stony Substances. *Transactions of the Royal Irish Academy,* 5, 51-87.

Lyell, Charles, 1830-1833. *Principles of Geology.* 3 vols. London. Further editions to 1875.

Playfair, John, 1802. *Illustrations of the Huttonian Theory of the Earth.* Edinburgh.

Playfair, John, 1805. Biographical Account of the Late Dr. James Hutton. *Transactions of the Royal Society of Edinburgh,* 5, iii, 39-99.

Powell, John Wesley, 1875. *Exploration of the Colorado River of the West.* Washington.

Powell, John Wesley, 1876. *Report on the Geology of the Eastern Portion of the Uinta Mountains*. Washington.

Ramsay, Andrew C., 1846. On the Denudation of South Wales and the Adjacent Counties of England. *Memoirs of the Geological Survey of Great Britain*, 1, 297-335.

Ramsay, Andrew C., 1863. *The Physical Geology and Geography of Great Britain*. London.

Ritter, Carl, 1865. *Comparative Geography*. Trans. William L. Gage. Philadelphia. Orig. pub. in German, 1852.

Scrope, George P., 1825. *Considerations on Volcanos*. London. Further editions 1862, 1872.

Scrope, George P., 1827. *The Geology and Extinct Volcanos of Central France*. London. Second edition, 1858.

[Scrope, George P.,] 1830. Review of Lyell's Principles, I. *Quarterly Review*, 43, 410-469.

Tinkler, Keith J., 1985. *A Short History of Geomorphology*. London and Sydney.

5

The Turning of the Worm - early nineteenth century concepts of soil in Britain - the development of ideas and ideas of development, 1834 - 1843

Brian T. Bunting

"Now mark me well - it is provided in the essence of things, that from any fruition of success, no matter what, shall come forth something to make a greater struggle necessary."

W. Whitman: *Song of the Open Road.*

Despite all warnings about singing the praises of former times, one may regard the fourth decade of the nineteenth century, especially the period 1833-1838, as marked by several seminal texts about soil and dramatically new ideas about its separation from geology, from several important publications, most notably those of John Morton (1781-1864, editions dating from 1838 to 1843), of Anon (1834 most probably John Wright, 1770?-1844), (see Figures 1 & 2 for the full titles of these works); of Thomson (1773-1852, 1838) and, of Daubeney (1795-1867, 1847); also, in 1837, a first statement from Charles Darwin (1809-1882). The *English*, later the *Royal Agricultural Society*, was also founded in 1838 and the *British Association for the Advancement of Science* first met in 1831 (Russell 1966).

The writings of Morton and Wright(?), were clearly influenced by Sir Humphry Davy (1778-1829, 1813), and by the geologist, William Smith; and one may assume their acquaintance with the geological maps published in the third decade of the nineteenth century. Both Smith and Morton lived near Bristol, and Morton was a Fellow of the Geological Society. As was the custom of the time, quotation and acknowledgement of the work of others is noticeable only by its absence. Both Morton and Wright were well-travelled in England and Scotland, setting up or sponsoring several example farms. Morton certainly shows an advanced and original level of understanding of geological stratigraphy. Both

THE

NATURE AND PROPERTY

OF

SOILS;

THEIR CONNEXION WITH THE

GEOLOGICAL FORMATION ON WHICH THEY REST;

THE BEST MEANS OF PERMANENTLY INCREASING

THEIR PRODUCTIVENESS,

AND ON THE

RENTS AND PROFITS OF AGRICULTURE.

TO WHICH IS ADDED AN ACCOUNT OF

THE PROCEEDINGS AT WHITFIELD EXAMPLE-FARM;

THE SYSTEM OF AGRICULTURE

RECOMMENDED ON THE ESTATES OF THE EARL OF DUCIE;

A LETTER TO MR. PUSEY'S TENANTS;

AND THE MODE OF

CULTIVATION ADOPTED ON STINCHCOMBE FARM.

BY JOHN MORTON.

Fourth Edition Enlarged.

LONDON:

JAMES RIDGWAY, PICCADILLY.

MDCCCXLIII.

Figure 1

The title page of Morton's text (1843 edition).

were firmly rooted in a practical approach to agriculture and the need to increase yields by manuring and land improvement, which they saw as the sole means of increasing the profitability of agriculture for, by 1837, "the farming industry had passed through a quarter of a century of misfortune," (Ernle and Hall 1936). Chemical methods of increasing crop yields were but little known before 1830; drainage, manuring, marling and mixing were more favoured.

The two writers have been largely neglected since 1860, for the simple reason that their recommendations are, and were, too complex and costly, of labour and time in the case of Wright, and in Morton's case, too demanding of the limited mechanical equipment of the time, both at the testing and farm level. However it is ironic that Wright's "Universal Compost" is somewhat analogous to sewage sludge, using urban waste and industrial wastes, while Morton's ideas of soil mixing have implicit appeal to landscape gardeners. The problem of so many cultivators being holders of short leases of land, especially in the remoter areas, in the 19th century, was a major obstacle to long-term improvement. Ideas of physical improvement were soon to give way to measures involving the use of chemical fertilisers, prompted by the work of Rothamsted and the Lawes Trust (Russell 1966).

Morton and Wright recognized the two great uniformities of nature: those of succession in time and those of the co-existence of different types; the latter very clearly expressed by Morton in causal laws and definitions (Hull 1974), while Wright was more of a proselyte, keen to establish the use of his universal compost and thus change unresponsive soils in drastic fashion. The two were objective scientists and observers, only invoking the Deity when the evidence was slim, the implication uncertain, or the understanding and collaboration of the reader desirable. Perhaps too, and without regret, one may point out that both writers were largely ignorant of, or apathetic to, the main geological controversies of the time.

Originators and precursors

Morton and Wright, perhaps William Cobbett (1763-1835, 1830), on whom see Moffatt (1985), derived little from the Board of Agriculture reports from the decades at the turn of the eighteenth century, which, with a few exceptions, [for example - John Farey (1776-1826, 1811-1815), Keith (1752-1823, 1811), Vancouver (1794), Young (1741-1820, 1771), rarely considered spatial variation of soil texture, and contained soil maps which were derived directly from geological sources, predating Smith. Morton, born a Scot, occasionally revisited places mentioned in these reports and expanded the information through interviews. He was particularly fond of Northamptonshire, where his namesake (Morton 1712) had worked over a century before. He worked intensively in Norfolk, through his largely political association with J. Trimmer (1847) and he was long resident in Gloucestershire. In some cases it is evident that they (Wright and Morton) have expanded and occasionally plagiarized pioneer material from earlier publications which had stood the test of time - the oldest that of Alexander Blackwell (?-1747, 1741), related to drainage of clay lands; followed by that of James Anderson (1739-1808, 1794) on bogs and swamps, Charles Clark (1792) on the use of lime and gypsum. Also the pioneer, but presumably not wholly original, input of William Ellis (1700?- 1758, 1764) whose writings had held their popularity, if not their credibility, for over a century.

DISSERTATION

ON THE

NATURE OF SOILS,

AND THE

PROPERTIES OF MANURE:

TO WHICH IS ADDED,

THE METHOD OF MAKING

A UNIVERSAL COMPOST,

TO SUPPLY THE PLACE OF DUNG,

WHERE THAT USEFUL AND NECESSARY ARTICLE
CANNOT BE OBTAINED:

THE WHOLE FORMING A BODY OF INTELLIGENCE CALCULATED
TO ESTABLISH FACTS THAT MAY BE APPLIED FOR
THE BENEFIT OF MANKIND.

WITH

A SYNOPSIS

OF

THE SCIENCE OF AGRICULTURE,

PRACTICALLY DELINEATED;

POINTING OUT THE NECESSARY THINGS TO BE TAKEN INTO CONSIDERATION
IN THE MANAGEMENT OF A FARM.

LONDON:

PRINTED FOR SHERWOOD, GILBERT AND PIPER,
PATERNOSTER-ROW.

1833.

Figure 2

The title page of an anonymous work, published in 1833. John Wright, a London publisher, issued a pamphlet with a very similar title in 1810 and internal evidence also supports the idea that he is the author of this text.

These, and many other texts are listed by Fussell (1950), while developments in the experimental aspects of agriculture in the early nineteenth century, and later, are treated by Russell (1966).

Morton and Wright eschew the grand Baroque manner of eighteenth century topographic writing, for instance, that exemplified by William Bray (1733-1812, 1783). The remark by Brent (1981) may be extended to them - "Darwin may have stepped out of the Biblical straitjacket rather earlier than most people have imagined," (i.e. by 1831), "and moved away from a fundamental acceptance of Genesis." In any case, no such luxury was available to Wright and Morton, for they often refer to matters teleological, presumably because their readership was used to it, and they were anxious to make their message as acceptable as it was comprehensible.

The rationale of land improvement through change of texture and social change

As was common for the time, both texts have titles of heavenly length (see Figures 1 & 2). Both claimed credibility through observation, experiment and reason (Kuhn 1970). Both urged that improvement of soil was possible through large-scale amendments of texture accompanied by chemical and organic additions. Both, in postscripts to their main work, addressed the rents and profitability of farming, - two parts of the "Cash, Corn and Catholic controversies which bedevilled Peel" (Briggs 1967). Morton, though only providing one geological base map as a soil map, (of the Whitfield Example Farm, Cromhall, Gloucestershire), and thus was a precursor of the soil survey, maintained that: "each farmer should know the nature, distribution and geological origin of the soil materials on his farm ... and prevent the error in adopting one system of culture for all kinds of soil." (M.97) Thus he could use that detailed knowledge to his advantage by paying attention to the textural requirements of crops and the response of manure, lime and other fertilisers on each of his soil classes, and "at once to see what materials are superabundant, and what are wanting." (M.96)

This message is relevant to earth sciences today for the surfaces of many long-settled landscapes contain materials derived from the addition of aberrant materials not related to the subsoil.

Wright was convinced that soil was alterable on a wholesale scale, a measure which he felt required an equally wholesale change of attitude on the part of most cultivators:

> "Discoveries of magnitude are rarely presented to the public eye for the good of society in general -- it is a misfortune which accompanies many attempts to enlarge the circle of science, that men are with difficulty persuaded to deviate from the accustomed...the injudicious Agriculturalist blunders on in the beaten track of his ancestors, being in some of his operations perfectly right, without knowing why he is so; and in others egregiously wrong, yet unable to detect the cause of his error." (W.2-3)

The universal compost - alteration of surface soils

His avowed intent was to urge all farmers to use whatever reasonable materials were available to make a replacement or supplement for manure, and to improve all land with a "universal compost." He found in this regard that:

> "by chemically examining the component parts of dung, a unctuous fat principle, an alkaline salt and a certain proportion of carbonaceous matter (are found), which Nature has united by putrefaction."..(the surrogate "universal compost") has these qualities, united by a natural process, (which) render it suited to all types of soil and crops. (W.3)

His basically simple recipe for this "universal compost" to replace "dung" is rarely clearly stated and is subject to many modifications, but in its essentials it seems to be (W.131):

2 parts (112 lbs.) mineral alkali, (muriate of soda), (viz. common salt)
1 part (50 lbs.) vegetable alkali, viz. English potash (the inferior Russian or American may be used in its stead)
1 part (36 pounds) (4 gallons) whale oil (oleous, or fat, substance)
1 part (50 lbs.) (one bushel) quick lime

Salt was a common ingredient of fertilizer, for taxes on it were relatively low and whale oil was a commonly available, under-used, commodity. Wright had several aims in promoting widespread use of this efficacious compound: the improvement of land, the generation of local service industries, retention of labour on the farm and increases of trade and profit. The "minor considerations" he adjoins are that "at least ten millions of acres could be so treated" in Britain (of a total farmed area of 50 million acres) requiring about 910,000 tons of the above ingredients, and he calculated that "five hundred million acres if extended to the Continent of Europe, could be so treated", and this likewise at roughly 200 lbs. per acre. As substitutes for, or additives to, the compost to be spread on the soil surface, Wright cited "ditch cleansings, peat earth, soap ash, night-soil" - ("the richest of all manure" (W.67))-"oil-cake, sea-weed, mud, wood and coal ashes."..."bone or malt dust, soot, salt, sea-sand, manure shells, offal, burnt clay, street-sweepings and road scrapings."..."refuse of manufacturers, (instead of being suffered to run into rivers rendering their waters unwholesome"!)..."lime kiln refuse, gypsum, greaves, wooden rags, curier's shavings, furrier's dippings and leather cuttings"!

For such ideas, perhaps fortunately, Wright (and Morton with his mineral additions) had a relatively limited market - the larger landowners, some reclaimers of land and the larger enclosures, and those tenant farmers who had transferred the burden of rates to their land owners, or in the words of Cobbett, (quoted in Jones and Pool 1959), had "substituted the old-fashioned substantial way of living with a gimcrack magnificence." Nine-tenths of all farmed land was under lease, many farm labourers had fled to the towns or sought poor relief and by 1821 many farmers were bankrupt (Crosby 1977). Thus Morton travelled, Cobbett spoke, and Wright advertised, because they sought to neutralize the effect of the Corn Laws and of Speenhamland on rural life and diet and wished to circumvent the prevailing and convenient Malthusian precept that "in the long run it was in the interests of the poor that the receipt of relief be made as disagreeable as possible...and a spartan diet be enforced." (Crosby 1977).

Soil characterisation and the differences of soil and rock

Wright's ideas on soil advance the subject in significant ways, or make a clearer synthesis of established ideas. Many ideas are original, but have not received credit because he applies everyday parlance to observations and ideas which only later came to be expressed in scientific terms. He often accepted then- current, but now forgotten terms for land types, such as those presented in another study (Bunting 1975). He clarified and expanded the existing textural classification of soils and related it to varying degrees of decomposition of organic wastes.

> "A soil inherits a proper texture...But if nature has not provided this happy mixture, assistance by application of an alternative" (rather than an alimentary) "manure is possible." (W.28)
>
> To this he adds "sandy, gravelly hungry soils are most benefitted by argillaceous and calcareous marls as alterations of the fat and oily principle." (W.423, 103- 104) and "Properly speaking, the upper stratum is a vegetable mould...mixed with large portions of clay, chalk or sand. Hence proceeds the differences of soil." (39)

Wright defined soils as "formed by combination of two or more primitive earths, more or less united with other materials, mineral, vegetable and animal." It is therefore (W.26):

> "evident that they must vary very much, both in qualities and proportion of ingredients...In some districts, one sort of material is abundant, consequently it enters largely into the composition of the soil; in others it is deficient."

This abundance or dominance was regarded as having four results, while others related to origin: (42-44, 99-105)

- i "Calcareous or siliceous matter, often in great abundance" - chalk soils or loams.
- ii "Argillaceous and heavy loams giving a material difference" - clay soils, or clay.
- iii "Some soils abound much more with animal and vegetable matter." (some peats and vegetable moulds).
- iv "Sandy soils and gravelly loams; ferruginous loams."
- v "Alluvious soils" or "warp."
- vi "Boggy, Peaty and Heathy soils."

Soil, too, was regarded as materially independent of the rock beneath, in "appearance, origin and properties" related implicitly to climate, organisms, and other factors, usually represented by state of cultivation or of processes, in turn indicated by color - "For The All-Wise Disposer has ordained that the whole earth should be covered with plants, but every soil is not equally fit for the same plant." (38)

> "we may perceive the necessity of attending the quantity of rain that falls, and also the climate in order to form a true judgement of the soil." (30)

> "The soil at the surface...is much better calculated for the purposes of vegetation than the under-stratum, the soil ... in which plants grow is a vegetable mould ...intimately mixed with primitive earths (and) contains oleaginous, alkaline and carbonaceous principles which are the food of plants...having been impregnated with ... rain water and gaseous principles of the atmosphere also tending to promote their fertility."(28)

These and many other statements give an alternative and earlier view of the soils as affected by factors, pre-dating the much-vaunted claims of late-nineteenth century Russian writers.

Geography of soils: factors of formation

A pioneer attempt to represent and explain the "most fertile soils of Europe" is derived and tabulated from his account (W.29):

Location	**Latitude** *North*	**Rainfall** *inches*	**Percent** *Silex*	*Argil*	*Calx*	**AUTHORITY**
TURIN	45°04'	40	77-79	9-14	5-12	Gilbert
UPSAL	59°52'	24	56	14	30	Bergmann
PARIS	49°50'	20	46-56	11-17	37	Tillett
LONDON	51°31'	23	50-70	20-30	10-20	?

This data is interpreted somewhat cautiously (30-32):

> "the soil in drier climates will be more retentive, with greater clay and lime contents, and in a moister climate (it) is more open and porous, correctable by alternative manures (and) on the achievement of the ideal texture, there is then need of alimental foods (which function in) an oxygenated soil" or "are impregnated with the vital principle of the atmosphere" (which)" renders it productive of abundance."

He observed that

> "it is also certain that all water, whether it flows within or upon the surface of the earth contains some earthy particles in solution."

However he is here clearly referring to salts and carbonate and "pyritous substances" rather than to clay migration, though it is possible he assumed both particulate and solute transport. He suggests a class of "Vitriolic soils" impregnated with "sulphuric acid and iron" -

> "frequently abundant about coal-mines (where) it is scarcely ever worth the trouble of cultivating for grain." (33-35)

Soil analysis

Directions for performing seven methods of analyzing, or at least testing, soil by experiment are outlined, presumably elaborated from Davy, with some refinements (34-38):

1 "Quantity of water",
2 Proportion of sand and clay,
3 Calcareous earth (by filtrating with muriatic acid, weighing the dried precipitate).
4 "Calx in marl" (loss in weight of total earth after acid digestion),
5 Presence of metallic or earthy salts,
6 "To know if the salt contained be metallic or aluminous" - "by infusion of galls" and,
7 Presence of neutral salt.

Soil 'touch' and colour

Though Liebig visited Britain in 1837 (Russell 1966), Wright had already deemed these "precedaneous methods" as "laboreous" and

> "for all the ordinary purposes of agriculture, a sufficient degree of accuracy.may in general be obtained from external appearance and from colour..the touch also serves to direct the judgment as to the quality; for when due proportions of clay and sand are blended, the soil will not adhere much to the fingers in handling." (37-38)

His use of colour relates to red, white, grey, yellow and blue, the latter associated with wet compact clays. He noted the darkening of more indurated sands and the diversity of organic content of chalk soils.

Perception, right-emphasis and unity in soil description

"Properly speaking," Wright deemed it:

> "unnecessary to enter into philosophical minutiae, in defining the nature and properties of soils For the sake of perspicuity, we follow the popular method, and divide them into Clay, Chalk, Sand, Loam, Boggy and Heathy soil." (40)

This is, in fact, a simplification of the 1665 Georgical Committee of the Royal Society which formulated eleven "kinds of soyls of England", quoted in Russell, (1966), who states that "this brilliant outburst of activity was followed by a long period of groping for new ideas that always eluded capture."

Wright discusses the ideas of Davy (106-107) in fertilizing these soils, the appropriate textural additives to each and the role of his universal compost in their improvement and yet, using manures instead of fallow was still not acceptable to farmers, even if they had heard of the practice. However, not subject to editorial processes either, Wright was adept at, or oblivious to, contradiction:

> "no solid progress can be made in the art of Husbandry without an attentive application to the principles of Philosophy and Chymistry.which directs him in the proportions of vegetable and animal matter to be employed" (in the universal compost) (15) and practically: "by exchanging the soil of one spot in a field for that of another" (117). "From these the Agriculturalist, by scientifically examining the nature of the soil and the properties of manure, he may learn to apply the latter to the benefit of the former." (15)
>
> He deemed "the application of lime to a calcareous soil...is rather injurious" (123) and warned "when to check the putrefactive process, so as to prevent the fertilizing powers becoming effete." (101-102)

Morton, likewise, was guarded in his assessment of the effect of lime, (vide infra) Just as Darwin is claimed to have derived "a sense of unity of nature" from the writings of Baron von Humboldt (Brent 1981), especially the *Narrative of Travels to the Equinoctial Regions*, (Ehwald 1960 Wilde, 1963), so was Wright imbued with this idea so typical of the time:

> "we cannot help being struck with the wise ordination of Nature: but what makes it all the more astonishing is, that all orders of vegetables are produced from so small a number of natural substances; gas, or air, water and carbon, to which may be added caloric and light," though the results "convince us of the unbounded comprehension of the Divine Mind" from "causes too minute and intricate for our feeble comprehension." (58-59)

Yet, reasserting the obligation of the early scientist to advise:

> "by a judicious mixture of the nutriment of vegetable substances, the proper food of plants can be imparted to the soil." (121)

Wright had a profound grasp on the unity of the subject:

> "Notwithstanding the varied appearance of the earth under our feet, and that of the furrows of the field, whose diversified strata present substances of every texture and shade; yet the whole is composed of but few earths." (116)

Soil and natural processes

The author of this 1833 treatise represents a transition from teleology to objectivity. He had not, like Sedgwick, "begun in the conviction that the truth of divine revelation could never be falsified by scientific truth" (Brent 1981), rather that what he could not understand of "the mysterious operations of Nature (that) take place unseen to human eyes" may be "accredited to the Wisdom of the Creator" who was not entirely interchangeable with "Providence", who/which:

> "has neglected no region intended for the habitation of man; for if, in hot climates the heat meliorates the soil, it is no less...than by the frost in cold; yet frost has no effect upon sand or dry earth." (93)

His allegiance, hence, is to natural forces, wherever possible, even though far superior to those allotted to man:

> "Frost...upon wet earth, puts every particle out of place, a process at once simple and powerful, and far superior to the plough, reaching the minutest parts and so too, to any operation that could be invented by the mind, and performed by the hand of man." (92-93)

Nor, for the record, could he tolerate:

> "The antiquated and unphilosophical notion, of leaving the ground to rest, (it) is too absurd to require refutation: all that is necessary is to interchange the succession of crops,...as a garden is planted without cessation." (98)

But this, like the "infinite service of snow" (58) and the "congenial influence of the sun's rays" cause

> "various exhalations from the earth, from the oceans and streams;...and these "ascending in the upper regions of the atmosphere...are carried about by the wind until condensed, they fall upon the earth ...affording food for man and beast." (65)

All these processes are:

> "beautiful instances of omniscient wisdom; and the Philosopher cannot but notice, with profound admiration, the Creator's providential care over the minutest of his works." (59)

He summarizes his case:

> "the most fertile soils contain the greatest quantity of the food of those vegetables that nourish man...the essentials requisite to a fruitful soil, are; a due mixture of the simple earths, fully charged with the food of plants; and the texture such, as to enable it to admit and retain so much water as is necessary to vegetation, and no more."(84)

To this he linked an idea derived from the emerging views of human physiology:

> "If we analyze the property of soils, and also the nature of Manures, we shall find that the earth is the stomach of plants in which their food is concocted, and naturally prepared; and that the finest and most subtle parts of its soil is their chyle, from which they receive their nourishment." (17)

From these statements we see a pioneer attempt to characterize the soil as an independent, vital, living entity, interacting with and interdependent upon climate and plants, capable of change and being changed by man's action, perhaps on a scale too considerable for most operators of the time in regard to Wright's proposals, and yet expressing a long-lost desire for a wholesome combination of improvement of land, use of waste materials and solicitude for society and its environment.

John Morton and the geological approach

Morton had a more thorough knowledge of the geological work of his time than did Wright, and more respect for it than had Darwin in 1835:

> "Geology is a capital science to begin, as it requires nothing but a little reading, thinking and hammering." (*Letter* to W.D. Fox, from Lima, July, 1, 1835, quoted in Burkhardt and Smith, 1985).

Morton's text, in its enlarged fourth edition of 1843, has 68 sections totalling 234 pages, with the individual farm reports extending to page 432. Sections 4 to 27 (pages 3 to 95) consider the rocks and soils of Great Britain individually, starting with the youngest; before outlining a classification of soils (Section 28 to 35), then examining the current ideas of plant growth and climatic action (Sections 36 to 40), mineralogy and soil texture (Sections 41 to 47), an outline of the "Best Constituted soil" (48), and methods of land improvement (49-60), the concluding sections discussing rotations and the role of land, labour and capital in the profits of agriculture. Some of his ideas on soil classification, mapping and land improvement have been outlined previously by Bunting (1964 a, b, c,) and Russell (1966), quoted in Yaalon (1988).

Morton's view of a soil body independent of solid geology

Though one could suppose from the title, and Morton's initial definition of soil, that he was solely interpreting soil as derived from rock: - "the surface of the earth partakes of the nature and colour of the subsoil or rock on which it rests," (M.1) yet Morton cites many departures from this idea. "Lime, potash, and iron...are acted on by the atmosphere, and the rock is decomposed; some of it into fine impalpable matter"..."with the addition of vegetable and animal matter, in every state of decay intimately mixed with it;"..."Iron becomes oxidized and gives a redder colour to the soil than that...exhibited by subsoil." (1-2)

Morton gives many examples of the variability of soils within a given rock type, according to the changes within the lateral extent of its outcrop; to changes related to variability of stoniness and he also frequently cites examples of more successful cultivation on materials of varied texture with stones, than on stoneless soils:

> "as with the more extensive Oolite, the variation of the rocks gave rise to well-mixed textures and productive soils but worthless if one component dominated, the soil was best when mixed with finely fragmented rock," (but was) "*dead* or *sleepy* land where these were absent, in a close brownish soil which forms a close impervious crust, agglutinized together by the rain...and vegetation ceases. To look at, of a good quality, and considerable depth, but unproductive and worthless." (60)

He frequently maintains that soils have differing properties according to the varying dip of the rock strata. The soils on Gray Wacke, and clay slate are:

> "of a thin shellotty nature, loose fragments between solid rock and soil, embedded in a reddish or grey shivery substance formed by decomposition. If the lamina of the clay slate are parallel with the horizon the soil is a free tender loam, if much inclined it is a greedy soil...In the valleys the decomposition is more complete...formed into a brownish clay that is of little value, being cold and wet." (90-91)

Morton also showed that there is, frequently, a varying response to, and stability of, soil structures after rain, within otherwise uniform materials:

> "bluish or blackish when wet; brown, pale or yellowish or grey, near the surface, when dry ... in dry weather opens into perpendicular cracks for a considerable depth ... with horizontal layers of nodules."..."The soil is slightly calcareous, sticks to the plough like pitch, and chokes if ploughed when wet." (20-21)

Departures on a regional scale from the idea of a close link between soils and subjacent rocks relate to features attributable to erosion and slope wash - the intermingling and mixing of materials at the margin of geological outcrops such as at the Greensand and Chalk boundary, where the soils are:

> "so blended together as to render it difficult to perceive the line that separates them...the sand increases as we recede from the chalk, and the calcareous matter increases as we approach it."..."The soil is greenish mottled, friable, rich and productive." "If sand abounds a deep loam is formed, in Kent forming the best garden land, but, owing to the mode of culture, for the soil is very weak crumbling"..."The black sand and white silvery sand are the worst soils, and naturally produce nothing but Heath...In Dorset a fox land occurs - rich, tender loam of a brown colour"...(presumably that of the fox). (38-39,40)

The possibility that soil profiles relate to, and are developed in, residues from overlying formations, especially in the south of England on the Chalk, and in East Anglia, is frequently cited.

Morton gives no direct indication of either acceptance or rejection of "diluvial" ideas, though several times he clearly states that the soils in the northerly parts of British outcrops are different to those in the south, either in depth, stoniness, compactness or variability, implying the perception, but not the knowledge, that glaciation caused this regional contrast. For example in the New Red Sandstone areas:

> "soils so indurated, tolerably compact, with clay beds for mixing, with varieties of red argillaceous marl, unctuous in the south-west, very responsive to liming...but in the sandy tracts of Nottingham and Yorkshires...poor, barren, soft, light sand." (71-72)

and, indicating fluvioglacial activity, or else ancient-valley gravels:

> "calcareous gravel, not larger than the size of beans, from Lincolnshire into Wilts., mostly resting on clay rather than the edge of shelley

oolite...evidently formed from the washing out of the valleys in the north-west of these (Oolite) hills." (60)

He frequently refers to different occurrences of 'Diluvial' materials:

"There is a diluvial deposit on some parts of (the Oak Tree, or Weald Clay), composed of a flinty iron sand and gravel"..."There is a great portion of the upper chalk, which is covered either with a thin coating of sand (35) or of vegetable mould" (36). Of the "Oxford or Clunch Clay" it formed "the rich pasture land of Dancy in Wilts" on "brownish clay loam", (54) to the "close heavy compact clay of Northampton shire" which "soil is not only difficult to work, but is the most expensive of all clays to cultivate" (54-55)..."On the surface, there are frequently beds of gravel, formed of very small rounded gravel, sometimes agglutinated together with a calcareous cement." (56)

Factors of soil formation

In attempting to explain the character of soils he invokes what we now term other factors of soil formation - climate, in both the annual, regional and daily aspects of wetting and drying, of frost action, leaching and erosion. His attempts to show the input of plants and soil fauna are less frequent, but he cites the importance of earthworms, of roots, of crop residues, though more in relation to crop production than soil character. Surprisingly he frequently cites the activity of ants on clay soils, presumably the areas long unfarmed, and certainly at variance with modern ideas of their dominant habitat.

His clearest statement on the factorial approach to soil formation relates to his qualified acceptance of the idea of an optimal combination of sand, clay and lime in a "best constituted soil" which "proportions depend entirely on climate, situation, the nature of the subsoil and other local circumstances." (118)

Morton had a very close appreciation of the role of climate and of seasonal weather. He referred to the variation of rainfall within Britain, the changes in humidity away from the coast, the great alteration to climate consequent on land reclamation, as in the Fens, and the different problems of sowing and harvest, and the timing thereof, on sands and clays (201-202). He commented on the effects of altitude and of the danger of assessing soil quality as poor "if cultivation is neglected or improperly or carelessly executed."

Many comments relate to a perception of the importance of landforms or relief, and of an appreciation of the role of biotic activity in a distinct surface "mould", not necessarily a "vegetable mould" as expressed by previous authors.

Morton cites the "low, level aspect of the Greensand"; the "smooth rounded, never rugged aspect...of the chalk hills"... "where the district is broken into irregular parts"..."intersected by deep winding valleys...generally without water in summer."

He contrasts (21) the "low, uneven gentle-waving surface, with sufficient slope for drainage" of the London Clay, having "small risings, no abrupt deviations"; with the "external character" of the "Plastic Clay" which "is rather hilly in some places" where the "sands and clay are alternate" (25) and "when beds of gravel exist...there are large reservoirs for water." He also maintains that thin layers of clay are "more easily improved" but "some new impulse must be given to agricultural speculations, before the cold wet clay soils will ever attain that degree of improvement which they are capable of" (146).

Mention has been made of Wright's intent to increase the organic matter of soils by every means available. Morton was more cautious - "Manure applied to soil increases its vegetative powers, but the way it acts is not well understood" (132) and he discusses "the use of cold manure for light soils and hot for clays" (169).

He did not appear to know of, or at least use, Darwin's idea of the development of a surface mould through the activity of earthworms, burying layers of cinders and burnt marl to a depth of 4 inches in 15 years (Darwin 1837) (See Figure 3 below).

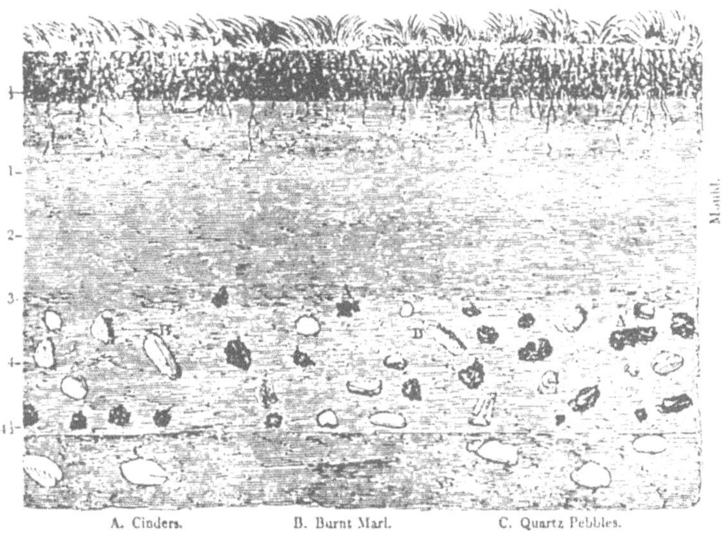

Figure 3

Charles Darwin's representation, in 1837, of a soil profile, with burial of cinders to depths of 4 inches by the action of earthworms in 15 years.

Of great interest to geomorphology are the references to soil formation processes and the precursors of concepts which may be of relevance to the history of biogeochemistry. Examples will be cited to show that Morton had a thorough grasp of a theory of leaching, of salts, lime and of particulate matter. He described the mottling associated with wet soils, the decomposition of roots and of straw, and the role of storm water in the eroding of some clays, but not others.

An outline of Morton's observations on the nature of soils on the major geologic materials will illustrate many of these interpretations.

Soils on recent deposits

Morton devotes four sections of his text to soils on post-Tertiary deposits. He explains many local differences in alluvial soils and their use and shows a qualified acceptance of a Universal Deluge. He essays an outline of many different forms of Diluvium, especially those "resting on the Chalk Formation", and queries the correctness of the use of the name for many of the sandy areas of East Anglia. He saw that much of the earthy matter of the Diluvial deposits of southern England was of the same nature as some of the beds of Plastic Clay, (late Eocene Woolwich Sands and Reading Clay for example) which had been altered in situ to lie as Diluvium over older geological materials:

> "The Plastic Clay forms...every variety of soil owing to the rapid succession of sand and clay and other materials it contains"..."There are materials at hand to mix a good friable loam and pulverizing it by manure" though the "only difficulty is that the soil is easily carried off by the stream, which forms deep gullies when the water is in any quantity." (27-29 passim).

Soils on pre-Tertiary clays

Morton had similar comments to make on the other major clays of Britain - the Gault (Section 12), the Oak Tree or Weald clay (13), the Oxford Clunch or Fen Clay (16), the Blue Lias (18) and the Graywacke and Clay Slate (25).

He contrasts the "solid compact mass of impervious clay of the Gault" with the "irregular...slaty appearance" of most of the others. He shows that the surface soil within the Gault outcrop derives calcareous matter by slope wash from the adjacent Chalk hills, but that:

> "the subsoil does not effervesce with acid"..."It is so impervious that no water is retained within it, and of course there are no springs. It is kept in a wet surface by surface water. The surface has a mixture of very small rolled and angular flints. It has marl in the Oxford to Bedford areas (Vale of White Horse), and is very productive; it is thin and cold in Cambridge and Huntingdon shires, (where) it is the most worthless district. When slightly wet (it) is sticky as glue, and frost makes it as porous as a honeycomb." (41-43)

Of the Weald Clay he considered that:

> "the surface is very uniform",..."it is a close compact substance"..."no particle can be perceived by the touch"...(it has)..."a darker fawn or hazel colour, a pale white, yellowish sickly appearance, cutting like a piece of soap"..."When once soaked when wet, it requires a long time to dry, as the atmosphere has little or no effect upon it." (46-48)

> "The Oxford or Clunch Clay rises into low round-headed knolls"..."is dark bluish"..."turns brown by exposure to the atmosphere and becomes strong and adhesive. The frost reducing it to fine powder and though it looks like a good loam, is very thin and poor."..."This formation is so close that no springs are found in it, it is wet from the coral rag above it." (53-55)

Soils of the limestone areas

Morton, like Cobbett, (1830 and 1855) had a close knowledge of the limestone tracts of England and was well aware of the many non-calcareous soils of these elevated areas. The calcareous soils on chalk are described as:

> "very thin, soft, weak and full of fibres...a light hazel mould or flinty chalky mould, dry and friable called white land...all contain flints in more or less abundance". "The lower chalk has a fine chalky loam or white malm, without flints"..."a most productive soil of the finest quality." (34)

Contrasted with this were:

> the "rolled chalk marls" of East Anglia, "rubbly"..."as if it had undergone the wearing process of moving water" (37).

In the northern areas of the Chalk, the climatic conditions "of cold and exposure to east winds" were a severe limit on productivity. Paradoxically, the use of bones is recommended as the best improvement of Chalk Soils.

The soils on the Inferior Oolite were regarded as:

> "calcareous, ferruginous and micacious"..."a soil of good depth, friable, soft sand and deep brownish in colour, dry and healthy, draining off all superfluous water." (61-62)

Soils of sandstone formations

Morton had many insights into these soils, pointing out the admixture of materials at the margins of the formations and the additions from weathered, newer, overlying materials, as he had done for the limestones. He refers to the "green earth of chlorite" of the Cretaceous sandstones and recommends deep ploughing of most sandy soils, with or without an additive mixing of other finer material - chalk marl or clunch clay - to improve the texture:

> "to enable receiving or transmitting of moisture more freely, so it is neither so easily injured by wet weather or excessive droughts."

The Iron or Hastings Sand is regarded as capable of similar improvement:

> "a fine silicious sand, mixed with clay and a large portion of ferruginous ochre, a yellowish fawn coloured or brown sandy loam which is very weak." (48-49)

This is very different to his assessment of soils of the New Red Sandstone:

> The "red, rich friable, marly loams" are frequent in the non-sandy parts of its outcrop: "slippery and greasy when wet, a soapy feel when dry, and often covered by a thin Diluvial gravel, rounded, with boulders belonging to the most primitive of rocks and...much furrowed into hill and dale, and exhibits a variety of beautiful undulations, consisting of little flats and gentle swells." (68- 69)

Being a Scot, Morton has some penetrating comments on the varieties of soil found on basalt, regarding it as:

> "easily decomposed from the iron and potash which they contain...Reddish, brownish or greyish soils...Friable, well-drained"...(which under proper management) "become of great fruitfulness." (93-95)

Association of soils and subsoils: the basis of soil classification

Adopting this geological outline, Morton showed "an intimate connection between soil and subsoil", which was "useful in forming a classification advantageous to the agriculturalist":

> "It identified their peculiar properties, the mode of culture best calculated for the kinds of crops produced. It could identify the superabundant materials and direct us to materials to produce the alteration of texture needed,...prevent us using injurious materials and direct the agriculturalist to a more minute study of the nature and property of soil."..."Clay, lime and silex" (were the) "distinguishing features in any association of soils." (96-97)

The Associations he suggested (98-103) for his classification were three:

I "*Aluminous soils (clays) with associated silicious matter in an impalpable state*"

1 "Non-calcareous aluminous soils"
2 "Calcareous aluminous soils" (deficient in silicious matter)

II. "*Calcareous soils" (some clay, no silicious matter)*

1 "Impalpable soils" (marls, and oolitic clay)
2 "Calcareous gravelly soils" (Upper chalk and shelly Oolites)
3 "Silicious, calcareous, fragmented soils" (older limestones and Coral Rag)

III. "*Silicious soils*"

1 "Dry sands", "loose", (with little aluminous and no calcareous matter)
2 "Gravelly strong soils" with some clay, (most "Diluvia" on acid clays and older sands)
3 "Fragmented soils" (on graywacke and some basalts)
4 "Calcareous fine sands and marls", (with silex and clay) - "of the first quality"

This classification bears a close resemblance to the existing ideas of "primitive earths," with the additional subdivisions according to lime status and content of coarse materials and with some reference to soil consistence. In detail, it also implies in situ and lateral mixing of different geological materials. It also excludes, at this secondary level, the effects, because locally varied, of wetness, organic matter content and depth, except where implied. Yet Morton clearly cites these as tertiary influences, appreciable to each farm owner. Morton quoted

Davy, at least in the limits of sand - "seven-eighths of silicious matter." He also considered that "an excess of sand is much less injurious than an excess of clay." (112)

The nature of clay, comparison with sand and the use of lime

Davy's definition of a clay soil- "at least one-sixth part"- is adopted and clay was considered as:

> "tenacious, compact, adhesive, its particles have a strong attraction for each other and for moisture, it retains it with the greatest obstinacy;" (113) and "it retards the progress of decomposition in vegetable matter." (128)..."unmixed with silex, (clay) is barren, it forms an unctuous, clammy soil, white, yellow, grey brown or reddish in colour." (131)

For Morton, soil formed of "pure carbonate of lime was also barren and unfruitful." This substance in soil "is either fine, impalpable, or hard rubbly shape." (114)

> "Calcareous loams retain carbonaceous matter" both "natural and that applied as manure" (as opposed to) "silicious sandy soils which soon decompose manure",(hence) "hungry soils." While "the remains of mould...forms a soft, light black substance; the cause of blackness in garden mould which has long been in cultivation: the French have given to this substance the name of Humic acid." (115)

Morton had very clear ideas on liming, unlike Ellis (1700?- 1758), 1764)- who interpreted the treatment of sour clay lands by liming:

> "calcination sets free and enables to act a balsamic alkaline salt that is coagulated in the crude stone or chalk and till the acid barren quality is evaporated by fire, the salts in them are of little or no signification to the land."

Morton, having quoted Davy on lime and liming, posed these questions:

> "Is lime only a stimulant exerting its influence on something that is already in the soil? and if so does it exhaust that something? or is it an enriching manure?"..."From pretty long experience we are inclined to think it acts both as an alternative and as a stimulant; operating in the one case as a medicine...in the other, as arousing...the vegetable powers...contained in the soil...which would have remained dormant without it."..."Lime also acts powerfully on any irony matter and on the gravel sands and clays..." (173-174)

Morton was also well aware of the differences in moisture retention in the main associations of his classification scheme:

> "Deep sandy or silty soils when there is water in the subsoil within three feet of the surface, are less affected by long drought than strong clays in the like situation. This is owing to the capillary attraction..Light sandy soils

have been greatly improved thereby, by new systems of cultivation, while the poor clay remains as they were a century ago". (132-135)

A "best-constituted soil" and impediments thereto

Morton, building on the three-fold classification and the varying proportions of clay, sand and lime in each of his classes, essayed an approach to the idea of "the best constituted soil", defined as:

> "That in which the earthy materials, the moisture, and manure, are properly associated; and on which the decomposable vegetable or animal matter does not exceed one-fourth of the weight of the earthy constituents."..."The soil being therefore, merely the reservoir of water, air and heat, and decomposing organic matter"..."yet it may be rendered either fertile or sterile by giving it the power of storing up these elements for use in much greater quantity than before."..."thus forming a natural laboratory." (117-118)

He thought that the proportions should be nearly equal parts of sand, clay and lime, though he quotes Kirwan (probably Richard Kirwan (1733-1812), 1801), to the contrary:

> "the proportion where rain to the depth of twenty-six inches falls per annum, is fifty six per cent of sand, fourteen of clay, and thirty of calcareous matter." (118)

Somewhat taken aback by this honestly-held illusion, Morton adds:

> "But these proportions depend entirely on the climate, the situation, the nature of the subsoil, and other local circumstances: More silicious sand is required (if) the soil (is) wet; and more clay, if dry." (118-119)

which represents a first approximation to the factorial study of soils. He favours "50 to 70 per cent silicious, 20 to 40 of clay and from 10 to 20 of calcareous matter."

He continues by examining the role of drainage:

> "The soil which is best adapted for retaining and transmitting, in all circumstances of wet and dry weather, the necessary quantity of moisture to growing plants, may be reckoned the best and most productive." (However) "The energies of the soil are frequently held in bondage by some pernicious quality inherent in it, or imported to it, which if neutralized or extracted, the soil would become productive."..."If clay is in excess, it may be remedied by application of sand, chalk marl or light manures. Peaty soils may be corrected by burning, by addition of sand or any thing heavy. Lime destroys sulphate of iron"..."None of these applications will have a desired effect unless a perfect subsoil drainage removes superfluous moisture". (119-120, 133-135)

Similarly the effect of "stagnant water" caused a soil to acquire this "pernicious quality" for it "melts down the particles which compose it."

Feasibility of major changes to soil texture and composition

Morton was enthusiastic about the many possibilities for changing the texture and lime content of soils, justifying the expense in terms of future benefits:

> "To alter the nature and properties of the constituents of any soil may be more expensive than to manure it; but the effect of the former will be lasting, while the latter is transitory; the one permanently improves the nature and quality of the soil, the other only imports a temporary excitement." (130)

From 1840 onwards (Jones and Pool 1959) the cost of labour on the farm was only exceeded by its paucity and the relatively massive measures inherent in both Wright's and Morton's methods were no longer possible. As well, the use of chemical fertilisers was increasing as Liebig's ideas, and the results of experimental trials, became more widely known (Russell 1966).

Most agricultural writers of their day and the writers of the Board of Agriculture recounted observations - urging a methodology by example. Both Wright and Morton travelled widely, the one to universalize the application of a compost, the other to instruct farmers to study their soils, especially their cultivability and their potential if managed wisely. Their view of soil was not as a static mass, but as a changing and changeable dynamic entity, only partly related to the rock beneath, but mixed by internal processes, with additives from superposed now removed geological materials, subject to gains and losses by lateral slope processes, to change by moisture and frost, to change by cultivation and additives, and functioning as a support for plants. Their knowledge of organic matter decay and plant nutrition was admittedly deficient, but there is an implicit longing for a finer microscopic view in their writings, and an appreciation of the need to work at a minute scale:

> "The processes of the small rootlets are so very minute, that no crude substance can pass through them; it can therefore only be taken up by them in the form of water or gas, and be absorbed by leaves...and we think the result of our researches will convince us that the food of plants or the principle of vegetable life, will appear to consist more in the condition and constitution of the soil, than in any single or compound specific."

Although Morton's apophthegm was that:

> "the soil on each geological formation is composed of the same materials as the subsoil on which it rests, yet we do not wish to infer that the soil over the whole of each formation is of the same value"..."the quality of the soils is infinitely varied, and the productiveness either arises spontaneously, (or) has to be drawn from it by culture."

Thus, to modify Yourcenar (1984), Wright and Morton expanded and refined our view of soil as an independent part of nature to an even greater extent than they proposed its modification.

Conclusion

The development of ideas relating to soil character and composition in Britain in the fourth decade of the nineteenth century is best exemplified in the work of Morton, and an anonymous precursor, presumably J. Wright, who both presented many clear ideas of the relation of rock type, drainage, soil texture and soil productivity, which ideas contain a clear indication of their appreciation of surface geomorphic processes and the dynamic interaction of climate, biota, and cultivation in the differentiation of surface soils from geological substrates.

That their work fell from favour and was long neglected may be due to the impracticability of the soil improvement methods proposed - of large scale mixing of physical components, or the massive additions of a dubious but Universal Compost, respectively. The nature of field observations clearly shows a grasp of many modern concepts -soil structure, profile dynamics, frost action and of the origin of many recent deposits and of the associated slope processes. Both expressed the desire for a soil classification of practical utility and of a factorial approach to soil formation and land capability assessment, for just as "a poor soil is dear at any price" (W.122) and "the richest land produces the sweetest pile of grass" (in the old-fashioned sense) (W.11) so:

> "By an improvement in nature, no class of individuals will be injured, but all degrees of society more or less benefitted: the landed interest...the agricultural...and the mechanic and laborers thereby enabled to...bring up their families with decency and credit." (W.6-7)

References

Anderson, James, 1794. *A Practical Treatise on Draining Bogs and Swampy Grounds*. London.

Blackwell, Alexander, 1741, A New Method of Improving Cold, Wet and Barren Land, particularly Clayey Ground. Walltove, London.

Bray, William, 1783. *Sketch of a Tour into Derbyshire and Yorkshire*. White, London.

Brent, P.L., 1981. *Charles Darwin*. Heinemann, London.

Briggs, A., 1967. *The Age of Improvement*. Longmans Green, London.

Bunting, B.T., 1964a. John Morton (1781-1864): a neglected Pioneer of Soil Science. *The Geographical Journal*, 130, 1, 116-119.

Bunting, B.T., 1964b. Pioneers of Soil Science - A British View. *Soil Science*, 97, 358-59.

Bunting, B.T., 1964c. John Morton, F.G.S. (1781-1864) an Early British Pioneer of Soil and Land Classification. *Transactions of the 8th International Congress of Soil Science*, Bucharest, 981-87.

Bunting, B.T., 1975.The Language of Site. South Pennine dialect terms and their use in geomorphology: An approach towards the terminology of site." in *Environment, Man and Economic Change. Essays presented to Prof. S.H. Beaver*. Longman, London.

Burkhardt, F. and Smith, S. (eds.), 1985. *Correspondence of Charles Darwin*. volume 1. 1821-36. Cambridge University Press.

Clark, Charles, 1792. *Treatise on the Earth called Gypsum; with an Account of its Extraordinary Effect as Manure*. London.

Cobbett, William, 1830. *Rural Rides*. New edition, 1855, with notes by Pitt Cobbett, London.

Crosby, T.L., 1977. *English Farmers and the Politics of Protection*, 1815-1892. Hassock,, London.

Darwin, Charles Robert 1837. On the Formation of Mould. *Transactions of the Geological Society of London*, 5, p. 505 - 509, also briefly cited in Darwin on Humus and the Earthworm. Faber, London, 1966.

Daubeney, Charles G.B., 1847. On the Distinction between the Dormant and the Active Ingredients of the Soil., *Journal of the Royal Agricultural Society*, VII, 237-44.

Davy, Sir Humphry, 1813. *Elements of Agricultural Chemistry*. Longman, London.

Ehwald, E., 1960. Alexander von Humboldt and V.V. Dokuchayev. *Albrecht Thaer Arkiv*, 4, 8, 561-82.

Ellis, William, 1764. *The Practice of Farming and Husbandry in all Sorts of Soils*. 2nd Ed. Dublin.

Ernle, Lord; Hall, Sir A.D. (ed.), 1936. *English Farming, Past and Present*. Longmans, Green & Co., Ltd., London

Farey, John, 1811-1815. *General View of the Agriculture and Minerals of Derbyshire*. 3 volumes, London.

Fussell, G.E., 1950. *More Old English Farming Books, 1731-1793*. Crosby Lockwood, London.

Hull, D.L., 1974. *Darwin and his Critics*. Harvard University Press

Jones, G.P. and Pool, A.G., 1959. *A Hundred Years of Economic Development in Great Britain, 1840-1940*. Duckworth, London.

Keith, George Skene, 1811. *A General View of the Agriculture of Aberdeenshire*.

Kirwan, Richard, 1801. The Manures most Advantageously applied to Soils and the Causes of their Beneficial Effects. *Transactions Royal Society of Dublin*, 2, 2.

Kuhn, T.S., 1970. *The Structure of Scientific Revolutions*. University of Chicago Press.

Moffatt, A.J., 1985. William Cobbett: Politician and Soil Scientist. *The Geographical Journal*, 151, 3, 351-55.

Morton, J., 1712. *The Natural History of Northampton-shire*. London.

Morton, John, 1843. *The Nature and Property of Soils*. Ridgway, London. (Fourth Edition Enlarged).

Russell, Sir E. John, 1966. *A History of Agricultural Science in Great Britain*. Allen and Unwin, London.

Thomson, T., 1838. *Chemistry of Organic Bodies*. London.

Trimmer, Joshua, 1847. On the Geology of Norfolk as illustrating the laws of the distribution of soils. *Journal of the Royal Agricultural Society*, VII, 444-485.

Vancouver, Charles, 1794. *General View of the Agriculture of the County of Cambridge*. London.

Wilde, S.A., 1963. In Memory of the Founder of Pedology, Baron von Humboldt. *Soil Science*, 96, 151-2.

Wright, John (?), 1833. *A Dissertation on the Nature of Soils and the Properties of Manure*. Sherwood, London.

Yaalon, D.H., 1988. Forerunners and Founders of Pedology. *Newsletter, Working Group on the History of Soil Science*, International Soil Science Society, Washington, 4-5 .

Young, Arthur, 1771. *The Farmer's Tour Through the East of England*. London.

Yourcenar, M., 1984. *The Dark Brain of Piranesi and other Essays*. Translation by R. Howard. Farrar, Straus, Giroux, New York.

6

James David Forbes on the Mer de Glace in 1842: early quantification in glaciology

Frank F. Cunningham

James David Forbes' visits to the Mer de Glace in 1827, 1832, and 1839 had been merely day trips familiar to tourists. In 1827 he and his brother Charles, accompanied by three guides, made their way up glacier from the Montanvert, keeping to its western side and reaching possibly as far as l'Angle. In 1832 with a single guide, Forbes was a little more adventurous, crossing the Mer de Glace to its eastern flank and proceeding to le Jardin de Talèfre, an outcrop in the middle of the Talèfre Glacier commanding a splendid view, and already a popular excursion. In 1839 Forbes repeated this excursion in the company of his brother John.

How are to be explained the facts that in spite of extensive Alpine journeys up to 1841, journeys during which he traversed some of the least known parts of that range, crossed several snow and glacier passes and, *en route*, conducted many experiments (meteorological and geological but never glaciological), Forbes evinced no scientific interest whatsoever in glaciers, yet in 1842 emerged as the leading British glaciologist? In a word the answer is Agassiz, and it is impossible to understand the subsequent thrust of Forbes' researches without examining the relationship between the two scientists.

By 1840 Jean Louis Rudolphe Agassiz (1807-1873), already famous as a biologist, was from his fieldwork and publications (especially his *Études sur les Glaciers* (Agassiz 1840a) which appeared in that year) regarded as the foremost of that group of Swiss scientists who had brought glaciology to Europe-wide prominence. Agassiz' threefold enquiries - the study of existing glaciers, examination of adjacent areas from which it was known they had retreated, and the survey of much more extensive areas with similar landforms - have formed the basis of subsequent research. Invited to the meeting of the *British Association for the Advancement of Science* in Glasgow in 1840, Agassiz sought to promote his ideas there and to follow up his friend Buckland's suggestion that they look for evidence of former glaciation in Britain. Neither in a paper read to the Geological Society of London by Buckland (Agassiz 1840b) before Agassiz arrived in Britain, nor in his own presentations in Glasgow before his glaciological tour (Agassiz 1840c), nor in an address he gave after it to the

Geological Society (Agassiz 1841), did Agassiz convince many of his hearers that their country had once been glaciated. Forbes, a founder member of the British Association who played a prominent part in its Glasgow meeting, met Agassiz and Buckland by chance in the Highlands while they were on their tour. Agassiz invited Forbes to join him the following summer to be shown his glaciological experiments and findings from his bivouac on the Unteraar Glacier in the Bernese Oberland. Agassiz judged Forbes to have such standing among British scientists that should he, also well-known for his Alpine travels, be converted to Agassiz' views, others would take them more seriously. Forbes accepted the invitation. It was to change his life.

Following a productive but truncated second tour of the Dauphiné Alps (during which, significantly, he continued to show no interest whatsoever in glaciology) Forbes, together with his friend Heath, had to hurry to reach the Grimsel Hospice on August 8, 1841, as arranged with Agassiz. Agassiz and a team of helpers used the Hospice as a base, but field work was carried out from a rough shelter in the lee of a large boulder (the famous 'Hôtel des Neuchâtelois' visited by many notables, Swiss and others) some 8 miles to the east near the head of the medial moraine of the Unteraar Glacier. F.J. Hugi (1830) had earlier worked at a base almost at the same place. Forbes spent almost all of the following three weeks being instructed by Agassiz, examining several other glaciers, and climbing the Jungfrau with him. On Agassiz' recommendation, Forbes proceeded (with Heath for part of the time) to inspect glaciers about Zermatt and Saas-Fee. On his journey home Forbes called on Agassiz and some of his entourage at Neuchâtel where Agassiz was an esteemed professor at the local university.

Forbes proved an apt but infuriating pupil. On his first field day with Agassiz (August 9) they walked to the Hôtel des Neuchâtelois, and soon Forbes was disputing there with his mentor about 'the ice-structure,' narrow, parallel, vertical laminations occurring in alternate bands of hard, blue ice and soft, white ice. These, Agassiz explained were superficial, temporary (they had not been so widespread the previous summer he averred), and, from their parallelism with the medial moraine, must be related to it. Forbes quickly pointed out, however, that because 'the ice-structure' could be seen in crevasses (and equally in a bore-hole Agassiz had sunk!) it could not be superficial; that because the structure could be matched on opposite flanks of a crevasse it could hardly be temporary; and, far from it being associated only with the medial moraine, a walk across the complete width of the glacier showed it to be widespread.

From a visit (without Agassiz) to the Rhône Glacier, Forbes became convinced that the disposition of the 'ice-structure' (which he liken to foliation in metamorphic rocks) in various parts of a glacier reflected its total constitution (Forbes 1842a). The dispute between Forbes and Agassiz about the nature and significance of the 'ice-structure,' as well as who should claim its discovery (neither of them was the first in this and Agassiz knew it), became the principal subject of an increasingly acrimonious correspondence between Forbes' 1841 and 1842 Alpine tours. The two scientists never met again and all correspondence ceased when in 1842, *en route* to Chamonix, Forbes attempted to call on Agassiz in Neuchâtel but was rebuffed.

Forbes was unconvinced by Agassiz on other issues. As a cautious empiricist he was reluctant to concede Agassiz' 'Ice Age,' and was even wary of some supposed glacial evidence left by earlier advances of Alpine glaciers because water could create similar forms. He was, however, altogether enthused about studying the behavior of existing glaciers, especially after the research he pursued to produce a masterly article on the then state of glaciology (Forbes 1842b) convinced him that there existed no reliable evidence to prove or refute any theory about glacier motion. Two such theories were widely supported at this time, the Gravitational Theory popularized by the revered H. B. de Saussure (1779-96) (but earlier formulated by Gruner (1760)) and the Dilatational Theory espoused by Jean de Charpentier (1841) and Agassiz (but advanced much earlier by Scheuchzer (1723)). According to the former theory the supply of snow to the upper reaches of a glacier, converted by pressure into ice, forced the mass downhill, gravity being assisted by terrestrial heat which melted and lubricated the lowest layers. Supporters of the Dilatational Theory disagreed, supposing instead that interstitial water, in crevasses and within capillary structure in ice, froze and expanded during the night causing a glacier to extend along the path of least resistance - downhill. Such supporters contended that during the winter, when there might be no melting during daytime, no movement of ice would occur.

Well aware from his research that any evidence or surmise about glacier movement was either wildly contradictory or unproven, (Agassiz' estimate of the speed of the Unteraar Glacier was twice that of Hugi) Forbes nevertheless went to his field work in 1842 with several *a priori* notions. He believed that measurements would reveal that a glacier behaved somewhat like a river, moving fastest in the center and slowest along the flanks (Agassiz, as well as most Alpine peasants and guides, thought the opposite was true), that glacier speed would accelerate downhill, and that in negotiating a corner the ice on the inner side of a bed would move slower than on the outer side. More importantly, Forbes was convinced that the major problems of ice motion being 'entirely a matter of mechanics,' appropriate experiments could solve the main issues in a few days.[1] To these experiments and the equipment necessary to carry them out he gave his formidable organizational ability well before his 1842 Tour.

Forbes on the Mer de Glace, 1842

1842 was Forbes' *annus mirabilis*, not only from the originality and importance of his glaciological enquiries but also because it was, on his own admission, the happiest summer of his life. Abandoning earlier ideas that he might co-operate with Agassiz, Forbes characteristically decided to carry out his experiments on glacier motion solo. With his sharp deductive mind, his familiarity in handling experiments and instruments and with meticulous recording, his Alpine travelling experience, his recent tutelage under Agassiz, and his grasp of what the leading contemporary glaciologists thought and had done, he was unusually well equipped for the task. He enjoyed comparing his solo venture with Agassiz' large

[1] The list of instruments and equipment that Forbes took to Chamonix in 1842 in order to make his map, measure glacier motion, and take a variety of meteorological observations is lengthy indeed. He seems only to have borrowed a single item, a ten meter chain to recheck his base line. He made all other linear observations by steel tape and never employed the metric system.

entourage, but even a Forbes required someone to hold the other end of his measuring tape or to carry a survey pole to appropriate points. He was extraordinarily lucky, through the good offices of the Curé at Chamonix, to have recommended to him on his arrival there on 24th June, the guide Auguste Balmat. Balmat was not only a skillful mountaineer but a man of remarkable intelligence and integrity with a special familiarity with the Mer de Glace (when a boy he had been a herder at the Montanvert). Balmat had taught himself English. He quickly became Forbes' co-worker.

Although Forbes had planned his 1842 Tour so as to include the examination of glaciers on the Italian flank of Mont Blanc and a traverse of the Pennine Alps, his most significant glaciological work, the best of his career and, arguably, the most seminal in the history of glaciology, was accomplished during three periods on the Mer de Glace (24 June - 1 July; 23 July - 11 August; 16 - 19 September), more particularly during the first of these. He had selected the Mer de Glace partly from earlier knowledge of it, partly because it promised to offer a great variety of circumstances, but also because the main building at the Montanvert, close to the glacier's west flank, was a sturdy stone structure with a permanent tenant, the guide David Couttet. Forbes was wise enough to prize a storm-proof base where he could store equipment and to value a sound table on which he could write, map, and record, and thought his base much superior to Agassiz' bivouac on the Unteraar Glacier.

The glacier of his choice was not the easiest for his purposes. Some nine miles long it embraces several contrasting sections. The higher reaches are a complex basin bounded by an arcuate divide which is marked by a succession of peaks (e.g. Aiguille du Midi, Mont Blanc du Tacul, Aiguille du Géant) and cols which are the crests of glaciers arising from either flank. Of these cols the most important and the only true pass is the Col du Géant (the highest pass in the Alps). This upper basin, the Glacier du Géant, is cut off from the lower reaches by an extensive ice-fall (les Seracs) leading to the Glacier du Tacul (a name sometimes applied to all the upper glacier). This reach is joined by tributary glaciers from the Chamonix Aiguilles on the west and from the ridge of Les Périades on the east. Where the Aiguille du Tacul descends to the ice the Glacier du Tacul is joined by the Glacier de Leschaux deriving from the precipices of the Grandes Jorasses. In Forbes' time (but not now) the Glacier de Talèfre combined with the Leschaux by a steep ice fall (now bare rock). From the Tacul-Leschaux junction the glacier forms a single tongue of which the upper part is the true Mer de Glace and the lowest, steepest reach, the Glacier des Bois. In Forbes' day the Glacier des Bois extended to the Chamonix Valley but the banana-shaped extremity has since disappeared. On both the Mer de Glace and the Glacier des Bois there is a marked difference between the eastern parts and the remainder. The greater supply from the Tacul-Géant catchment area is continued as a cleaner stretch with splendidly exhibited 'ogives,' occupying some two thirds of the total width. The supply from the Leschaux arm is compressed against the east flank which is consequently higher, more crevassed, but sullied by the number of moraines, (Figure 1).

Forbes was initially apprehensive about the problems the Mer de Glace presented to the surveyor, with its steep, heavily-crevassed lower section (Glacier

Figure 1

Mer de Glace from the North, Airphoto. Mittel holzer c.1920.

des Bois), its winding course and steep, sometimes precipitous flanks which together rendered triangulation difficult, its large range of altitude (over 7,000 feet between the snout and the Col du Géant), its liability to bad weather, and the virtual isolation of the main Géant Glacier by the difficult ice-fall at its lower reaches. In addition the surface of the glacier is almost everywhere rough and complex topographically, the contrast in this respect with such a glacier as the much-visited Athabasca Glacier in Alberta (upon which buses can operate) being remarkable.

It is always of great interest to compare Forbes' accounts of his travels and experiments in his field notes compiled on the spot[2] with the later published versions. His 1842 work appears in his finest book, *Travels through the Alps of Savoy* ... (Forbes 1843) where all is topically arranged and expressed with his customary lucidity. From his field note book to his 1842 Alpine tour, (unfortunately only his notes up to 1st Aug. have survived) however, we can visualise Forbes and Balmat tramping up, down, and across the glacier, setting up experiments and making daily records of their progress. At the same time Forbes was establishing the seven principal and seven subsidiary trigonometrical stations from which he would construct his map of the Mer de Glace. This, on a scale of 1:10,000 based largely on theodolite observations but in difficult positions on compass bearings, the first reliable map made of any glacier, was a remarkable achievement which would need and deserves separate consideration. Its completion allowed Forbes to locate precisely the places where his glacier measurements took place.

On 25 June Forbes and Balmat reconnoitered the Mer de Glace proper that is the area between the Montanvert and the Tacul, and Forbes decided on the following locations for his experiments, l'Angle, the outer edge of the glacier in its western swing under the Aiguille de Charmoz and where the ice abutted on a steep cliff; an area of glacier near the Montanvert (some 60 yards below to the south); the Pierre Platte, a gigantic boulder on the Lechaux Glacier between Tacul and the Couvercle (Figure 2).

Work began on the 26th at l'Angle. From there, only half-an-hour's walk from the Montanvert, Forbes picked his way across the glacier for 250 feet where, with his blasting iron he bored a hole (A) in the ice. Centring his theodolite over this borehole, he first aligned the central state of its telescope (it had three) on the east side of the Montanvert building, then swung it leftward for 100° to where Balmat stood at Position A1 on the west flank of the glacier (see Figure 2). Forbes motioned Balmat, who was holding a white sheet of paper against the cliff adjoining the ice, to the exact spot. Balmat first made a pencil line on the rock, then later marked it with a chisel and red paint. Forbes also trigonometrically calculated the height of the glacier at Balmat's feet. On the following three days the pair repeated the procedures, Forbes centring his theodolite over the hole (which of course was on the move), aligning with the Montanvert building, traversing 100° left, and instructing Balmat where to make

[2] St. Andrews University Library, 1968. *An Index to the Correspondence and Papers of James David Forbes 1809-1868*. This valuable Index (prepared by R.N.Smart, Keeper of the Muniments) is an excellent guide to the voluminous correspondence of Forbes housed by the Library, with almost 5,000 incoming letters and over 2,500 copies of outgoing letters. The Index distinguishes (as this paper does not) between "*Pocketbooks*" in which Forbes entered his field notes and "*Journals*" in which he sometimes wrote them up in more finished style. Not all of these have survived.

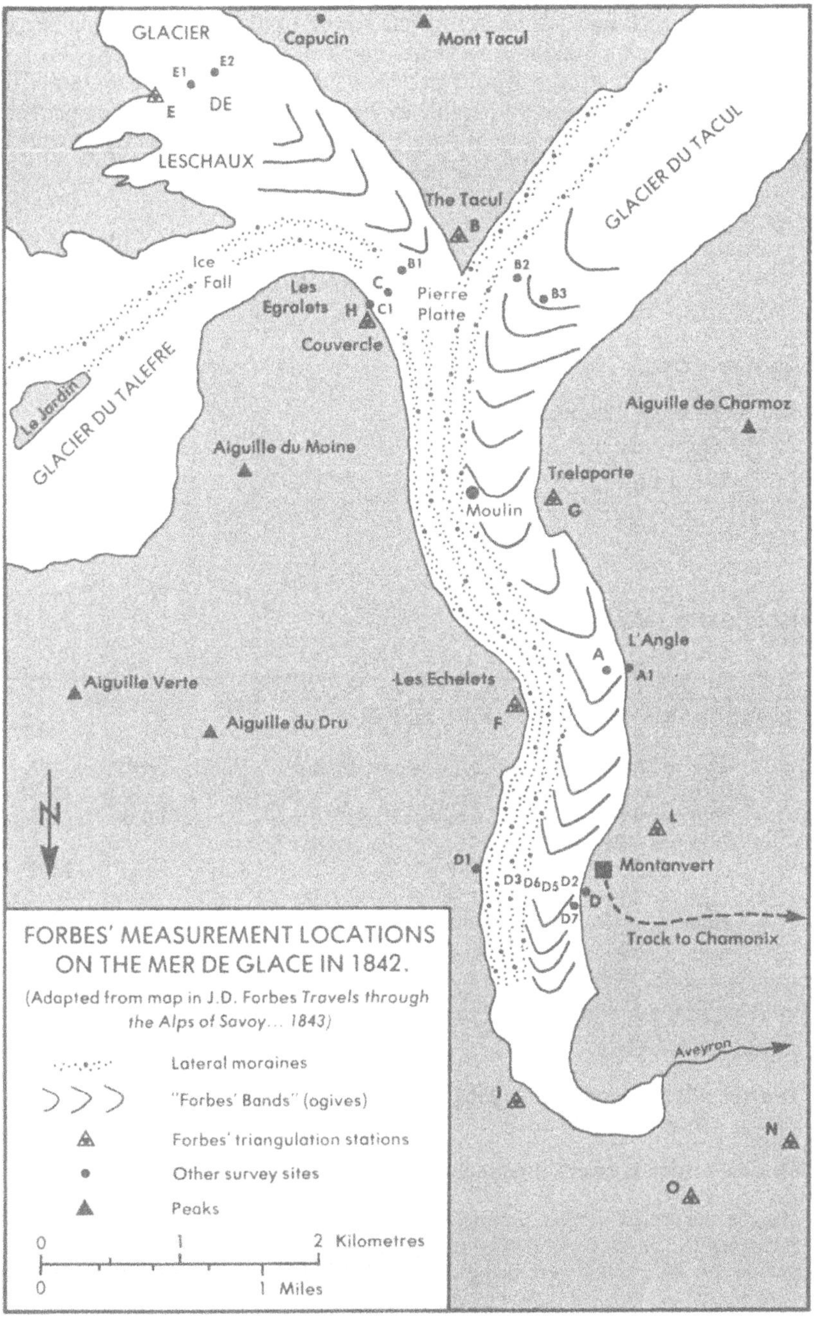
GLACIER
DE
LESCHAUX
E1
E2
E
Capucin
Mont Tacul
GLACIER DU TACUL
The Tacul
B
B1
B2
B3
Ice
Fall
Les
Egralets
H
C
C1
Pierre
Platte
Couvercle
Le Jardin
GLACIER DU TALEFRE
Aiguille de Charmoz
Aiguille du Moine
Trelaporte
G
Moulin
L'Angle
A
A1
Aiguille Verte
Les Echelets
F
Aiguille du Dru
N
L
D1
Montanvert
D3 D6 D5 D2
D
D7
Track to Chamonix
FORBES' MEASUREMENT LOCATIONS
ON THE MER DE GLACE IN 1842.
(Adapted from map in J.D. Forbes Travels through
the Alps of Savoy... 1843)
Lateral moraines
"Forbes' Bands" (ogives)
Forbes' triangulation stations
Other survey sites
Peaks
0
1
2 Kilometres
0
1 Miles
Aveyron
I
N
O

Figure 2

his next mark. We have to ask ourselves why Forbes, a skilled surveyor, would adopt such an indirect method of tracking the moving borehole. His purpose was to have on the rock face a semi-permanent record of glacier motion, and because the measurements turned out to be small (between 1 and 2 feet per day) his point was made, but the results would only have been absolutely accurate if the borehole had travelled exactly parallel to the cliff face, an unlikely coincidence. The results and the significance of these measurements are succinctly provided in Forbes' field notes, produced verbatim below.

Extract from Forbes' *Field Notebook 1842*

"Station A . L'Angle (30 June) p38

Motion since 6 PM yesterday		Inches
to 5 PM today	....	15.5
to 6 PM	....	17.4

---

p39

	ft.	in.
Level of glacier sunk since 26 June	1	4½

Motion of Station A (at l'Angle) since June 26

	26th	2½ PM to)		Inches		Daily	
	27th	4½)			16.5		15.2
to	28th	6 PM	....	17.4		16.3	
to	29th	6 AM	....	18.0)			
to	____	4½ PM	....		7.8)		17.5
to	____	6 PM	....	1.7)			
to	30th	6 AM	....	8.5)			
to	____	5 PM	....	7.0)		17.4	
to	____	6 PM	....	1.9)			

Whence I think it plain

1. That the Glacier motion is approximately regular and continuous even when the mass is exceedingly crevassed.

2. That the motion is nearly if not quite as great during the night as in the day.

It may be suspected 3. that the cause of the increased motion on the last 3 days compared to the first was owing to the heat of the weather which had only set in fine on the 24th; there were many clouds on the 27th, whilst the 28th, 29th & 30th were almost cloudless & very hot days."

The first two of these conclusions, to the effect that a glacier is continuously and regularly moving day and night, disposed of the Dilatational Theory. His measurements of the surface of the ice in contact with the rock at Station A1 showed a continuous shrinkage which he ascribed to the effect of summer heat, but was dissatisfied with his results.

Forbes' measurements near the Montanvert were carried out by more normal survey methods and had a different aim. He marked two points on opposite sides of the glacier, a painted cross D1 on a rock on the east flank, and D on a large boulder some 60 yards down from the Montanvert. These two points, he calculated from his map triangulation, were 2,898 feet apart, and along the line between them he blasted two boreholes in the ice, D2 some 300 feet from the boulder on the west flank, and D3, 795 feet further across and near the centre of the glacier (see Figure 2). Between 29th June and lst July (his last measurement during this first spell on the Mer de Glace), Forbes determined by theodolite observations from D that the hole at D2 had moved an average of 17.5 inches per day while that at D3 had moved 27.1 inches per day. He thus disposed of the widely held view that the flanks of a glacier move faster than its centre.

Forbes' third experiment, carried out between the Couvercle and the Tacul (the apex where Tacul and Leschaux glaciers join), had a third intention and required another different survey procedure. Anticipating that glacier motion would be found to increase downhill (the Gravitational Theory implied this), he sought a situation which might be compared with borehole A near l'Angle. He chose a giant boulder, the Pierre Platte, then resting on the ice some 400 feet from Couvercle and almost on the medial moraine deriving from the Glacier du Talèfre (since Forbes' time the Talèfre and Leschaux glaciers have shrunk and no longer join). Using two fixed points, his triangulation Station B on the Tacul and a new Station C1 on Couvercle, Forbes triangulated the Pierre Platte on two occasions (27th and 30 June), discovering that over the interim it had moved an average of 10.2 inches per day. As the corresponding rate for the borehole at A was 17 inches, even though it was nearer the glacier edge (250 against 400 feet for the boulder), he thought the point clearly made that the upper reaches of a glacier move slower. Boulders are, in fact, far from reliable recorders of glacier motion. They may move laterally from one 'stream' of ice to another where the speed is different; they may move to the edge of a glacier and become at least temporarily stranded (this happened to the Pierre Platte); or slip into a crevasse.

Nevertheless Forbes had, as early as July 1st when he set off for Italy, done remarkable work. His four conclusions, that glacier motion is regular and therefore incompatible with the Dilatational Theory, that the motion is almost as rapid during the night, that the centre of the glacier moves faster than its sides, and that motion was faster in warmer weather (a conclusion fatal to the Dilatational Theory), he communicated to his old teacher Professor Robert Jameson in Edinburgh, in a letter dated 4 July, for publication in the *New Philosophical Journal* there (Forbes 1842c). This date is of importance in that it preceded by some weeks Agassiz' earliest results which have often (wrongly) been accorded priority. Forbes maintained he suggested these experiments to Agassiz.

On July 1st, Forbes and Balmat left the Montanvert for Chamonix and thence made their way round the south end of the Mont Blanc Range (via the Col du Bonhomme) to Courmayeur. From there Forbes posted his historic *First*

Letter to Jameson, then proceeded to Turin while Balmat returned to the Montanvert with instructions to continue certain measurements and to keep all the boreholes near Station A and between Stations D & D1 cleared for subsequent use.

23rd July - 11th August 1842

After a fruitless journey to Turin where he had hoped to witness an eclipse of the sun, Forbes returned to Courmayeur. Thence he examined the Miage and Brenva Glaciers on the Italian flank of Mont Blanc, and on the 23rd July crossed the Col du Géant back to his Montanvert base, where the faithful Balmat awaited him. The Col du Géant, the highest true pass in the entire Alpine chain, was then regarded as a serious undertaking and, indeed, had not been crossed during the previous five years. Leaving Courmayeur at 1:30 AM with two guides (one from Chamonix, the other from Courmayeur) Forbes made the traverse in 14½ hours and apparently finished in better shape than his companions. At the Col they saw a few remnants of the cabin de Saussure had used during his epic stay in 1788 when over a period of 17 days he had conducted and unprecedented set of scientific experiments at high altitude. It must have been a moving experience not only for Forbes, who had the utmost admiration for de Saussure, but for his Chamonix guide, Jean Marie Couttet, whose father had been one of de Saussure's team of helpers. On the descent, perfect weather allowed Forbes comprehensive views of the entire basin of the Glacier de Géant and he observed several illuminating details - the overall 'clam shell' structure he thought so important, the rising of the ice surface where the glacier forces it way between the constricting spurs of the Rognon and La Noire, and below the famous Seracs (the ice-fall gave the party some trouble) a sequence of dirt bands which puzzled him but which he was soon to resolve.

During this second spell (as well as on his third in September) Forbes was unlucky with the weather. It is true that the 24th July was another splendid day, but the long exposure to ice glare the previous day had so inflamed his eyes that Forbes remained in the Montanvert hut until evening when he strolled up on to the Charmoz ridge about 1,000 feet above the Mer de Glace. In the subdued light the 'dirt bands' he had noticed the previous day showed up with great clarity. Hyperbolic in shape, regularly spaced, and demonstrating elegantly and unmistakably that glaciers move faster towards their centres, they otherwise puzzled Forbes so on 25th July, in spite of deteriorating weather, Forbes took Balmat up glacier to examine them. Forbes thought the 'dirt-bands' were surface expressions of layers arranged according to the 'clam shell' structure, but it was only later that he realized that the dirt entered the glacier in the crevasses of the ice-fall.[3]

Rain and cloud on 26th July preventing any work on the glacier, Forbes and Balmat descended to Chamonix and, because of torrential rain, remained there next day. Returning on the 28th, the pair proceeded up the east flank of the Mer de Glace, then crossed the Montanvert examining the D series of boreholes en route. During Forbes' absence in Italy Balmat had measured the advance of

---

[3] For a modern assessment, see Dawson 1988.

boreholes D2 and D3 only on the 16th of July. Since that date they were found to have moved 40 feet 3½ inches and 60 feet respectively. Forbes and Balmat next bored another hole (D4) between these two before proceeding up glacier to the important borehole at l'Angle which Balmat had also kept clear, and, on the same occasions, continued to measure the shrinkage of the glacier surface there. The borehole A, they found, had advanced 32 feet 10 inches since July 1st.

On July 28th Forbes and Balmat left their Montanvert base to establish a new Station (E) on a rocky promontory on the eastern edge of the Glacier de Leschaux. After marking it with red paint Forbes set up his theodolite and sighted on to the Capuchin, a conspicuous tower on the opposite side of the glacier. Balmat sank two more boreholes along this alignment, the first (E1) 200 feet in from the eastern edge of the glacier, the second (E2) in the middle of it. Forbes expected the former to move the slower, but in addition he anticipated that while E1 was similar in distance from the glacier flanked to A that, being further up the system, it would not move so quickly. No sooner has this preliminary work been done than a violent thunderstorm drove Forbes and Balmat back to their base. The bad weather continuing Forbes walked down to Chamonix to meet his friend Bernard Studer, the distinguished Swiss geologist with whom he had arranged to make an extensive tour of the Pennine Alps. On 31 July the weather remained impossible, but on August 1st Forbes was able to take Studer up to the Montanvert and from there to examine his boreholes along D-D1 and his Station A (both of which he had last measured on 28 July with Balmat). At the former, borehole D2 had in the interim moved by 5 feet 7 inches, D3 by 7 feet 3 inches, and the borehole D4 located between them had also moved 7 feet 3 inches, representing average daily rates of 16.2 and 21 inches. Studer was impressed both by the precision of Forbes' measurements and by the proof that glacier motion could be calculated over so short a time. Forbes had indeed measured motion at l'Angle in a period of only 1½ hours. The while Forbes was explaining not only his methods but his ideas, and on the way to l'Angle he expounded his interpretation of the 'ice structure,' its relation to his 'clam shell' conception of an entire glacier (Figure 3) and the newly noticed 'dirt bands.' The borehole at Station A was found to have moved 4 feet 6½ inches since 28 July, a diurnal average of 13.6 inches.

Back at the Montanvert, Forbes drew up a list of what he claimed as his original discoveries about glacier motion and, according to Forbes, Studer agreed with the list. Regrettably these claims begin (in French) but are not completed on p.98 of Forbes 1842 field journal, the last page which has survived, but certainly thereafter Studer championed Forbes against Agassiz' claims of priority in the measurement of glacier motion. Forbes maintained that he was the first to treat glacier motion as a problem of mechanical forces; that he had invented two experiments to distinguish between and test the 'theories of de Saussure and de Charpentier' (i.e. the Gravitational and Dilatational Theories); and that he was the first to determine the diurnal motion of a glacier by reference to three coordinates. These were, in fact, unnecessarily modest claims, and in his Second Letter to Robert Jameson (Forbes 1842d, posted from Courmayeur on 11th August) Forbes evaluated his achievements more amply. If there is overlap with his First Letter to Jameson (of July 4th) the contents are sufficiently important to merit restatement. By August 11th Forbes felt able to assert that he had disproved both

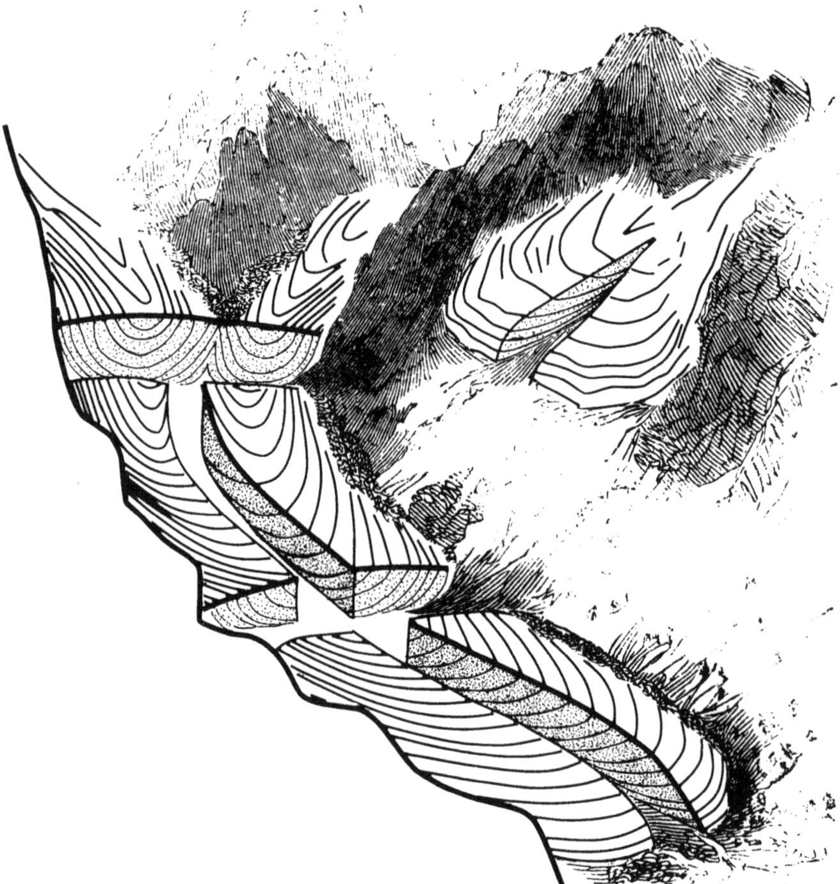

Figure 3

Idealised sections of a 'canal' glacier showing flow lines retarded by friction along its sides and base. Forbes believed that these flow lines influenced the disposal of the ice-structure (which followed them) and of crevasses (which developed at right angles to them). This diagram, enlarged from Fig. 3 in Forbes' *Travels through the Alps of Savoy... 1843,* is obviously based on an aerial view of the Mer de Glace from the N.E.

Gravitational and Dilatational Theories and to suppose that his results would enable him to formulate a better explanation; that he had shown "the continuous imperceptible motion of a glacier;" that he had proved motion to vary from day to day; that such variations occurred in all reaches of a glacier; that the centre moved faster than the sides and that the disproportion between these increased down glacier; that variations in motion affect movement in the centre most; that the greatest daily movement he had measured had been 27.1 inches; that he was now uncertain about differences in motion between the upper and lower parts of a glacier; and that he was convinced that there was some connection between glacier motion and 'the ice structure' to which he intended to give closer attention. This list, omitting as it does his discovery of the 'dirt bands,' was an impressive and legitimate set of claims.

Studer had left Forbes with the commitment that they would meet at the Great St. Bernard Hospice on August 12th. In the meantime, Forbes and Balmat continued their work on the Mer de Glace, but in the absence of the later parts of Forbes' field journal for 1842 we have to rely on the tables of measurements in his *Travels in the Alps of Savoy...* for an indication of what they did both for this period and for Forbes' third and last spell on the Mer de Glace that year (except for a single day of the latter). Below are the Tables from Forbes' *Travels ...* (pp. 139 - 140).

Experiments on the Motion of Ice

TABLE I

GLACIER MOTION

RECKONED IN EACH CASE FROM THE COMMENCEMENT OF THE OBSERVATION

Near Montanvert (1). D 2.			Near Montanvert (2). D 4.			Near Montanvert (3). D 6.			Near Montanvert (4). D 3.			L'Angle. A.		
1842.	English. Inches.	English. Feet.	1842.	English. In.	English. Ft.	1842.	English. In.	English. Ft.	1842.	English. In.	English. Ft.	1842.	English. In.	English. Ft.
June 29	0	0	July 28	0	0	Sept. 17	0	0	June 29	0	0	June 26	0	0
July 1	35·0	2·9	Aug. 1	84·0	7·0	,, 20	59·1	4·9	July 1	54·2	4·5	,, 27	15·2	1·2
,, 28	518·5	43·2	,, 9	278	23·2	,, 26	179·7	14·9	,, 28	774·2	64·5	,, 28	31·5	2·6
Aug. 1	585·5	48·8				,, 28	227·1	18·9	Aug. 1	861·2	71·7	,, 29	49	4·1
,, 9	715·5	59·6							Sept. 16	1962·7	163·6	,, 30	66·4	5·5
Sept. 16	1399·5	116·6	D 5.						,, 17	1986·4	165·5	July 28	460·9	38·4
,, 17	1416·4	118·0							,, 20	2047	170·5	Aug. 1	515·4	42·9
,, 18	1430·2	119·2	Sept. 17	0	[0				,, 26	2169	180·7	,, 9	633	52·7
,, 19	1443·3	120·3	,, 19	37·2	3·1				,, 28	2215	184·6	Sept. 16	1127	94
,, 20	1459·6	121·6	,, 20	57·5	4·8							,, 26	1238	103·2
,, 26	1544·8	128·6	,, 26	172·7	14·4									
,, 28	1583·8	132·0	,, 28	223·1	18·6									

Glac. de Léchaud. Pierre Plate. C.			Glacier de Léchaud. B 1.			Glacier du Géant. B 2.			Glacier du Géant. B 3.			Glacier de Léchaud. E 1.			Glacier de Léchaud. E 2.		
1842.	English. In.	English. Ft.	1842.	English. In.	English. Ft.	1842.	English. In.	English. Ft.	1842.	English. In.	English. Ft.	1842.	English. In.	English. Ft.	1842.	English. In.	English. Ft.
June 27	0	0	June 30	0	0	June 30	0	0	Aug. 4	0	0	July 29	0	0	July 29	8	0
,, 30	30·6	2·5	Aug. 2	355	29·6	Aug. 2	454	37·8	,, 6	36	3	Aug. 2	45	3·7	Aug. 2	54	4
Aug. 2	359	29·9	,, 6	395	32·9	,, 4	482	40·2	Sept. 17	564	47	,, 8	131	10·9	,, 8	152	12
Sept. 17	758	63·2	Sept. 17	803	66·9	,, 6	510·5	42·5				Sept. 25	672	56			
						Sept. 17	949	79·1									

TABLE II

MEAN DAILY MOTION

D 2.	Inches.
June 29 to 1 July	17·5
July 1 — 28 ,,	17·3
,, 28 — 1 Aug.	16·2
Aug. 1 — 9 ,,	16·6
,, 9 — 16 Sept.	18·0
Sept. 16 — 17 ,,	16·9
,, 17 — 18 ,,	13·8
,, 18 — 19 ,,	13·1
,, 19 — 20 ,,	16·3
,, 20 — 26 ,,	14·2
,, 26 — 28 ,,	19·5
D 4.	
July 28 — 1 Aug.	21·0
Aug. 1 — 9 ,,	24·7
D 5.	
Sept. 17 — 19 Sept.	18·6
,, 19 — 20 ,,	20·3
,, 20 — 26 ,,	19·2
,, 26 — 28 ,,	25·2
D 6.	
Sept. 17 — 20 Sept.	19·7
,, 20 — 26 ,,	20·1
,, 26 — 28 ,,	23·7

D 3.	Inches.
June 29 to 1 July	27·1
July 1 — 28 ,,	25·7
,, 28 — 1 Aug.	21·0
Aug. 1 — 16 Sept.	24·0
Sept. 16 — 17 ,,	23·7
,, 17 — 20 ,,	20·3
,, 20 — 26 ,,	20·4
,, 26 — 28 ,,	22·5
A.	
June 26 — 27 June	15·2
,, 27 — 28 ,,	16·3
,, 28 — 29 ,,	17·5
,, 29 — 30 ,,	17·4
,, 30 — 28 July	14·
July 28 — 1 Aug.	13·6
Aug. 1 — 9 ,,	15·4
,, 9 — 16 Sept.	13·0
Sept. 16 — 20 ,,	11·15
C.	
June 27 — 30 June	10·2
,, 30 — 2 Aug.	9·9
Aug. 2 — 17 Sept.	8·7

B 1.	Inches
June 30 to 2 Aug.	10·8
Aug. 2 — 6 ,,	10·0
,, 6 — 17 Sept.	9·7
B 2.	
June 30 — 2 Aug.	13·8
Aug. 2 — 4 ,,	14·0
,, 4 — 6 ,,	14·25
,, 6 — 17 Sept.	10·4
B 3.	
Aug. 4 — 6 Aug.	18·0
,, 6 — 17 Sept.	12·6
E 1.	
July 29 — 2 Aug.	11·3
Aug. 2 — 8 ,,	14·3
,, 8 — 25 Sept.	11·3
E 2.	
July 29 — 2 Aug.	13·5
Aug. 2 — 8 ,,	16·3

Between his second and third spells on the Mer de Glace, Forbes made his famous traverse of the Pennine Alps accompanied by Victor Tairraz and, for some of it, by Studer and his guide. This was Forbes' most ambitious mountain journey and it allowed him to cross several glaciers and view many more, experiences which only confirmed his ideas about glacier behaviour.

September 16th - 27th 1842

Forbes returned to Chamonix on September 8th, but was plagued by bad weather for his final spell of glacier work in 1842 and could not even start until the 16th. During his Pennine tour, Balmat had managed to keep all the boreholes open (except D4 in the stretch across glacier just below the Montanvert, and this was replaced by a new one D6), but the paucity of recordings in Forbes' Table I reflects the difficulties he and Balmat had in negotiating the Mer de Glace above their Montanvert base. After the 18th September an unseasonably early onset of wintry conditions made survey work difficult, sometimes impossible and dangerous. There was as much as 2 feet of new snow on the glacier surface and, at times, only Balmat's skill, as well as his familiarity with the terrain, kept them from falling into crevasses. At least on one occasion Forbes employed a second guide - David Couttet the tenant of the Montanvert hut. Even at l'Angle only 2 measurements of the progress of the borehole at A could be taken (on 16th and 20th September). Even when it was not actually snowing, wind-driven snow obscured the markers and reduced temperatures markedly. It was commonly only 20°F on the glacier and even in the Montanvert building only 34°F.

The fewness of measurements possible at his lower stations at this time was annoying to Forbes, but he knew he had ample evidence for his contentions. He was determined, however, to get a record of glacier motion from his higher Station E on the east flank of the Leschaux Glacier. He and Balmat twice slept out at Tacul in the vain hope of a morning break in the hostile weather. By chance, the words of his field journal for September 25th are known to us from their inclusion in the only biography of Forbes (by Shairp, Tait, and Adams-Reilly 1873), whose authors had access to all the 1842 journals and to others which have subsequently disappeared. From them we learn that on that day Forbes, accompanied by both Balmat and Couttet, started off from the Montanvert at 6:30 AM in a temporary clearance of weather. Within an hour, however, driving snow returned and continued for the remainder of the day. At times the trio, forcing their way up glacier, were compelled to shelter behind boulders. Both guides fell through the surface snow into waterholes below. When they at length reached Station E (Couttet thought they were fools to be there) Forbes set up his theodolite, Couttet protecting him and it with an umbrella while Balmat struggled across to E1, the nearest borehole (200 feet on to the ice), to find that it had moved 45 feet since August 8th. As the same borehole had moved 10 feet 11 inches between 29th July and August 8th, the total movement represented a progress of 56 feet in 48 days with an average daily motion of 11 feet 3 inches. On the descent Forbes attempted to sound a moulin on the west side of the lower Leschaux Glacier by lowering his geological hammer on a line. Unfortunately the line broke and Forbes thought his hammer lost but, remarkably, it was discovered *on the ice surface* below the Tacul on June 22nd 1858 by Forbes' friend Alfred Wills and his guide Auguste Balmat! This provided not only a long term measure of glacier motion but confirmed the contention of Alpine peasants that objects in glaciers rise to the top.

Following two more days spent refining data for his map, Forbes descended to Chamonix for the last time in 1842 and made his way in leisurely fashion back to Edinburgh. He was entitled to the pride he felt about his achievements both in the study of glacier motion and in cartography. Balmat, who had not accompanied Forbes in either his crossing of the Col du Géant or his traverse of the Pennine Alps because of his value in continuing measurements on the Mer de Glace solo, was asked to check through the coming winter the progress of the large boulder D7. These latter observations, transmitted to Forbes, showed its progress to be little different to that observed in summer and finally disposed of the Dilatational Theory. A hardy fellow indeed, Balmat nevertheless rated his work with and for Forbes in 1842 the most demanding of his career. Considering Forbes' tubercular weakness, his own performance had been equally remarkable.

During the remainder of 1842 and the first half of 1843 Forbes not only taught his usual courses at Edinburgh University but laboured at his records to produce his classic book, *Travels through the Alps of Savoy...* A model of clarity in thought, arrangement, and expression, Charles Kingsley described it as "an epic poem." It not only summarized Forbes' field evidence but included his observations from laboratory experiments conducted over the winter which attempted to replicate the behaviour of glacier ice in a variety of prepared materials. Forbes cames to the conclusion that such ice resembled a plastic substance, able to stretch or compress in accommodating its flow to bends in a glacier's course or to irregularities of its bed. When the stretching imposed was excessive then crevasses occurred, but pressure could anneal these (as in the areas below icefalls where his 'dirt bands' indicated debris accumulated in open crevasses above them). Noticing that sub-glacial streams (notably the Aveyron) flowed even in the coldest conditions, he supposed that here was always enough free water in the lower parts of a glacier to lubricate its passage.

A disputed conclusion

Forbes, had he known it, was now at the apex of his career. Unexpectedly to his siblings and friends Forbes, after a brief courtship, married Alicia Wauchope. In July 1843 the pair set off on a honeymoon which was to include an Alpine tour. He was struck down by a serious illness occasioned by a chill but involving his tubercular condition and perhaps a nervous breakdown. At Bonn the doctors despaired of his life but he recovered slowly, missing 2 entire sessions at Edinburgh and never achieving the health of 1842 again. To be sure he returned to the Alps and revisited the Mer de Glace with Balmat in 1843, 1844, and 1846, but his work on these occasions was mainly to confirm his theories of glacier motion and improve his map.

In 1851, following a strenuous solo tour of Norway, Forbes' health completely failed him. By this time, at the age of 42, he was the acknowledged leader among British glaciologists, but before his death (in 1868) his reputation was under a serious cloud from which, unjustly, it has never properly recovered.[4] He came under continued attack from Agassiz and his followers and was especially hurt by criticisms from John Tyndall (1860) and his influential supporter T.H. Huxley, "Darwin's bulldog." Tyndall, an excellent physicist (unusually, he graduated in Germany) and first-rate mountaineer, attacked Forbes on two main grounds. Firstly he thought Forbes' explanations of glacier movement, depending as they did mainly on gravity and lubrication at the base by free water, should be replaced by regelation. Tyndall's distinguished mentor Michael Faraday (also his predecessor as head of the Royal Institute in London) had conducted the earliest experiments of the property of ice to melt under pressure, and to reform when pressure is withdrawn (the latter is regelation), although the quantitative resolution of this process which glaciologists required was finally formulated by James Smith (1849), the brother of Lord Kelvin, a priority Tyndall was loth to admit. Although regelation is known to operate, as for example when the lower reaches of a glacier are being forced over a protrusion in the bed, Forbes' ideas were essentially sound. The Smith brother always supported them against Tyndall's and Agassiz' criticisms, but for a long while the tide of opinion was against Forbes. Tyndall cited as further evidence

4 An account of this controversy can be found in Rowlinson (1971).

the fact that glacier ice had no tensile strength and could not therefore exert force on its beds and flanks. Only in the 1880's did physicists show that this was entirely false and that it had very considerable tensile strength (Main 1887).

Tyndall's second line of attack, derived from Agassiz and his supporters, was that Forbes had knowingly laid claim to important ideas in glaciology which were, in fact, the work of others. In defiance of the facts, Tyndall maintained that Agassiz had been the first to measure glacier motion and that Forbes' notions of a glacier behaving like a slow river had been misappropriated from the writings of Bishop Rendu of Anneçy. In fact in his *Travels...* Forbes had been the first glaciologist to mention that work, had made a point of visiting the Bishop and preserved cordial relations with him. Forbes, more hurt by these smears than by the attacks on his scientific ideas, was in a poor position to defend himself. His health in ruins, no further work on glaciers was possible, but he produced an effective refutation of Tyndall's claim that Rendu had been slighted (Forbes 1860). Most regrettably, none of the three authors of the only biography (Shairp who succeeded Forbes at St. Andrews, Tait who succeeded him at Edinburgh, Adams-Reilly who continued his mapping of the Mont Blanc Range) was a competent glaciologist. Tyndall did not hesitate to attack their production (Tyndall 1873).

In Forbes' failing years he had also been left behind in the rapid development of the study of previous glaciation in Britain, an aspect to which, even when physically able to conduct enquiries, he had shown a strange aversion. Even after his remarkable research in 1845 into former ice action in the Coolins of Skye (Forbes 1846), which Archibald Geikie thought the most significant work of its kind so far done, Forbes evaded a real commitment to the subject. One wonders if Agassiz' priority in this field influenced him.

But let there be no doubts. James David Forbes was unmistakably the first to measure glacier motion, and his explanations of the nature of that motion and of the structure of glaciers were the bases for the sophisticated research which has more recently been accomplished.

References

Agassiz, L., 1840a. *Études sur les glaciers*. Neuchâtel.

Agassiz, L., 1840b. On the polished and striated surfaces of the rocks which form the beds of Glaciers in the Alps, *Proceedings of the Geological Society*, 3, 71, 321-322.

Agassiz, L., 1840c. On Glaciers and Boulders in Switzerland. This lengthy address in French is only preserved in short English summaries by *British Association Report*, Glasgow, 1984, Part 2, 113-114 and in *The Athenaeum*, 17 October 1840, No. 677, 824.

Agassiz, L., 1841. Glaciers and the evidence of their having once existed in Scotland, Ireland and England. *Proceedings of the Geological Society*, 3, 72, 327-332.

Charpentier, J. de., 1841. *Essai sur les glaciers et sur le terrain erratique du bassin du Rhône*. Laussanne.

Dawson, T.A.I., 1988. Dirt Ogives on the Mer de Glace, Chamonix, France. Abstract of unpublished Alfred Steers Dissertation Prize Winning Essay, 1987, *Geographical Journal*, 154 (2), 303-304.

Forbes, J.D., 1842a. On a remarkable Structure observed on the Ice of Glaciers (1841). *Edinburgh New Philosophical Journal*, XXXII, 84-91.

Forbes, J.D., 1842b. The Glacier Theory. *Edinburgh Review*, 80, 49-105. This reviewed eight publications - two by Agassiz, two by Charpentier, and one each by Hugi, Necker, Rendu, and Venetz.

Forbes, J.D., 1842c. First Letter on Glaciers. *Edinburgh New Philosophical Journal*, 33, 338-341.

Forbes, J.D., 1842d. Second Letter on Glaciers. *Edinburgh New Philosophical Journal*, 33, 341-344

Forbes, J.D., 1843. *Travels through the Alps of Savoy and other parts of the Pennine Chain with observations on the phenomena of Glaciers*. Edinburgh & London.

Forbes, J.D., 1846. Notes on the Topography and Geology of the Cuchullin Hills in Skye, and on the Traces of Ancient Glaciers which they present. (1845). *Edinburgh New Philosophical Journal*, XL, 1846, 76-79.

Forbes, J.D., 1860. *Reply to Professor Tyndall's Remarks in his Work 'On the Glaciers of the Alps,' relating to Rendu's Théorie des Glaciers*. Edinburgh.

Gruner, G.S., 1760. *Die Eisgebirge des Schweizerlandes*, Bern.

Hugi, F.J., 1830. *Naturhistorische Alpenreise*. Solothurn.

Main, J.F., 1887. Note on Some Experiments on the Viscosity of Ice. *Proceedings of the Royal Society of London*, 42, 329-330.

Rowlinson, J.S., 1971. The theory of glaciers. *Notes and Records of the Royal Society of London*, 26, 189-204.

Saussure, H.B. de, 1779-96. *Voyages dans les Alpes, précédés d'un essai sur l'histoire naturelle des environs de Genève*. Volume. 1, 1779, Neuchâtel; Volume. 2, 1786, Genève; Volumes. 3 & 4, 1796, Neuchâtel.

Scheuchzer, J.J., 1723. *Itinera per Helvetiæ Alpinas Regiones facta annis 1702-11*. Leyden.

Shairp, J.E.C., Tait, P.G., & Adams-Reilly, A., 1873. *Life and Letters of James David Forbes, F.R.S.*, London.

Smith, J., 1849. Theoretical Consideration of the Effect of Pressure in lowering the Freezing - Point of Water. *Proceedings of the Royal Society of Edinburgh*, 2, 204-205.

Tyndall, J., 1860. *Glaciers of the Alps*. London.

Tyndall, J., 1873. Principal Forbes and his Biographers. *Contemporary Review*, 22, 484-508.

7

"Extraordinary and terrifying metamorphosis" - on the seismic causes of slope instability

David Alexander

"The whole land surface, which suffered these concussions, shows an extraordinary and terrifying metamorphosis, hardly allowing those who dwell there to recognize the ancient mien of the place." Andrea Gallo, *Letters;* Messina, 1784.

The nature of earthquakes

As seismic activity is common over much of the Mediterranean basin, it is hardly surprising that many classical, Mediæval and Renaissance authors mentioned earth tremors. To Anaximenes (544 B.C.), for example, earthquakes resulted from huge cycles of desiccation and rehydration, that alternately split and swelled the earth's crust (Aristotle, *Meteorology*, II, vii). Lucretius, in De *Rerum Naturæ* (VI, 535-702), argued that subterranean landslides cause the earth to shake, and Seneca (in *Quæstiones Naturales*) believed that underground cavern collapse led to blasts of wind that caused the shaking (*cf.* Pliny the Elder, *Natural History*, II, 80). Plato (*Critias*, III, 111) stated that Atlantis had been engulfed during earthquakes, and Strabo (*Geography*, I, iii, 110) was somewhat responsible for the idea that earthquakes cause underground caverns that may eventually swallow up cities (Solbiati and Marcellini 1983). In each case, subsidence and mass movement consequent upon the earthquakes were magnified out of all proportion in the describing, often because the accounts of the ancient natural philosophers were based on derived sources.

In more modern times, authors from Petrarch (*Lettere senili*) to Kant (*Universal Natural History*) lamented the dangers and devastations of earthquakes. Whatever their causes, they seemed to demonstrate that the *terra firma* underfoot was, paradoxically, hollow. A full exposition of this was given in Athanasius Kircher's *Mundus Subterraneus* (Amsterdam 1664), which included detailed diagrams of the earth's interior (reproduced in Adams 1938, Chapter 12). A great fire at the centre of the earth communicated through branching fissures and passages with smaller fires not far below the crust. Water from the oceans percolated down to caverns beneath the mountains (*hydrophylacia*), where it was heated by the peripheral fires and distilled, as in an alembic. The water could

Lectures and Diſcourſes

OF

EARTHQUAKES,

AND

Subterraneous Eruptions.

EXPLICATING

The Cauſes of the Rugged and Uneven Face of the EARTH;

AND

What Reaſons may be given for the frequent finding of Shells and other Sea and Land Petrified Subſtances, ſcattered over the whole Terreſtrial Superficies.

Aaaa

Figure 1

The title page of Robert Hooke's book on earthquakes and eruptions (Written 1688, published posthumously 1705).

break through the surface as hot or cold springs and the fire as volcanic eruptions. Kircher had visited Mount Vesuvius when it was active and had travelled through Calabria in March 1636, when a major earthquake occurred there, and both phenomena had impressed him greatly. Through analogy with gunpowder, then widely used in warfare, he regarded earthquake shocks as a result of the explosion of sulphurous and nitrous gases pent-up in the hydrophylacia and heated by the same fires that distilled the underground waters. In essence, this explanation was a more sophisticated version of a prototype expounded by Anexagoras of Clazomenæ (500-428 B.C.) and many luminaries since (Geikie 1905, p. 13).

In 1761, John Mitchell considered this by then ancient hypothesis very carefully. Deeper fires, he suggested, caused greater tremors; but volcanoes, because they vented the subterranean vapours whose explosion led to the earth shaking, were relatively free from major seismic activity. Another theory, gleaned by Mitchell from the effects of the 1692 Port Royal, Jamaica, earthquake, was that the nearer to the mountains that the earthquake occurred, the greater the shaking that it caused (Mitchell 1761). In other words, mountains, as Kircher had supposed, are puffed-up blisters on the skin of the earth, and so are inherently unstable. Under this hypothesis, which stemmed from Ovid (*Metamorphosis*, 15), underground fires made caverns as they burnt their way through flammable rock material. The burning would eventually enlarge the caverns to the point of static instability, while contorting the strata; but before this percolating groundwater would flash and cause a spontaneous exhalation, with a rapid vibratory motion (the earthquake), which would split the strata along faults and collapse the roof of the cavern. The vapours would not necessarily exit to the atmosphere, as the cooling action of the percolating seawater created a solid rock barrier above the burning cavern, and the best route of escape was often lateral, over great distances along sub-horizontal strata (Mitchell 1761). In 1688 (published 1705), Robert Hooke came up with an immensely complicated treatise that seems to attribute most of geology to seismic causes (Figure 1). His ninth proposition states that the tops of the highest mountains have been underwater and their present elevation is due to earthquakes; and his tenth, "that the greatest part of the Inequality of the Earth's Surface may have proceeded from the Subversion and tumbling thereof by some preceding Earthquakes" (Hooke 1705, p. 291).

Catastrophism seems to have owed its persistence in no small measure to Mediterranean volcanism. Hooke took historic eruptions in Greek and Italian territories as his model for morphogenesis. Even Charles Lyell, who was as much a Grand Tourist as many of his most influential forebears, retailed the Lucretian and Senecan explanations of earthquakes based on fire, wind and subterranean cavern collapse (*Principles of Geology*, 11th Edition, 1883, II, Ch. 29). He drew heavily on the work of Sir William Hamilton, the author of *Campi Phlegræi* and an enthusiastic student of volcanic activity.

In 1745 the Leyden Jar was invented, and William Stukeley seems to have been the first to suggest electrical discharge as the cause of earthquakes (Stukeley, 1750, pp., 657-668). The odour of lightning bolts created in laboratory experiments convinced Stukeley that metalliferous particles were present in the underground vapours, which had been generated by the action of electrical energy on various ores and minerals, including phosphorus and arsenic. Electrical pulses could be transmitted through the earth along mineral veins and would cause sparks underground that ignited subterraneous vapours. As mineral veins were apparently more common in the roots of mountains, the generation of vapours would be greatest there (according to G. Isnard, *Mémoire sur les Tremblemens de Terre*, Paris, 1758), and this created the caverns, which caused earthquakes by collapsing when the force of pent-up gases was too much for them to sustain. This idea remained popular for at least fifty years. Even Robert Mallet, who did so much in the mid-nineteenth century to found seismology as an observational science, attributed earthquakes to underground steam explosions (Mallet 1862, Vol. 2). He also believed that the non-elastic component of seismic waves was responsible for orogeny (Mallet, 1846).

Thus theories of earthquake genesis that predate modern seismology can be classified broadly into vulcanist and electricist, with a variable amount of overlap concerning the details. Vulcanists such as Kircher and Mitchell believed that subterranean fire was the primary motor for earthquakes, while electricists such as Stukeley and Pietro Vannucci (*Discorso Istorico-Filosofico sopra il Tremuoto*, Cesena, 1789) preferred underground static discharges. Steam or gases could figure in either hypothesis, as could cavern collapse under the mountains. Hence, with so little progress in explaining the *cause* of earthquakes it is perhaps surprising that some of their *consequences* have fared better. Seismically-induced ground failure, which is after all a highly visible phenomenon, has accumulated a rich observational history and has frequently contributed to the understanding of earthquakes themselves.

Historical recognition of seismically-induced mass movement

The historical perception of seismic landsliding should be considered in the light of several issues: whether the phenomenon has been recognized regularly enough during history for it to constitute a theme in landscape studies; whether historical descriptions illuminate its causes and mechanisms; and whether the relationship between earthquakes and landslides has been perceived correctly. The following account will be limited to Italy and neighbouring areas.

Seismic landsliding was mentioned by several Classical authors. The elder Pliny (*Natural History*, II, lxxxi, 191) noted that in the fifth century B.C. Anaximander is reputed to have saved the Spartans from death with a timely and accurate prediction of seismically-generated mass movement on Mount Taigeto, but he did not explain the criteria used in the forecast. The loss of the city of Helice on the Peleponnesian coast in 373-2 B.C. may have been caused by seismic influences on ground stability (Marinatos 1960). Plato (*Critias*, III, 111) suggested that the soil had disappeared from the Acropolis under the combined onslaughts of earthquake and cloudburst (mistranslated as 'flood' in certain English editions). Titus Livy (*Ad Urbe Condita*, XII, v, 8), who described the central Italian campaigns of Hannibal in 217 B.C., referred to an earthquake in the mountains near Cortona that caused landslides which dammed rivers and destroyed villages; Ovid (*Metamorphosis*, 15) also referred to the alteration of drainage patterns during earth tremors.

By and large, Classical accounts treated seismic landsliding as a significant phenomenon that was either an incidental fact to be mentioned in the narrative of some other matter (soil erosion in the case of Plato, military history by Livy), or a symbolic metaphor portraying the end of the world, or of Atlantis (Plato, *Critias*, III; *Timæus*, 2b). In the Mediæval period, by contrast, seismic landsliding became more a part of narrative history: Ricobaldo of Ferrara, for example, chronicled such events (*Rerum Scriptorum Italicum*, IX, 138) in the central Italian city of Ancona in "Anno MCCLXIX Tunc magni terræmotus mons Anconæ scisso in mare dissolvit" (Almagià 1910, 327-328). Similar phenomena had been observed at Ancona in A.D. 558, and have occurred regularly since then, including in 1972. However, Mediæval sources must be treated with caution: Guidoboni (1986), who studied the Brescia earthquake of 1222 (estimated MCS intensity XI), noted that references in contemporary sources to "mountains collapsing" must be treated more as symbolic statements than as hazy descriptions of seismic effects.

Notions of causality were sometimes erroneous in the early accounts. For example, the official historian of the Venetian Republic, M.A. Coccio Sabellico (1436-1506) - who was by all accounts a credulous and superstitious man - claimed in his *Rerum Venetarum ad Urbe Condita Decades* (published in 1516-17 and 1718) that the strong earthquake of 1456 in southern Italy caused "many miraculous effects," including landslides in the mountains around Lake Garda, 675 km from the epicentre (Figliuolo 1985). In reality, there may instead have been a tenuous link with the earthquake of April 1457 at Gubbio in central Italy. It was left to Valerio Fænzi to state the causal relationship unequivocally, which he did in the influential work *De Montium Origine* (Venice, 1561): "nothing in all nature is more puissant, nothing of more mordant efficiency than an earthquake. It can scatter great tracts of earth..." While a contemporary account of the 24 April 1593 earthquake at Corleone in Sicily spoke unambiguously of "splitting and mutation of the soil" (in A. Mongitore, *Istoria cronologica dei terremoti in Sicilia*, Palermo, 1743), news of a major mudflow in Basilicata provoked by the earthquake swarm that began on 31 July 1561 found its way into Athanasius Kircher's *Mundus Subterraneus* (Amsterdam, 1664, I, p. 221) in very imprecise form.

The quality of accounts from the seventeenth century is extremely uneven. One of the more perceptive authors was Marco Antonio Mellio, who (in *De Terræmotu Aemiliano*, 1693) regarded the landslide of 11 April 1690 at Budrialto in Romagna as a delayed reaction to earthquakes in 1688 and 1689 (Almagià 1907, p. 267). On the other hand, D. Marcello Bonito, who was one of the most influential writers on earthquakes at this time, marred his compendium of seismic effects (*Terra Tremante, ovvero, Continuatione de' Terremoti*, Naples, 1691) with specious nonsense, catastrophism and sensationalism. A more scientific approach is to be found in Robert Hooke's lectures of 1688, which were published with a commentary in London in 1705 (Fig. 1). Hooke was an unabashed catastrophist, who created an entire system of tectonogenesis on the basis of what he presumed were seismic effects. He owed much to Athanasius Kircher's *Mundus Subterraneus*, but also to accounts that had little directly to do with earthquakes: "It is reported, faith *Childrey*, that in a Parish by the Sea-side, not far from *Axbridge* in *Somersetshire*, within these 50 years a Parcel of Land swell'd up like a Hill; but on a sudden clave asunder, and fell down into the Earth, and in the place of it remains a great Pool" (Hooke 1705, p. 303). Thus one could recognize the hollowness of the earth and the workings of its subterranean fires. The influence of Mediterranean volcanism is easy to identify in Hooke's writings, but volcanic activity, seismicity and slope failure are often too intermingled to be distinguished clearly from one another. Yet the theory is orderly and as comprehensible as it is comprehensive; his first genus of seismic effects comprised raising of the land level, raising of the sea bed above sea level, raising up mountains out of plains, and raising parts of the earth's surface in a manner that creates landslides. Adams (1938) regarded Hooke as the first Anglo-Saxon writer to draw attention to the detailed effects of earthquakes upon the land surface. Robert Hooke, however, was not without his detractors. Jean Etienne Guettard (1715-1786) had the 1733 landslide at Issoire in the Auvergne in mind when he refuted the notion that landslides are caused by subterranean winds or internal fires ("Des pluies at des averses," in *Mémoires sur différentes parties des Sciences et des Arts*, Paris, 1770, III, pp. 209-220).

Thus the theoretical debate had not been resolved, but the quality of field observations had nevertheless undergone a dramatic improvement. The eastern Sicily earthquake of 11 January 1693 (MCS intensity XI) produced land surface effects that are described with great clarity by a local man of letters, Vincenzo Bonaiuto, whose account appeared in English in the *Philosophical Transactions of the Royal Society* for 1693-4 (Bonaiuto 1693-4) and was later used as an example by Charles Lyell in the first edition of the *Principles of Geology* (1830). This is how Bonaiuto described a small rotational slump produced by the tremors:

> In the territory of *Sortini*, in a piece of ground half a mile long, but much narrower, the ground at several little interstices is sunk from the level, in some places two, in others three palms [52-78 cm], and ends in a very deep circular gulf or swallow.

Evidently, the earthquake also produced severe modifications of drainage, as Bonaiuto noted:

> Not far from the country of *Cassaro*, from the tops of two mountains, between which through a long valley ran a river, two very great rocks were loosened, which tumbling over against each other, met so exactly, as to close up the valley, and stop the current of the river, which not finding any subterraneous or side-passage, has filled up the valley to the top of the rocks that were thrown down, and runs over them, forming a lake three miles round of a considerable depth.

While Bonaiuto made some very perceptive observations on the differing performance of rock outcrops and alluvium as foundation materials (Hughes 1983), he sometimes lacked the ability to put a name to the phenomena he described:

> In the city of Noto [Noto Antica] is a street half a mile long, built of stone, which is at present settled into the ground, and quite hanging on one side, like a wall that inclines; and in another street before the ascent del Durbe, is an opening big enough to swallow a man and a horse.

Thus slumping and subsidence of the ground are not mentioned explicitly, even though the phenomena are described clearly.

The Lisbon earthquake of 1 November 1755 had a profound impact on Western philosophy for decades afterwards (Kendrick 1956), largely as a result of the collapse of buildings during the tremors and the subsequent fire and tsunami waves. However, the mass movements that it provoked did not escape attention. John Mitchell (1761) wrote that "the mountains were impetuously shaken, as it were, from their very foundations; and most of them opened at their summits, split and rent in a wonderful manner, and huge masses of them were thrown down into the subjacent vallies." Mitchell regarded the water that saturated unstable slopes as having condensed from the "subterraneous vapours" that caused earthquakes.

The Calabrian earthquakes of 1783

Between 5 February and 28 March 1783 southwestern Calabria experienced six catastrophic earthquakes (Fig. 2), which were followed in the succeeding months and years by at least 1197 perceptible aftershocks (Vivenzio 1788). The catastrophe damaged 202 villages and destroyed 182 others at a total cost of 31,250,000 ducats, and the death toll was computed variously as between 29,515, and 35,160 people (Placanica 1985). The tremors created an extraordinary range of effects in the young and heavily tectonized geological formations of the Calabrian arc: 12 major lakes and 203 minor lakes and ponds appeared, rivers ceased to flow for many hours and then returned with floodwaves; liquefaction, slumping, rockfall and mudflow were widespread on a scale unprecedented in written history (Cotecchia *et al.* 1969). As the events coincided with an outgrowth of scientific endeavor in Europe, they attracted interest and visits to the area by prominent natural philosophers, not merely from Naples, but from elsewhere in Europe. The 1755 Lisbon earthquake had caused the more cheerful arcadian view of nature to be replaced by a bleaker vision in which as Voltaire satirized (*Candide*, 1759), mankind received fearful and arbitrary punishment for its misdemeanours (Besterman, 1969). But the 1783 Calabrian disaster elicited a more positive response from contemporary thinkers, many of whom assembled compendia of observations gleaned from the impact and its aftermath, a series of "earthquake anthologies" to be perused by members of the European intelligentsia and scientific societies (Placanica 1985).

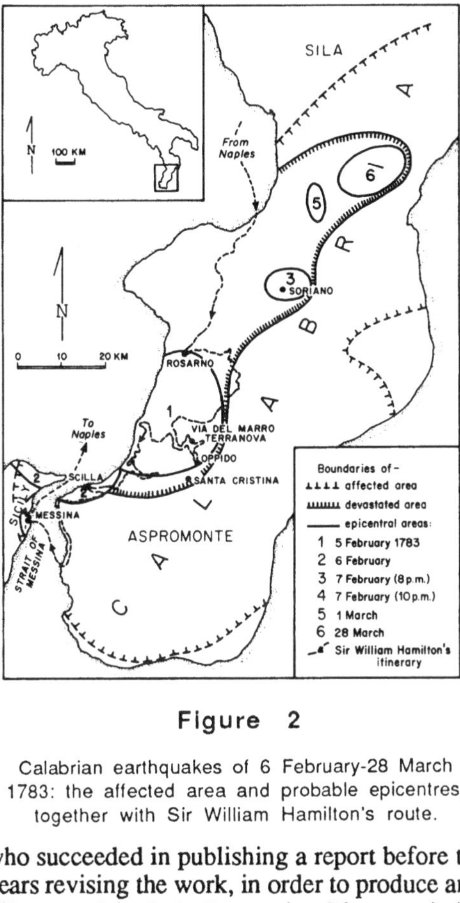

Figure 2

Calabrian earthquakes of 6 February-28 March 1783: the affected area and probable epicentres, together with Sir William Hamilton's route.

On 5 April 1783, barely a week after the last main shock had occurred, an expedition to the disaster area left Naples under the auspices of its *Reale Accademia delle Scienze e Belle Lettere*. At its head was the Secretary of the Academy, Michele Sarconi, who had under his leadership seven men of letters and three draftsmen, including the eminent Neapolitan architect Pompeo Schiantarelli. Their report, published in Naples in 1784, included 68 drawings (many of which depict seismic mass movements, Figure 3), which were etched and in some cases turned into gouaches by Pietro Fabris, a Neapolitan court painter.

The work also included the famous macroseismic survey and prototype seismic intensity classification by Padre Elisio della Concezione (Sarconi 17-84). The Vicario Generale at Naples sent to the disaster area his secretary Giovanni Vivenzio, who succeeded in publishing a report before the end of 1783. He later spent five years revising the work, in order to produce an authoritative catalogue of seismic effects and include Domenico Pignataro's "Earthquake Journal," a complete catalogue of perceptible aftershocks and their effects (Vivenzio 1788).

Britain's envoy to Naples, Sir William Hamilton, was absent on leave in February 1783, but he returned and set off for Calabria by boat on 2 May of that year, touring the disaster area on horseback from 6-17 May (Figure 2). His account was published by the Royal Society in London at the end of 1783 (Hamilton 1783) and appended further observations by a local nobleman, the Marquis Francesco Ippolito (Ippolito 1783). Other detailed records were compiled by Giuseppe Coccia, a local official, and Andrea Gallo, who was Professor of Philosophy and Mathematics at the Royal Lyceum of Messina. The former was motivated by his duty to report to his superiors in Naples (Coccia 1894), and the latter by his international connections and membership of the scientific societies of Naples and Bologna (Gallo 1784). Another investigation was carried out by the Frenchman Deodat Dolomieu (1750-1801). In February 1784 Dolomieu was *en route* for Mount Etna, being in the process of acquiring a detailed knowledge of the field relations of volcanic deposits, and he paused in Calabria long enough to survey the geological effects of the earthquake (Dolomieu 1784).

a Changes of the surface at Fra Ramondo, near Soriano, in Calabria.

b Fissures near Jerocarne, in Calabria, caused by the earthquake of 1783.

c Section of one of the circular hollows formed in the plain of Rosarno.

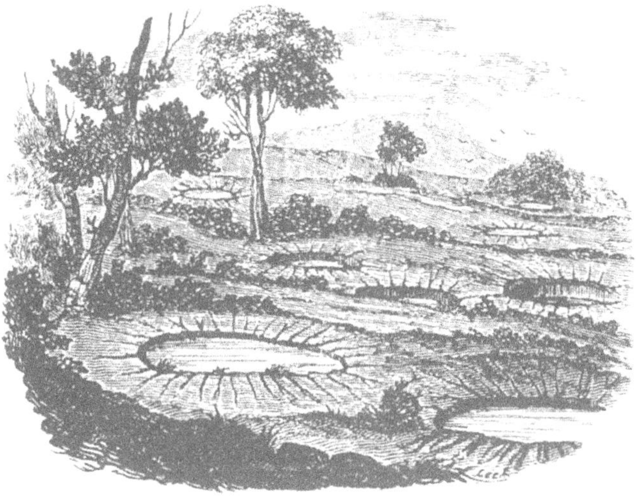

d Circular hollows in the plain of Rosarno, formed by the earthquake of 1783.

Figure 3

Ground failure phenomena resulting from the 1783 Calabrian earthquakes, (illustrations from Lyell, 1883, after Sarconi, 1784). Lyell's captions are given, modern interpretations are:

a	Rotational slumping at Fra Ramondo, near Soriano.
b.	Liquefaction subsidence at Gerocarne.
c. and d.	Sand boil formation on the plain of Rosarno.

In considering the work of these investigators, one should bear in mind which of the contemporary systems of explanation of earthquakes each of them espoused, as it is likely to have influenced both the observation and the explanation of seismic landslides. Hamilton, as befitted the author of *Campi Phlegræi*, was a vulcanist (Lyell, in Chapter 29 of the *Principles*, Vol. 2, was mindful to refute Hamilton's identification of tuff deposits in southern Calabria). Dolomieu was also a vulcanist, but Vivenzio was an electricist, as was Gallo. Sarconi and Dolomieu, however, concentrated more on naturalistic descriptions of the surface mutations that they encountered in Calabria, without needing to invoke static electricity or subterranean fire.

Although the contemporary debate of the vulcanists and electricists was both animated and sustained, it is not always easy to detect the standpoint of a writer from his descriptions of surficial phenomena. In the first place, the two theories overlapped considerably, especially with regard to the importance of combustions and explosions in underground cavities, and in the second the nature of surface effects was judged to be independent of the causes of earthquakes. For example, in both cases, upheaval or dislevelling of the ground was postulated: Dolomieu and Hamilton attributed it to volcanic fires; Cristofano Sarti (of the University of Pisa) and Francesco Ippolito regarded it as caused by spontaneous underground combustion resulting from thunderstorms in the atmosphere; and Gallo and Vivenzio found the spark in underground discharges of static electricity conducted by veins of metallic ore. With the exception of sand boils (as described below), contemporary investigators did not find their perception of mass movements clouded by ætiology, as the following sub-sections will show.

Slumps, slides and rockfalls

The accounts by Gallo, Sarconi and Vivenzio are rich with descriptions of rotational slumping in soft sediments, block glides, rockfalls, earthflows, mudflows and solifluction phenomena. For example, Sarconi's report (Sarconi 1784) includes illustrations (see Figure 3a, previous page) accompanied by written explanations of very widespread rotational slumping at Bozzano a Oppido in the Cumi Valley, translational sliding with a sub-vertical backwall at Terranova, an earthflow that exposed the stone lining of a well, also at Terranova, and a mass movement complex along the Via del Marro, showing the thrusting and chaotic upheaval that occurred in the toe zone. He described the classic form of rotational slumping thus (Sarconi 1784, p. 61):

> We reached the brook called Porcione. To our surprise, in the place of a wide plain, which used to be there, we found a ruinous upheaval of sand and gravel. The ancient soil was torn asunder, and, with opposite movements, part of it was deeply subsided and another part elevated, in such a way that a narrow, winding trench of water separated them...thus we saw a change, become common throughout the ancient plain, that is, a notable extension of the land.

Hamilton (1783, pp. 189-90) was much impressed by the slumping at Terranova, which he struggled hard to explain:

> Sometimes I met with a detached piece of the surface of the plain (of many acres in extent) with the large oaks and olive trees, with lupins or corn under them, growing as well, and in as good order at the bottom of the ravine, as their companions, from whom they were separated, do on their native soil in the plain, at least 500 feet higher, and at the distance of about three quarters of a mile. I met with whole vinyards in the same order in the bottom, that had likewise taken the same journey. As the banks of the ravine, from whence these pieces came, are now bare and perpendicular, I perceived that the upper soil was a reddish earth, and the under one a sandy white clay, very compact, and like a soft stone; the impulses that these huge masses received, either from the violent motion of the earth alone, or that assisted with the additional one of the volcanic exhalations set at liberty, seems to have acted with greater force on the lower and more compact stratum than on the upper cultivated crust: for I constantly observed, where these cultivated islands lay (for so they appeared to be on the barren bottom of the ravine) the wider stratum of compact clay had been driven some hundred yards further, and lay in confused blocks, and, as I observed, many of these blocks were of a cubical form. The undersoil having had a greater impulse, and leaving the upper in its flight, naturally accounts for the order in which the trees, vinyards, and vegetation, fell and remain at present in the bottom of the ravine. This curious fact, I thought, deserved to be recorded, but is not easily described by words.

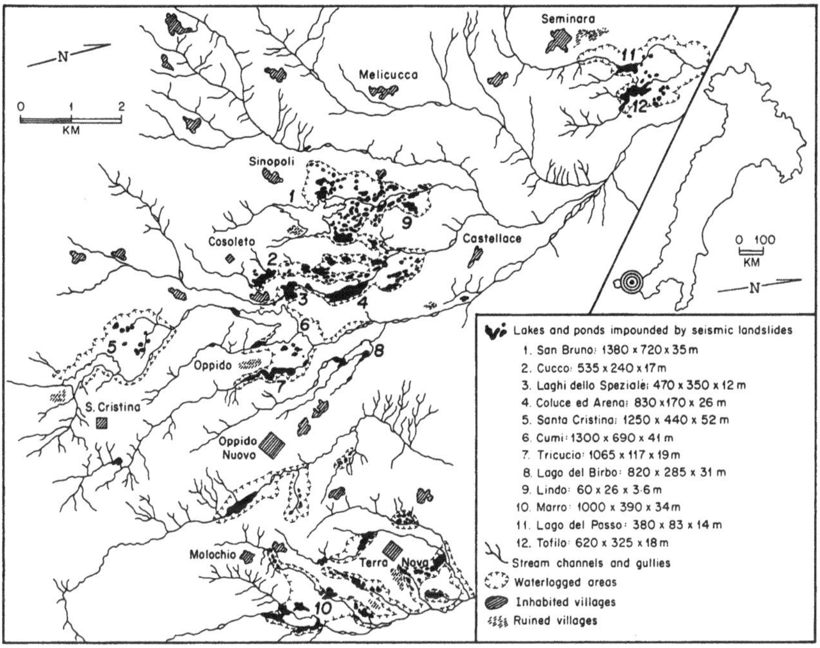

Figure 4

Lakes and ponds mapped in Calabria by Vivenzio (1788).

Dolomieu (1784) instead preferred a somewhat less visceral explanation, and likened the damage sustained by the Calabrian landscape to the disaggregation and dislevelling of cubes of damp sand lying on a tray that is vigorously shaken. Gallo (1784) seemed to discern the influence of underground fires in the slumping, and reported that smoke and heat were discernable at the headscarps. Both Vivenzio (1788) and Dolomieu (1784) deduced from the pattern and extensiveness of ground breakage at Oppido and Santa Cristina (both of which were situated on active faults) that these settlements were located close to the "centre of the explosion" - or in broad terms the earthquake focus.

It is salutary to reflect that knowledge of the earth's interior was minimal at the time, such that conjecture was often unbridled. For both the vulcanists and the electricists, explosive forces were at work close beneath the surface. According to the latter, lightning that penetrated the soil, or static discharges among the mineral veins, would ignite gases trapped in the underground caverns and the result would be both a heaving of the ground and a rumbling crash, accompanied by severe tremors. The same result could be achieved, according to the former, by underground fires associated with volcanic activity that never reached the surface, but this could also superheat steam in order to cause the explosion. Indeed, the flashing of groundwater in contact with subterranean fire remained in vogue as an explanation of seismicity throughout the "steam age" of the industrial revolution (see Mallet 1862, Vol. 2).

Contemporary accounts of the mass movements caused by the tremors were sufficiently detailed and rigorous to enable Cotecchia *et al.* (1969) to identify the affected locations and map the landslide deposits in the field. These authors noted that the earthquakes produced slope-failures that were exceptionally large and widespread, were often related to boundaries between soft sediments and harder rocks, were mainly roto-translational slumps and slides, and involved longer travel distances than mass movements in the same area that had not been set off by earth tremors. One of the largest failures occurred near Santa Cristina, in which a block of land 100 m thick descended 100 m, travelled laterally 500 m, left a trail of debris 1700 m long and dammed up the Torrente Lago. The lake, which Vivenzio reported as 1250 m long, 440 m wide and reaching a depth of 52 m (Figure 4, previous page), was not drained for 40 years (Cotecchia *et al.* 1969). The disruption to drainage caused by the earthquakes received comment from all authors, varying from the impressionistic tones of Giuseppe Coccia (1894): "formed a great lake... big enough, frankly, to accommodate a naval armada"; to the patient work of Vivenzio (1788), who measured the dimensions of and mapped all 215 lakes produced by the tremors.

Sarconi (1784 p. 307) detected groundwater: "emerging from the bosom of the earth"; wherever the slumping occurred. He was particularly fascinated by the headscarps of multiple regressive slumps and of flow-slides, but, although his descriptions are clear, they are not the most scientific in tone:

> To the yawning cracks in the ground were added many wide and deep lacerations, occurring along the top and sides of the hill. These clefts and cracks were of strangely raw earth, and the great masses that they defined could be seen to have been pushed from one place to another, or engulfed, or disintegrated, or mixed together with trees and plants, so that the horrible cataclysm led to the wandering of those self-same lands. (Sarconi 1784, p. 164)

Large-scale changes in the ground could be much more precisely observed and described, as was the fault scarp 15 km long that Dolomieu (1784) investigated where the plutonic rocks of the Sila-Aspromonte mountain chain abut the Plio-Pleistocene sediments of the coastal plain:

> It continued for almost the entire length of the plain, where the sediments abuting the granite at the bases of Mounts Caulone, Esopo, Sagra and Aspromonte, sliding on this solid core, very steeply, subsided. This has established a fissure many feet wide, and from 9 to 10 miles long, between the solid rock and sandy sediments; and this fissure runs, almost uninterrupted, from San Giorgio, following the piedmont, right behind Santa Cristina. Many sediments, thus slipping, have been carried very far from their primitive positions and have covered others sufficiently to make them disappear. (Dolomieu 1784)

Tectonic instability thus appears to have caused earthflows and other superficial spreading movements in poorly consolidated sedimentary formations.

One of the greatest and most catastrophic of the mass movements, and one which virtually all authors commented upon, was the abrupt collapse shortly after the February earthquake of a section nearly 2 km long of the coastal cliffs at

Monte Montesina, near Scilla. The cliffs had been weakened by the tremors and also by the Strait of Messina tsunami that followed them, and their collapse caused a water wave that overwhelmed the Prince of Scilla and 2473 of his subjects, who had sought refuge on the beach nearby. Contemporary accounts, however, tended to identify this shock wave with the tsunami, such that cause and effect were confused. Yet Sarconi (1784) was perspicacious enough to discover that fissuring in the headscarp of the cliffs at Scilla had begun in 1780, three years before the tremors.

Mudflows

Hamilton (1783, p. 191) could not resist likening the mudflows produced by the earthquakes to lava flows he had observed on Vesuvius:

> I observed also that many confused heaps of the loose soil detached by the earthquake from the plains on each side of the ravine, had actually run like a volcanic lava (having probably been assisted by the heavy rain) and produced many effects greatly resembling those of lava during their course down a great part of the ravine.

Sarconi (1784, p. 105), by contrast, emphasized the importance of saturation with groundwater in creating fast mudflows. When he investigated major mudflows in clays thinly interbedded with sand and marl strata he was amazed to find that the movements had not entirely ceased three months after the earthquakes that triggered them.

Compaction and liquefaction in sands

Dolomieu (1784) noted that:

> The general effect of the earthquake on the sandy-clayey terrains of lowland Calabria... was to increase their density by lowering their volume, that is compact them, and to turn cliffs into escarpments of gentler slope.

Liquefaction effects were widespread during the tremors and many were described and drawn in meticulous detail by Sarconi and his associates. At Ierocane the illustrators sketched radial cracking produced by subsurface liquefaction (Figure 3b), and on the Plain of Rosarno they drew sand boils (Figures 3c and d). Both Sarconi (1784) and Dolomieu (1784) reported eye-witness accounts of sand-boil occurrence. According to the former (p. 105): "A large quantity of water was erupted and forced out of holes in the ground." Ippolito (1783), however, took these to be the holes through which subsurface lightning bolts issued (as noted above, he was an 'electricist' rather than a 'vulcanist' with regard to the causes of earthquakes). In sum, one cannot help but feel that a good portion of Andrea Gallo's "extraordinary and terrifying metamorphosis" was the result of break-up of the ground during spontaneous liquefaction of saturated sand layers at depth.

Earthquake motions

Finally, in the writings of both Hamilton and his correspondent Francesco Ippolito there is a hint that the type of earthquake motion may have influenced the form of mass movement produced. Ippolito (who must have felt the consecutive passage of P, S, L and R waves) classified earthquake motions into *oscillatorio*, *orizzontale* and *vorticoso*, which Hamilton translated - rather loosely - as *beatings, pulsations* and *whirlings*. In sequence (to use another of Hamilton's translations) undulation preceded shaking, which preceded vorticose motion. In some earthquakes one of these motions predominated over the others, and this may, one supposes, have determined whether slumping, mudflows or sandboils were created. Details that might have clarified this assumption further were not presented by either writer, and no attempt was made to explain the different types of motion in terms of the fundamental cause of earthquakes.

Lyell's retrospective view of the Calabrian earthquakes

Charles Lyell attacked the Grand Tour of Italy with rather more sense of purpose than many of his contemporaries. Hence his *Principles of Geology*, which retained its position as the dominant authority for fifty years after it was first published in 1830, made extensive use of Italian examples. One chapter (Lyell 1883, Vol. 2, Ch. 29) was largely devoted to a re-analysis of the events of 1783 in Calabria, and includes eleven illustrations gleaned directly from Sarconi (1784) and redrawn to a slightly inferior standard (Figure 3). Most of Lyell's account is a remarkably uncritical rehash of the late-eighteenth century authors, with translations that are faithful neither in metaphrase nor in spirit. But there are some more inspired passages, for instance, in which he noted that the dislevelment of land by fault block movement is unlikely to have been great.

Like those who visited Calabria in the 1780s, Lyell was fascinated by the fissuring of argillaceous terrain during mass movement. He observed that the fissures often closed up, sometimes tightly, when slumps were mobilized. But his best contribution is to be found in the summary to the section of *Principles* concerned with the Calabrian earthquakes. In this, Lyell revealed that he had thought profoundly about the relationship between orogeny and erosion, with respect to their effects on the circular interaction between stream incision and sideslope collapse (Lyell 1883, Vol. 2, pp. 133-134):

> If our space permitted, we might fill a volume with local details of landslips, which the different authors above alluded to supply, showing to how great an extent the power of rivers to widen valleys is increased where earthquakes are of periodic occurrence. A geologist can never fully understand the manner in which valleys have been formed until he duly appreciates the part which subterranean movements repeated at long intervals play in combination with rivers, during that lapse of ages which must always be required for the elevation of a country to the height of many hundreds of feet above the level of the sea.
>
> Time must be allowed in the intervals between distinct convulsions for running water to clear away the ruins caused by landslips, otherwise the fallen masses will serve as buttresses, and prevent the succeeding earthquake

> from exerting its full power. The sides of the valley must be again cut away by the stream, and made to form precipices and overhanging cliffs, before the next shock can take effect in the same manner.

In the winter of 1828, before completing the *Principles*, Lyell toured southern Italy. In southeast Sicily he had been impressed concurrently by the immense edifice of Mount Etna, signifying the rapid and cataclysmic genesis of rock material, and the 800 metres depth of limestone visible near Noto, representing slow, uniform accumulation of strata. One way in which he reconciled the two was to apply relative dating to the fossil shells occurring in limestone beds interleaved with or overlain by lavas. Thus Italian examples, both received and observed, were formative influences on Lyell's view of the earth: his conception of uniformity rested in part on the repetitiveness of earthquakes, eruptions and landslides, producing slow change cumulatively under the modulating influence of aqueous erosion (Wilson 1972, pp. 232-291). The matter is summed up succinctly at the end of Book III of the *Principles* (Lyell 1883, Vol. 2, p. 245):

> In the preservation of the average proportion of land and sea, the igneous agents exert a conservative power, restoring the unevenness of the surface, which the levelling power of water in motion would tend to destroy.

Robert Mallet and the Basilicata earthquake of 1857

The Irish engineer and geologist Robert Mallet (1810-1881) was in his time one of the most eminent authorities on earthquakes and their effects and a man who did much to transform seismology from rank speculation into an exact science. He invented a seismograph (Mallet 1846) and used it to conduct tests of seismic wave transmission in various rock formations (Mallet 1861). He also compiled maps of seismic activity around the world, using a vast catalogue of earthquakes that he had painstakingly created (Mallet 1853-5). When on 16 December 1857 a catastrophic earthquake occurred in Basilicata, central-southern Italy, he mounted an expedition to the disaster area and acquired the support of the *Royal Society of London*. Mallet, who took his lead from John Mitchell's writings of a century earlier, believed that the spontaneous condensation of pressurized steam underground gave rise to earthquakes (Mallet 1846, 1850). But although he understood neither faulting nor crustal stresses, he nevertheless made great advances in the study of seismic wave motion and in the construction of isoseismal lines on the basis of surface effects, which he documented with great care.

Mallet's journey into the more remote parts of the southern Apennines (Figure 5) was distinguished by much hardship and adventure. He described earth fissures, rotational slumps and rockfalls at numerous locations, with sufficient accuracy and detail for it to be clear that many of the phenomena recurred in quite the same manner - and in some cases the same location - during the catastrophic earthquake of 23 November 1980, at which I was present (Figures 6 and 7). For example, at Auletta Mallet observed a fresh crack in the soil which was up to

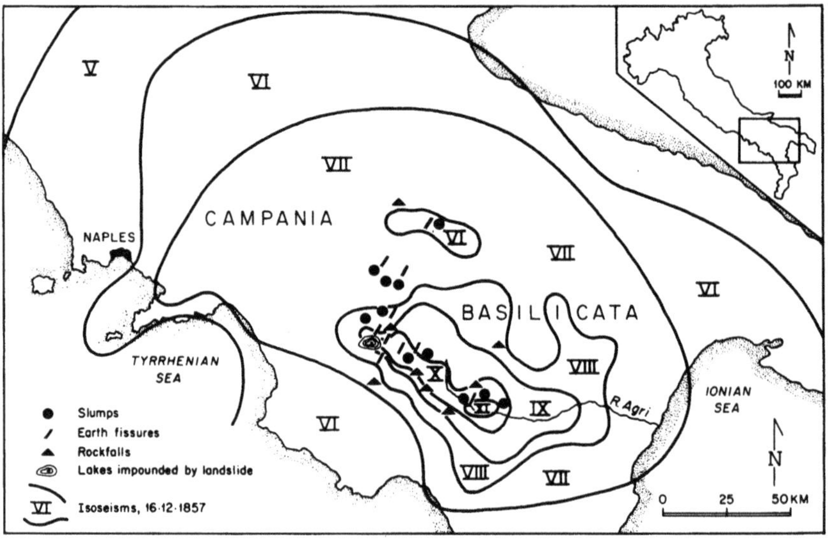

a

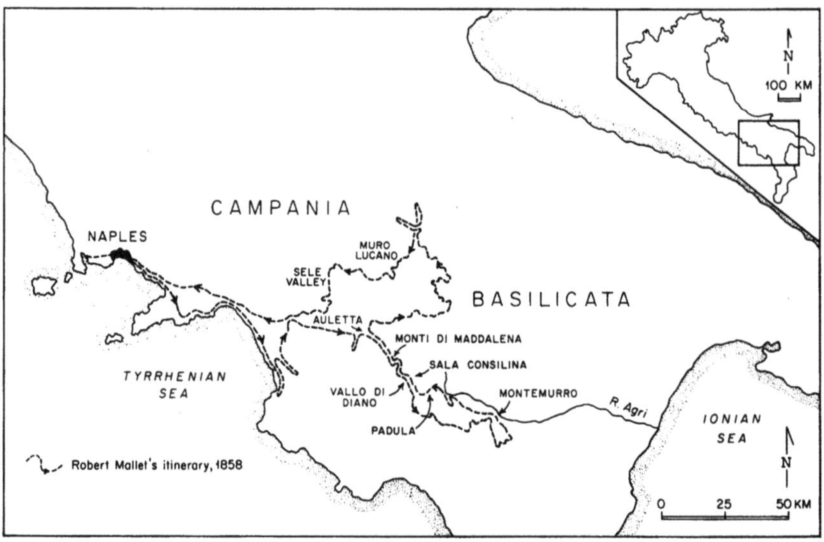

b

Figure 5

The Basilicata earthquake of 16 December 1857:
a. reconstructed isoseisms & location of principle mass movements;
b. Robert Mallet's itinerary.

0.5 m wide, 7.5 deep and several hundred metres long. Beneath it a shear plane was carried into the ground at 25° to the horizontal. Mallet concluded (1862, Vol. 1, p. 264) that:

> it was manifest that the fissure was the evidence of a great earth slip, and had resulted, not from any direct rending asunder of the ground or rocks beneath it, but that the clay masses had when shaken violently upon the inclined beds of rock upon which they were superposed, slid down bodily by gravity, and parted from each other at these fissures,...the fissure evidently had in no case run down plumb into the soil, but sloped in the same direction, but with less of an angle to the vertical that the hillside on which it was found.

Slumping was repeated at Auletta during the 1980 earthquake (Alexander 1981) and in the Agri Valley (Cherubini *et al.* 1981), where 122 years previously Mallet (1862) had observed that spurs between bends of the incising River Agri had provided the extra relief (and pronounced stream undercutting) for the 1857 earthquake to cause multiple regressive sliding of Miocene flysch on a series of curved shear planes (Figure 8). Close to Muro Lucano, Mallet (1862) recorded mudflows that gradually convulsed 200 hectares of land and which have undergone sporadic movement ever since (Alexander 1981). Nearby, he identified gigantic lateral spreads, which carried limestone blocks up to 25 m long and 4.5 m thick down a 12-14° slope in Plio-Pleistocene clays.

Mallet regarded earth fissures as not capable of being produced by the dilatory effects of seismic waves, which he thought too small in amplitude for this, but as a consequence of innate slope instability. Nevertheless, he felt that earthquakes would tend to cause fissuring that was transverse to the passage of waves out from the epicentre. He also found the seismic landslides of Basilicata to have been most commonly located at discontinuities between argillaceous sediments and harder rocks such as limestones and sandstones. In the Vallo di Diano he was puzzled by fault breakage that left a 30-metre surface trace in limestones (*cf.* Bollettinari and Panizza 1981, Westaway and Jackson 1984). As his preferred model of earthquake generation did not encompass slip-faulting, he was left without a ready explanation for this phenomenon.

At Padula Mallet encountered sizeable rockfalls on the limestone massif of the Monti di Maddalena and ascribed them to the inertia of the rock mass, coupled with the influence of fracture patterns, during the shaking. His description (1862, Vol. 1, p. 353) is veracious and also would require little modification to be applied to the rockfalls of 1980 in the Sele Valley, a little further north (Agnesi *et al.* 1982, 1983):

> The massive fragments with newly broken and glittering white surfaces encumbered the slope, or, after having traced their descending paths in lines of torn rock and furrowed detritus, block up the bed of the torrent below, which brawls between the immense fragments of the ill-discernable vertical strata, separating at joints, etc., and are mere cases of loss of equilibrium by the shock; but one most remarkable case I observed, in which an enormous mass of solid rock that had stood up as a sort of blunt aiguille from the steep face of the slope, not quite vertical, but rather overhanging (about 15° by the eye), to the downward side, had been broken clean off at its base, and again breaking into three massive pieces had slid down the rocky slope, and now occupies a place about 150 feet below, having crossed and wholly torn away the mule path in their progress.

Figure 6

Landsliding near San Fele in Basilicata caused by the 23 November 1980 earthquake.

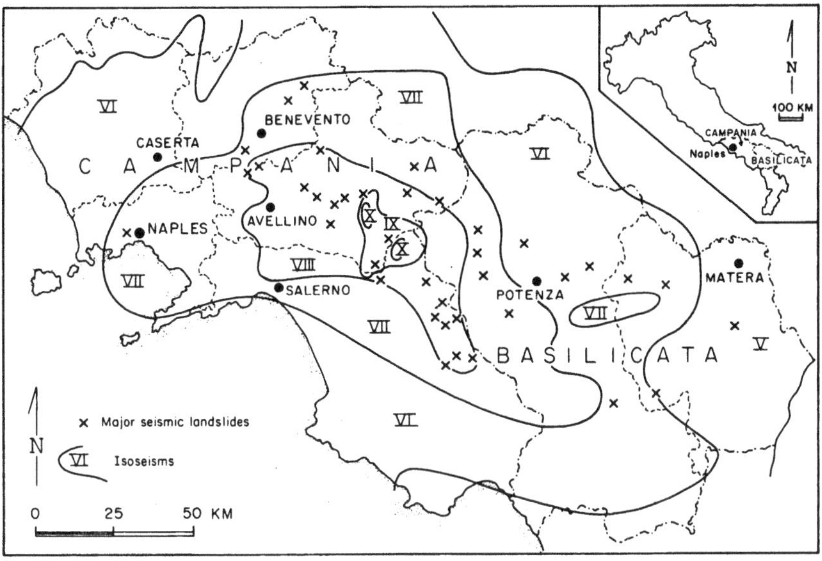

Figure 7
Campania-Basilicata earthquake of 23 November 1980: isoseisms and location of major landslides.

Local authors had described very similar rocksliding nearby at Sala Consilina during the earthquake of 1 February 1826, although with less precision (Almagià 1910).

Mallet (1850) had read the works of those who investigated the Calabrian disaster of 1783, and his thinking on the geological effects of earthquakes was much influenced by the Neapolitan author Francesco Grimaldi, who died in 1784, right after his visit to Calabria. Like Lyell, he was fascinated by Sarconi's engraving of radial cracking of the ground consequent upon subsurface liquefaction. Lyell (1883) took the cracks to evince bulging of the ground, whereas Mallet (1846) ascribed them rather elaborately to cyclical phase differences in the waves as they passed through the subsoil. Mallet also attributed the fountains observed in sand boils to sudden compression of underground water pockets as the seismic waves passed through them.

Rather strangely, he failed to observe the slumping that occurred at Montemurro, which was the closest settlement to the epicentre of the 1857 earthquake. A distinguished local historian, Giacomo Racioppi (in *Memorie sui terremoti di Basilicata del 1857*), ascribed the slumps to over-steepening of slopes by gully incision during winter rains. Almagià (1910) used the example of Montemurro to show that most places where seismically generated landslides have been recorded have had equivalent movements with non-seismic causes (at Montemurro these took place in 1844, 1884, and 1907). Mallet, however, arrived at Montemurro at the dead of night and was deeply impressed - indeed, somewhat disorientated - by the damage he encountered, most of the population of 5000 having perished in the tremors.

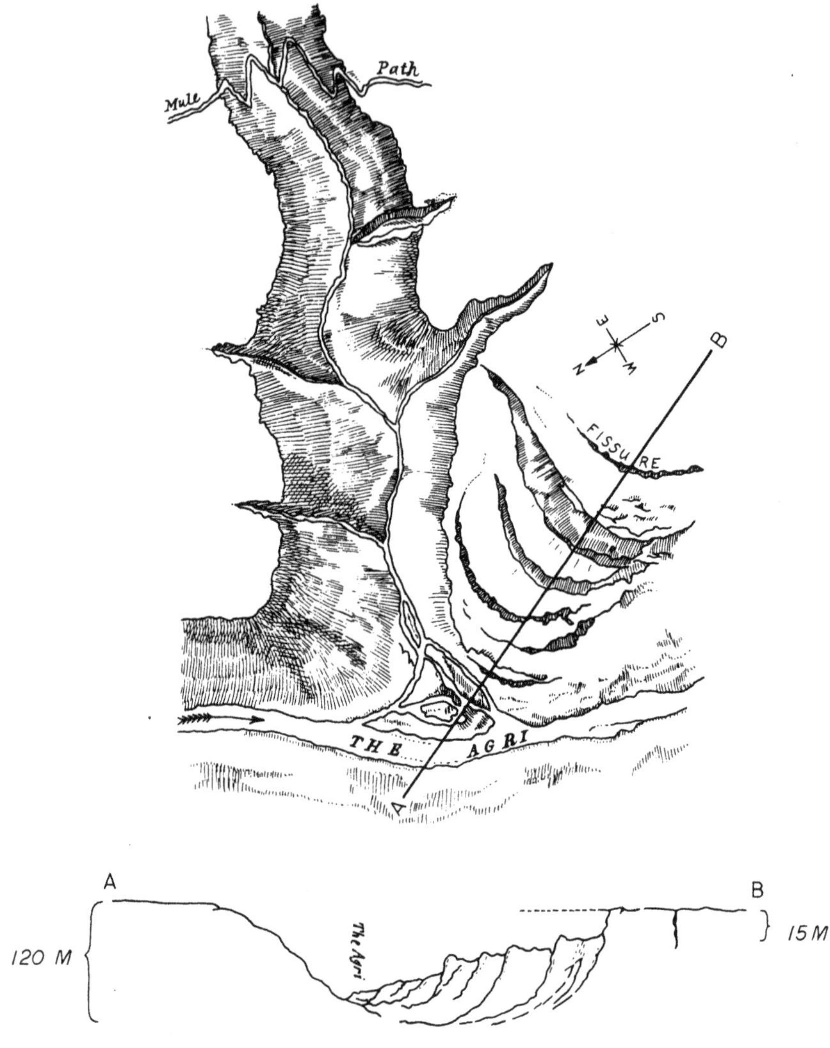

Figure 8

Mallet's diagram of multiple retrogressive slumping in the Agri Valley, Basilicata (Mallet, 1862, volume 2, p. 15).

In his field descriptions of seismically-generated mass movements, Robert Mallet was principally concerned to record all his impressions of earthquake effects. Indeed, he intended explicitly that his 1862 monograph be not merely a compendium of observations but also a travelogue that recorded his impressions of the remote hinterlands of southern Italy (Mallet 1862, Vol. 1). Perhaps, though, his main scientific interest in the landslides derived from their role, like the damage to buildings, in macroseismic survey - the pinpointing of epicentres, isoseismal lines and directions of wave travel. To this end, it was necessary that he record them as comprehensively as possible, with special reference to their position, aspect and local conditions, which thus might help explain localized seismic wave motions.

The twentieth century

In the early 1900s the Barese geographer Roberto Almagià carried out a national survey of landslides in Italy, publishing his findings on the North and Centre in 1907 and on the Mezzogiorno in 1910. His first volume included a long, case-by-case examination of the extent to which historical landslides in Italy have been synchronous with earth tremors (Almagià 1907, pp. 263-73). He judged the evidence to be equivocal, and also rated pore-water pressure (i.e. - heavy precipitation) a much more important cause of mass movements than earthquakes, no doubt justly. Where seismic landsliding had definitely occurred, one could invariably expect a long period of weathering and perhaps steepening of the slope to have preceded it. No particular information could be gleaned from historical accounts that might impute microseismicity, and so Almagià (1910, pp. 327-338) was unable to link this to slope instability.

Roberto Almagià's predilection for the non-seismic causes of landsliding undoubtedly influenced other researchers, as his works were read in Italy for decades. Investigations of slope failure in earthquakes thus became local and rather more circumstantial or descriptive than analytical. The reawakening of interest in such phenomena probably began in the Americas when the results of field investigations that described landslides were published after earthquakes in Montana in 1959 (Witkind *et al.* 1962), Chile in 1960 (Weischett 1963), Peru in 1970 (Plafker *et al.* 1971), and Guatemala in 1976 (Harp *et al.* 1981). Similar field studies were conducted in Calabria after the 1978 Ferruzzano earthquake (Bottari *et al.* 1981-1982), in Campania in 1980 (Agnesi *et al.* 1982, 1983) and in Basilicata in 1980 (Cherubini *et al.* 1981), to mention only three of many recent endeavours. There has also been a reawakening of interest in historical accounts of seismic landslides (Cotecchia *et al.* 1969, Hughes 1983, Alexander 1983).

Conclusion

Given the varied fortunes of geology over the last two and a half millennia, seismically-induced mass movement has received fairly consistent treatment, although it has never been a major issue. Instead, it has been regarded mainly as an exceptional phenomenon, and hence one to be reported descriptively, especially if it occurred on a grand scale, as it seems frequently to have done. The causes were more often neglected than investigated; but, if they were adduced, they were taken to be the endogenous fires or vapours, or, if they were sought in the field, were found to be more commonplace functions of water, rock mass and gravity.

The sheer scale of ground failure provoked in Calabria in 1783 makes this event stand out in European scientific history. Robert Hooke and John Mitchell provided the seismological basis for investigations, if one were needed by contemporary writers, but the fieldwork stood alone in furnishing a wealth of examples of slumps, mudflows, surface water impoundments and rockfalls. Thus the reports of Sarconi, Vivenzio and Hamilton were destined for posterity, not obscurity (thanks in no small measure to the publishing activities of the early scientific societies). In the light of descriptions of slope failure contained in their works, the writings of Mallet and Racioppi on similar phenomena in 1857 seem almost repetitious. But the latter authors, Mallet especially, wrote more objectively and perceptively, such that their observations bear direct comparison with slope failures caused by the tremors of 1980. Finally, macroseismic survey of Italian communities in the last few years (Postpischl 1985) has shown how useful historical accounts can be to modern applied seismo-geomorphology.

References

Note: Classical and some later historical works are cited in the text alone.

Adams, F.D., 1938. *The birth and development of the geological sciences.* Baltimore, Williams and Wilkins.

Agnesi, V. *et al.*, 1982. Osservazioni preliminari sui fenomeni dinstabilità dei versanti indotti dal sisma del 1980 nell'alta valle del Sele. *Geologia Applicata e Idrogeologia*, 17, 79-93.

Agnesi, V. *et al.*, 1983. Elementi tipologici e morfologici dei fenomeni di instabilità dei versanti indotti dal sisma del 1980 (alta valle del Sele). *Geologia Applicata e Idrogeologia*, 18(1), 309-41.

Alexander, D.E., 1983. "God's Handyworke in Wonders" - landslide dynamics and natural hazard implications of a sixteenth century disaster. *Professional Geographer*, 35(3), 314-23.

Almagià, R., 1907. Studi geografici sulle frane in Italia. I. Nord ecentro. *Memorie della Società Geografica Italiana*, 13.

Almagià, R., 1910. Studi geografici sulle frane in Italia. II. Sud. *Memorie della Società Geografica Italiana*, 14.

Besterman, T., 1969. *Voltaire.* New York, Harcourt, Brice and World.

Bollettinari, G. and **Panizza, M.**, 1981. Una "faglia di superficie" presso San Gregorio Magno in occasione del sisma del 23-XI-1980 in Irpinia. *Rendiconti della Società Geologica Italaliana*, 4, 135-6.

Bonaiuto, V., 1693-4. An account of the earthquake in Sicilia, on the Ninth and Eleventh of January, 1692/3. *Philosophical Transactions of the Royal Society of London*, 18 (207), 2-10.

Bottari, A., LoGuidici, E., Nicoletti, P.G. and **Sorriso-Valvo, M.**, 1981-2. The Ferruzzano earthquake of 1978: macroseismic effects and slope stability conditions in southern Calabria (Italy). *Revue de Géographie Physique et Géologie Dynamique*, 23(1), 73-84

Cherubini, C., Guerricchio, A. and **Melidoro, G.**, 1981. Un fenomeno di scivolamento profondo delle argille grigioazzurre Plio-Calabriane nella valle del T. Sauro (Lucania) prodotto dal terremoto del 23 november 1980 - nota preliminare. *Rendiconti della Società Geologica Ital.aliana*, 4, 155-9.

Coccia, G., 1894. Relazione al Maresciallo Pignatelli per la distrutta città di Santa Cristina col tremuoto del 5 febbraio 1783. *Rivista di Storia Calabrese*, 227-31.

Cotecchia, V., Travaglini, G. and **Melidoro, G.**, 1969. I movimenti franosi egli sconvolgimenti della rete idrografica prodotti in Calabria dal terremoto del 1783. *Geologia Applicata e Idrogeologia*, 1-24.

Dolomieu, G.D. De, 1784. *Mémoire sur les tremblemens de terre de la Calabre pendant l'année 1783*. Roma, Fulgoni.

Figliuolo, B., 1985. Il terremoto napoletano del 1456: il mito. *Terremoti e storia, quaderni storici, nuova serie*, 60(3), 771-801.

Gallo, A., 1784. Lettera storico-fisica de' Terremoti di Calabria scritta lì 7 giugno 1783 dal signor G. a Monsieur H. in Parigi. In *Collected Letters*, Messina, Di Stefano.

Geikie, A., 1905. *The founders of geology* (2nd Edn.). London, Macmillan.

Guidoboni, E., 1986. The earthquake of December 25, 1222: analysis of a myth. *Geologia Applicata e Idrogeologia*, 21(3), 413-24.

Hamilton, W., 1783. An account of the Earthquakes, which happened in Italy, from February to May 1783. *Philosophical Transactions of the Royal Society of London*, 73(1), 169-208.

Harp, E.L., Wilson, R.C. and **Wieczorek, G.F.**, 1981. Landslides from the February 4, 1976, Guatemala earthquake. *United States Geological Survey Professional Paper*, 1204A.

Hooke, R., 1705. Lectures and Discourses of Earthquakes and Subterraneous Eruptions. In *The posthumous works of Robert Hooke*. London, Richard Waller, 279-450.

Hughes, R., 1983. Historic disasters. *Disasters*, 7(3), 161-3.

Ippolito, F., 1783. Letter to Sir William Hamilton. *Philosophical Transactions of the Royal Society of London*, 73(1), i-vii (trans), 209-16 (in Italian).

Kendrick, T.D., 1956. *The Lisbon Earthquake*. London, Methuen.

Lyell, C., 1883. *Principles of geology* (11th Edn.). London, Chapman and Hall, Vol. 2, Ch. 29.

Mallet, R., 1846. On the dynamics of earthquakes. *Transactions of the Royal Irish Academy*, 21(1), 51-113.

Mallet, R., 1850. First report on the facts of earthquake phenomena. *British Association for the Advancement of Science*, 20th Mtg., 1-89.

Mallet, R., 1853-5. Third report on the facts of earthquake phenomena. *British Association for the Advancement of Science*, 22nd Mtg., 1-176; 23rd Mtg., 117-212; 24th Mtg., 1-326.

Marinatos, S.N., 1960. Helice: submerged town of Classical Greece. *Archæology*, 13 (3), 186-93.

Mitchell, J., 1761. Conjectures concerning the cause, and Observations upon the Phenomena of Earthquakes. *Philosophical Transactions of the Royal Society of London*, 51(2), 566-634.

Placanica, A., 1985. *Il filosofo e la catastrofe*. Torino, Einaudi.

Plafker, G., Erickson, G.E. and **Concha, F.J.**, 1971. Geological aspects of the May 31 1970 Peru earthquake. *Bulletin of the Seismological Society of America*, 61, 543-78.

Postpischl, D., ed., 1985. *Atlas of isoseismal maps of Italian earthquakes*. Roma, Progetto Finalizzato Geodinamica, Consiglio Nazionale delle Ricerche.

Sarconi, M., 1784. *Osservazioni fatte nelle Calabrie e nella frontiera di Valdemone sui fenomeni del tramoto del 1783 e sulla Geografia fisica di quelle regioni*. Napoli, Campo, e la Reale Accademia delle Scienze e Belle Lettere.

Solbiati, R. and **Marcellini, A.**, 1983. *Terremoto e societa*. Milano, Garzanti.

Stukeley, W., 1750. On the causes of earthquakes. *Philosophical Transactions of the Royal Society of London*, 52 (497), 641-5, 657-68, 731-50.

Vivenzio, G., 1788. *Istoria de' tremuoti avvenuti nella Provincia della Calabria Ulteriore, e nella città di Messina nell'anno 1783* (2nd Edn.). Napoli, Stamperia Reale, 2 Vols.

Weischett, W., 1963. Further observations of the geologic and geomorphic changes resulting from the catastrophic earthquake of May, 1960, in Chile. *Bulletin of the Seismological Society of America*, 53, 1237-57.

Westaway, R. and **Jackson, J.,** 1984. Surface-faulting in the southern Italian Campania-Basilicata earthquake of 23 November 1980. *Nature*, 319, 436-8.

Wilson, L. G., 1972. *Charles Lyell, the Years to 1841: the Revolution in Geology*. New Haven, Connecticut, Yale University Press.

Witkind, J.J. ***et al.*****,** 1962. Geologic features of the earthquake at Hebgen Lake, Montana, August 17, 1959. *Bulletin of the Seismological Society of America*, 52, 163-80.

8

From colonial science to scientific independence: Australian reef geomorphology in the nineteenth century

D. R. Stoddart

Of all the great phenomena of nature explored by Europeans during the voyages of discovery of the late eighteenth century and later, coral reefs were perhaps both the most exotic and the most enigmatic. Massive topographic structures apparently built by the simplest of skeleton-secreting organisms, the questions they posed were geological, geomorphological, zoological, botanical, and philosophical. Reefs came to the attention of the European imagination at a time when older interpretations of nature in terms of the great chain of being were beginning to give way to a more rational and functional understanding of the nature of things. But the sciences which would attempt to solve the mysteries of the reefs had themselves largely to be defined. Coral reefs too were themselves located on the farther limits of the scientific oikumene, generally on alien shores populated by heathens, cannibals, pirates, and (increasingly after 1788) convicted desperadoes of many kinds.

There is thus a double theme in our understanding of how reefs themselves came to be understood. On the one hand there is the rise of objective knowledge, as topographies, faunas and floras became known. This was achieved, at least in outline, over a surprisingly short space of time: it was but seventy years from the collecting and describing of Banks on Cook's first expedition to the theoretical speculations of Darwin and Dana. But there is also the transition between the kind of science carried out in the brief, expensive, and necessarily superficial forays of great expeditions - 'imperial science,' as MacLeod (1988) terms it - and the slow and fumbling beginnings of studies in the reef lands themselves, so often carried out by those exiled, often by financial considerations, from metropolitan science, and often almost by definition backward, parochial and derivative.

In this paper I explore how the Great Barrier Reef and other Australian reefs were gradually and belatedly incorporated into the canon of accepted knowledge. I shall consider why it was that the largest reefs in the world made so little impact on scientific thinking about reef phenomenon, and I shall discuss the interplay of those social as well as scientific factors which determined the pace at which the reefs became known.

Describing

Cook's entry to the Great Barrier Reef Province on his first voyage was less than auspicious: as the *Endeavour* passed Sandy Cape on the night of 22nd May 1770 Cook's clerk, lying in a drunken stupor, had both of his ears cut off, apparently by the ship's company's only American, Midshipman Magra, whose name is now commemorated by a sand cay in latitude 11 ° 51½' S (Beaglehole, 1955, p. 323). Twenty days later the ship ran aground on what is now Endeavour Reef (15° 45' S), a circumstance that even Cook naturally found 'alarming.' Small wonder that he had little time for speculation on the nature and origin of the reefs that threatened him.

It was not, of course, immediately apparent to Cook as he passed through the Bunker and Capricorn Groups on passage northward that he was entering an enormous lagoon enclosed by the Great Barrier Reef. On 7 August, from the masthead of the repaired and refloated vessel, he 'could see nothing but breakers all the way from the South round by the East as far as NW, extending out to sea as far as we could see': the Barrier was discovered (Beaglehole 1955, p. 369). Cook found his exit through it from the summit of Lizard Island, only narrowly to escape a second grounding on the seaward side:

> All the dangers we had escaped were little in comparison of being thrown upon this Reef where the ship must be dashed to pieces in a Moment. A reef such as is here spoke of is scarcely known in Europe, it is a wall of Coral Rock rising all most perpendicular out of the unfathomable Ocean, always overflowing at high-water generally 7 to 8 feet and dry in places at low-water; the large waves of the vast Ocean meeting with so sudden a resistance make a most terrible surf breaking mountains high especially as in our case when the general trade wind blows directly upon it (Beaglehole,1955, p. 378).

And there, other than the observation that sand patches accumulate at the leeward ends of lagoonal reefs (Beaglehole 1955, p. 369), Cook leaves it: sober, pragmatic, cautious, unproblematic.

It is perhaps more surprising that Cook's naturalists - Banks, Solander, Parkinson - took no greater interest in the coral reefs. As for Banks himself, in spite of his subsequent reputation as the 'autocrat of the philosophers' when President of the *Royal Society* (Cameron 1952, Beaglehole 1962, p. 20) rather savagely comments that Banks' occasional reference to the 'great chain of being' was 'the nearest to a philosophical or general scientific notion that Banks ever had.' Banks and his fellow naturalists were content to collect the plants and describe the aborigines during their time in Australia, and their interest, perhaps unsurprisingly, was overwhelmingly terrestrial rather than marine. Even in Tahiti, where they first encountered coral reefs, their priorities were the same (Beaglehole 1962, pp. 252-386).

Banks also climbed with Cook to the summit of Lizard Island, and 'saw plainly the Grand Reef still extending itself Parallel with the shore at about the distance of 3 leagues from us or 8 from the main.' He seemed to recognise its singularity, but in his summary systematic description of the east coast of Australia (Beaglehole,1962, II, pp. 111-137), Banks makes no mention of the reefs or their origins, other than as sources of fish, molluscs, crabs, and turtles.

Comparing

Banks, Solander and Parkinson applied a taxonomic expertise to the description of plants and animals, but their conceptual apparatus did not extend to the description of the landforms. This deficiency was supplied on Cook's second voyage by Johann Reinhold Forster, who never, however, had the opportunity to see the Australian reefs. In the Tuamotu Archipelago, however, he speculated on the surface features of atolls:

> [At Takapoto] Both Isles have the longer sides ... exposed to the East-ward, which probably was the first origin of these curious Isles: for their being some little Elevation or inequality in the bottom of the Sea the coral animals began to raise their large rocky branching out habitations; & as soon as the Animals came nearer to the Surface of the Sea, they extended their branches more & more North, in order to shelter themselves against the Surf of the Sea, & at last this work formed a Lagoon in the middle. On the ledge of Coralrock, the Sea threw some Seaweeds, Shells, & Sand, which with the dung of the aquatic fowl formed a small Quantity of Soil, on which some few seeds carried hither from the higher Islands & the Continents by Seas, wind, & birds began to grow. The leaves of the plants increased the Soil, & thus it became at last a habitation for Man (Hoare 1982, p. 494; entry for 18 April 1774).

In Fiji both Cook and Forster were struck by the widespread occurrence of uplifted reef limestones, and even the circumspect Cook speculated on their implications:

> If these Coral rocks were first formed in the Sea by animals, how came they thrown up, to such a height? has this Island been raised by an Earth quake or has the sea receded from it. Some Philosophers have attempted to account for the formation of low isles such as are in this Sea, but I do not know if any thing has been said of the high Islands or such as I have been speaking (Beaglehole, 1961, 438).

In his treatises on *The theory of the Formation of Isles* and *On the diminution of the Sea and Water*, Forster (1778, pp. 145-159) classifies islands into sea-level reef islands, raised limestones reefs, and volcanos, but allows no further speculation: 'We are the first to collect a great many facts, and by no means to form systems from a few particulars' (Forster 1778, p. 146).

Matthew Flinders, forty years after Cook, explored the reefs of northeastern and northern Australia in *H.M.S. Investigator* though his naturalist, Robert Brown, showed no interest in them. Though also a professional hydrographer, Flinders showed a keener interest in the reefs than either Cook or Banks. In 21°S he:

> went upon the reef with a party of gentlemen; and the water being very clear round the edges, a new creation, as it was to us, but imitative of the old, was there presented to our view. We had wheat sheaves, mushrooms, stags horns, cabbage leaves, and a variety of other forms, glowing under water with vivid tints of every shade betwixt green, purple, brown, and white; equalling in and excelling in grandeur the most fantastic parterre of the

> curious florist. These were different species of coral and fungus, growing, as it were, out of the solid rock (Flinders 1814, 87-88).

He described the reef in some detail, including the first account of reef blocks, 'which as some distance resembled the round heads of negroes' (Flinders 1814, p. 81).

But Flinders' view of reef growth was as unproblematic as that of Cook, perhaps inevitably so, given that the crucial fact of the limitation to shallow water of the depths in which reef-building corals could grow would not be demonstrated by the French naturalists Quoy and Gaimard, with Freycinet's expedition in *Uranie* and *Physicienne*, for another twenty years (Quoy and Gaimard 1824).

Flinders attributed the ability of corals to form reefs to instinct:

> It seems to me, that when the animalcules which form the corals at the bottom of the ocean, cease to live, their structures adhere to each other, by virtue either of the glutinous remains within, or of some property in salt water; and the interstices being gradually filled up with sand and broken pieces of coral washed by the sea, which also adhere, a mass of rock is at length formed. Future races of these animalcules erect their habitations upon the rising bank, and die in their turn to increase, but principally to elevate, this monument of their wonderful labours. The care taken to work perpendicularly in the early stages, would mark a surprising instinct in these diminutive creatures. Their wall of coral, for the most part in situations where the winds are constant, being arrived at the surface, affords a shelter, to leeward of which the infant colonies may be safely sent forth; and to this their instinctive foresight it seems to be owing, that the windward side of a reef exposed to the open sea, is generally, if not always the highest part, and rises almost perpendicular, sometime from the depth of 200, and perhaps more fathoms (Flinders 1814, Volume 2, p. 115).

The publication of Flinders' results was in any case long delayed, both by the loss of his ship on Wreck Reef in the Coral Sea and then by his imprisonment by the French for 6½ years in Mauritius. He gave currency to the name 'Australia' in his charts (Flinders 1814b), named the 'Corallian Sea' (Flinders, 1814a, p. 314), and introduced the proper name 'Barrier Reefs' (Flinders 1814a, p. 101), which on his chart became for the first time 'The Great Barrier Reefs' (Flinders 1814b, Ingleton 1979, p. 71).

But he afforded no broader view, and indeed no formulation of questions which might provoke answers about the nature and origins of the coral reefs. It was a pattern repeated by such subsequent Australian hydrographic surveyors as Philip Parker King, J.C. Wickham and John Lort Stokes in the years up to 1839. It is perhaps unreasonable to expect these professional men, engaged in arduous and difficult work, so to exceed their orders as to concern themselves with purely geological and biological questions. But the consequence was that the comparative vision of Forster - classifying, generalising, and thus seeking the causes of things - failed to illuminate the nature of the Australian reefs.

Synthesizing

Contemporary with Flinders' surveys in 1801-1803 was the French expedition commanded by Nicolas Baudin in *Géographe* and *Naturaliste* ships as appropriately named as any of Cook's. Indeed, *Investigator* and *Géographe* met in Encounter Bay, near modern Adelaide, on 8 April 1802. The French expedition worked mainly in southern and western Australia, and in the East Indies. Its membership included four zoologists, three botanists, four horticulturalists, two astronomers, five artists, and a naturalist who served as anthropologist and oceanographer. But it was not a happy group, and only four completed the voyage, most of the rest dying or deserting. Baudin, insecure, resented and disliked his chief naturalist, François Peron. Their journals, says Spate (1979 p. 93) 'constitute a lamentable record of human pettiness, ... [and] their admirable qualities were pitiably spoilt by human vanity.' Nevertheless, Peron became the first systematic geomorphologist to work on Australian reefs. It is remarkable not only that his name appears in no standard history of geomorphology but that his contribution to reef studies is completely misread by Böttger (1890, p. 267) and was ignored by Günther (1910, p. 36).

Peron was particularly impressed by the evidence of recent high stands of the sea. Indeed, for him,

> L'un des plus beaux resultats des recherches géologiques modernes, l'un des plus incontestables aussi, c'est la certitude du séjour de la mer a de grandes élévations au-dessus de son niveau actuel (Peron 1816, p. 165).

His memoir on the subject emphasized the evidence given by fossil marine shells, raised and lithified marine deposits (he speculated interestingly on the nature and origins of the cements), and elevated corals (which he realised were confined to tropical latitudes). His outline of the evidence for higher sea levels was based not only on his own observations in Australia and Timor, but on an analysis of the extent descriptive literature, carefully tabulated, with authorities, on no less than 243 distinct circum-tropical locations (Peron 1816, pp. 184-192). In this work Peron supplies the factual basis for generalisation that Forster had required. It is of particular interest that Peron is less concerned with the description of individual sites investigated by himself or others, but rather with the analysis and interpretation of classes of geomorphic phenomena through the comparative method.

This discussion of sea-level change leads to his final essay *Considération générales sur la formation des Iles et des Montagnes madreporiques* (Peron 1816, pp. 179-183).

Two questions are posed. Are the raised limestones and fossils evidence of former higher sea-levels? Given their similarity to modern reefs and faunas the answer must necessarily be so. How then have they reached their present elevations? - by volcanic upheaval or by a fall in sea-level itself? Given the frequent lack of evidence for volcanicity associated with the raised reefs, Peron argues for the latter. The raised reefs are for him 'image et produit a-la-fois du calme de la nature' (Peron 1816, p. 180), and he approvingly quotes the Comte de Fleurieu's conclusion, in his account of the circumnavigation by the *Solide* in 1790-1792, that

> elles sont le produit des siècles; que l'ouvrage n'en est pas termine; qu'il doit s'y faire un accroissment graduel; mais qu'une longue succession de temps est nécessaire pour que cet accroissment soit rendu sensible (Claret de Fleurieu 1798-1800, 3, p. 324).

It is thus that Peron for the first time lays open the whole subject of reef formation to theoretical enquiry, utilizing the tools of 'l'observation, l'éxpérience, le raisonnement et l'analogie' (Peron 1816, p. 179).

It was a method which itself prompted further consequent questions, and which takes him to the limits of explanation. Thus his final two questions - where has all the sea water gone? and where has all the calcium carbonate in the reefs come from? - he could not answer from observation or experience, and was forced to be content to leave them to the future rather than resort to imagination (Peron 1816, p. 181).

It was not until the voyage of the *Beagle* thirty years after Peron's that dynamic theory was brought to bear on these questions. Until then, Peron's treatment remained innovative and remarkable, and evidence of the keenest geomorphic insight. Peron's other activities were equally astonishing. His Australian faunistic collections comprised 18,414 specimens of 3872 species, of which 2592 were new to the Paris Museum. Cuvier claimed that Peron's work had yielded ten times more new species of animals than had Cook's second voyage. And like Forster too he was interested in the human race - the physical anthropology, culture, artefacts and languages of the peoples he studied (Wallace 1984, pp. 150-152). Yet Forster's work is remembered and Peron's largely forgotten. Indeed, Peron's work made little impact at the time, and he had no influence on future reef studies. He himself died from tuberculosis in December 1810, with only the first volume of his narrative published, and it took a further six years for his reef memoirs to appear. And there were doubtless reasons of national chauvinism as well as linguistic incapacity to account for the fact that Peron's work passed unrecorded by Lyell, Darwin and Dana.

But there was also the fact that the 'coral reef problem' from the beginning was articulated in terms of the nature and origins of atolls in the open ocean, and that this issue was firmly located from at least the time of Bougainville and Cook in the tropical Pacific (the descriptions of the great atolls of the Maldives in the Indian Ocean by Ibn Batuta in the fourteenth century and by Pyrard de Laval in the sixteenth belonged to a quite different tradition and were uniformly overlooked).

Hence in spite of the fact that Flinders' 'Great Barrier Reefs' with a length of 1200 miles, are the biggest in the world, and that there are many other diverse reef types around Australia's shores, the focus of interest shifted decisively to the South Pacific. Many of these studies supplied much new information, but their conceptual framework remained that of Forster, and indeed their conclusions often went little beyond his essay on *The theory of the formation of isles* (Forster 1778). Thus Adelbert von Chamisso discussed the Marshall Islands following the voyage of the *Rurick* under Otto von Kotzebue in 1815-1818 (Chamisso 1821). F.W. Beechey, in command of *H.M.S. Blossom*, worked through the Tuamotu atolls in 1825-1828, and contributed an extended discussion of their features and origins (Beechey 1831, 186-194). And even the authors of 'narratives of missionary enterprises in the South Sea islands' felt free to join the debate, and not surprisingly proved equally unoriginal (Williams 1837 pp. 19-36). Attempts were even made by Edward Belcher and Cadwallader Ringgold in the Tuamotus in 1840-1841 to bore through coral limestone to the underlying foundations, in the Tuamotus, but they failed because of conceptual naivety as well as technical inadequacy (Stoddart 1989).

Theorizing

The terms of the debate changed radically following Charles Darwin's voyage around the world in H.M.S. Beagle in 1831-1836, and the announcement of his new theory of coral reef formation to the *Geological Society of London* in 1837 (Darwin 1838, see also Darwin 1839, 1842, Stoddart *ed.* 1962). Darwin defined the dynamic processes responsible for the morphological development of reefs over geologic time: his theory called for the slow transformation of fringing reefs to barrier reefs and then atolls by the slow subsidence (at rates at which coral upgrowth could keep pace) of the reef foundations. The theory was applied in the first instance to the transformation of reef-fringed Pacific volcanos (such as Tahiti), to barrier-encircled islands (such as Moorea or Borabora), to atolls (such as those in the Tuamotus), but Darwin was prepared to generalize it to encompass the reefs of continental shores as well as those of the open ocean basins (Darwin 1842).

Darwin's direct reef experience was but limited, and indeed he had formulated the essentials of his theory in South America before ever seeing a coral reef (Stoddart 1976, p. 204). He had seen atolls in the Tuamotus from on board ship and had observed the fringing reefs of Tahiti and the barrier reef of Moorea in 1835; the following year he visited Cocos-Keeling Atoll in the eastern Indian Ocean, the only atoll he ever studied personally. Between Tahiti and Cocos-Keeling he visited Sydney and Western Australia, but the voyage had gone on for several years, his mind was on other things, and FitzRoy's orders precluded a view of Australian reefs.

Late in 1839 the *United States Exploring Expedition* also came to Sydney, with James Dwight Dana serving as scientist in charge of geology and invertebrate zoology. By his own account it was in Sydney that Dana first learned of Darwin's coral reef theory, after he himself had worked through the reefs of the eastern and central Pacific, and with slight modifications he adopted it wholeheartedly. But Dana too never saw an Australian coral reef, and his experience was firmly rooted in the open Pacific.

Thus when Darwin published his *Structure and distribution of coral reefs* in 1842 and Dana his *On coral reefs and islands* in 1853, both works were centred on the problem of open-ocean atolls. Darwin gives but one sentence to the Great Barrier reef proper (1842, reprint 1890 p. 93), with his information derived from Flinders and King. He gives a paragraph on the reefs of western Australia, derived from the surveys of King and his old shipmate Wickham, but withholds judgement on their status. The additional sources on whom Darwin relies, such as Fairfax Moresby in the Indian Ocean and F.W. Beechey in the eastern Pacific, are both to do with open-ocean atolls. Though Dana's experience was far wider than that of Darwin, he too almost completely ignores the Great Barrier Reef and other Australian reefs in his book, and locates the coral reef controversy within the ocean basins rather than on continental shores.

Soon after Darwin's *Structure and distribution of coral reefs* was published in 1842, *H.M.S. Fly* sailed for Australia to carry out extended surveys on the Great Barrier Reef. She carried two scientific men, John MacGillivary as zoological collector, and Joseph Beete Jukes (1811-1869) as geologist; the ship's surgeons attended to botany. Jukes had been a student of Adam Sedgwick at Cambridge University, and as such was the first trained professional geologist to work on Australian coral reefs (Stoddart 1988).

Three factors shaped Jukes' work on coral reefs. The first was the influence, both personal and scientific, of Sedgwick. Sedgwick was a pragmatist, devoted to the elucidation of the testimony of the rocks by meticulous field observation and the marshalling of empirical evidence. Jukes had worked with him in the field in England as well as being instructed by him at university, and he carried Sedgwick's intellectual attitudes with him through his life. It is interesting to note that the young Darwin too had geologised with Sedgwick as a student, though he did not attend his lectures, and that Darwin's own early geological work on board the *Beagle* likewise reflected Sedgwick's emphasis on collecting and recording in the field.

Second, Jukes was profoundly influenced by Lyell's *Principles of Geology*, the first volume of which was published in the year he went to Cambridge and the others while he studied there. Lyell's geology, with its emphasis on theory, hypothesis and conjecture was less than congenial to Sedgwick, but his uniformitarian approach supplied a ready tool with which to interpret field observations.

And finally there was Darwin's theory, explaining the different forms of reefs in terms of progressive slow modification consequent on the subsidence of reef foundations. Since Darwin believed that vertical movements of the crust were of broad extent, this also implied general regional coherence in the patterns of distribution of different reef types. Jukes himself discussed the theory with Darwin before he left England, and took a copy of *The structure and distribution of coral reefs* with him to Australia.

The *Fly* arrived in Sydney in October 1842. Jukes found there another of Sedgwick's pupils, W.B. Clarke, who was in the process of becoming one of the leading geologists in New South Wales. Clarke had also been urged by Sedgwick to discuss the new coral reef theory with Darwin before leaving for Australia, and he had a further opportunity to consider the matter when soon after he came to Sydney the *United States Exploring Expedition* sailed in, with James Dwight Dana as geologist. Dana already had extensive experience of reefs in the central and east Pacific, and it was in Sydney that he learned for the first time of Darwin's theory. He became an immediate convert, adding fresh arguments to Darwin's central thesis, drawing attention to the progressive development of reefs along linear volcanic chains, and presenting fresh evidence for the regional coherence of tectonic style. His subsequent work in Fiji was based on Darwin's theory. Clarke knew Darwin, as well as being a pupil of Sedgwick, and Dana and he must inevitably have discussed coral reefs during their joint geological fieldwork in New South Wales. Thus when Jukes arrived he could not only discuss Darwin's views with Clarke but would also know of Dana's support and his independent evidence for the subsidence theory.

The result of Jukes' work on the Great Barrier Reef were therefore not surprising, for the terms of the coral reef debate had now changed irreversibly.

Up to this time all studied of Australian reefs had been essentially descriptive, or, as in the case of Peron, classificatory. Jukes made them theoretical, but without losing sight of Sedgwick's empirical method, on which all his work was based. This was indeed easier than it seemed: Jukes indeed was forced to make a distinction between empirical evidence and theoretical conclusion, and neither he nor Sedgwick could have been unaware of the irony involved.

Paradoxically, as I have shown elsewhere (Stoddart 1988), Jukes' evidence for subsidence was restricted to an examination of the gross form of the reefs themselves. The presumed process of subsidence was so slow that it was undetectable by observation, and the drilling technologies then available were quite inadequate to demonstrate the great thickness of shallow-water reef limestones which subsidence theory predicted (Stoddart 1989). Darwin's theory thus could not be tested directly: the evidence for it was inferential, and indeed was confined to the phenomena it was intended to explain. The theory simply provided the most plausible explanation of reef forms, but required a considerable imaginative leap to be acceptable. And as Jukes found out on the Great Barrier Reef and others (beginning with Forster and Peron) had done elsewhere, such direct evidence of vertical movements as there was on coral reefs generally indicated recent elevation rather than subsidence. To accept Darwin's theory as he did, therefore, Jukes had to minimize the significance of the field evidence immediately available, in a manner that Sedgwick could never have approved.

Jukes' wholehearted acceptance of the subsidence theory yielded only an unproblematic and unverifiable model of the structure of the Great Barrier Reef (Figure 1 next page). He evaded the implications of the direct evidence of recent uplift by arguing that it was doubtless only of limited duration and had been preceded by prolonged subsidence (Jukes 1847, Volume 2, pp. 344-346).

Darwin was flattered that, like Dana before him, Jukes had apparently offered independent confirmation of the subsidence theory from an area which Darwin had not studied. Jukes' memoir undoubtedly helped to establish Darwin's as the dominant explanation of reef morphology over the middle decades of the nineteenth century. But Sedgwick might have taken pride in the fact that it is Jukes' meticulous fieldwork on the topography of reefs and islands, on beachrock and on phosphorites, rather than his structural interpretations, which have value still.

By applying theoretical constructs to the Great Barrier Reef, Jukes thus rendered obsolete earlier attempts to explain the coral reefs of Australia. But instead of being the precursor of a new generation of theoretical studies, Jukes' work had no immediate outcome. Thomas Henry Huxley, naturalist during *H.M.S. Rattlesnake's* cruise along the Barrier Reef in 1848-1849, was a functional morphologist who completely failed to apprehend the reefs through which he sailed, his focus firmly fixed on the anatomy of the medusæ With the completion of the hydrographic surveys of Australia's tropical coasts by King, Wickham and Stokes, interest in and access to the reefs both waned. Jukes' conclusions and Darwin's authority apparently brought the issue to a close. It was many years before Australian science renewed an interest in the reefs, and this itself was contingent on the emergence of Australian scientific institutions and of an indigenous scientific elite.

Diagram to represent an imaginary section of the Great Barrier reef. The proportions are enormously distorted, the perpendicular scale being fifteen to twenty times greater than the horizontal one.

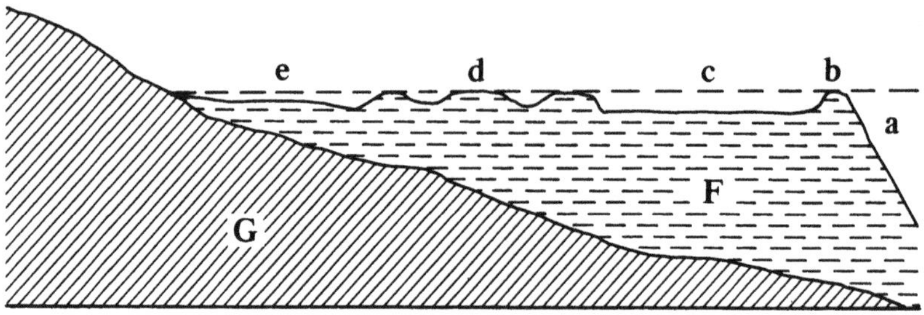

a. Sea outside the barrier, generally unfathomable.
b. The actual barrier.
c. Clear channel inside the barrier, generally about 15 to 20 fathoms deep.
d. The inner reefs.
e. Shoal channel between the inner reefs and the shore.
F. The great buttress of calcareous rock, formed of coral and the detritus of corals and shells.
G. The main land, formed of granites and other similar rocks.

Figure 1

Section of the Great Barrier Reef (redrawn from Jukes 1847, volume 1, p. 333)

Localizing

After Jukes the Great Barrier reef remained largely untouched for over half a century, and other Australian reef provinces proved no more attractive. Vigorous reef studies were pursued elsewhere in the world - by Semper in Palau, Walther in the Red Sea, Gardiner in the Indian Ocean, Agassiz in the West Indies and the Pacific, and many others. But the Australian reefs continued to be too remote and apparently conceptually unenticing to attract attention.

New South Wales itself was scarcely the optimal locus for initiatives in scientific exploration of the reefs. Such energies as there were were directed to the interior of the continent and not to the seas. The convict system persisted until 1840, and the country's first university, at Sydney, was not founded until 1850. Its Department of Geology was established in 1866, but it always had a strong practical interest, devoted to field mapping and to mining, especially after its most distinguished head, T.W. Edgeworth David, arrived from England in 1882. The University of Queensland at Brisbane was not founded until 1910, and its Department of Geology also showed but peripheral interest in reefs, even though its first head was H.C. Richards, a prime mover in the establishment of the Great Barrier Reef Committee in 1922 (Hill 1982).

Perhaps more immediately important was the foundation of the Australian Museum in Sydney. Here Charles Hedley, himself a malacologist, made the enlargement of the collections his chief aim, organizing the *Chevert* expeditions to north Queensland for this purpose. Though primarily a taxonomist, and somewhat at odds in consequence with Richards, who was unsympathetic to the filling of museums for its own sake, Hedley was responsible, with the young geologist from Sydney, T. Griffith Taylor, for perhaps the first indigenous scientific paper produced on Australian reefs (rather than reef biota), in 1908.

When the impetus finally came for the expansion of Australian science into reef studies, it took a paradoxical form, in the three expeditions to Funafuti Atoll in the Ellice Islands between 1896 and 1898. The intention of these expeditions was to test Darwin's theory directly by boring through an atoll rim in a attempt to reach the reef foundations. The first, led from London by W.J. Sollas, was in terms of this intention a failure, but Hedley, one of its members, made enormous collections which resulted in a series of faunistic papers in Volume 3 of the *Memoirs* of the Australian Museum in 1896. The second and third expeditions, though sponsored from the Royal Society in London, were largely in Australian hands. The second was led by Edgeworth David, and the third by the Australian A.E. Finckh; it was on the latter that the boring was carried to a depth of 1100 feet (335 metres). MacLeod (1988) has rightly reflected on the way in which these expeditions symbolized a rejection by Australian men of science of domination by a colonial model of science organized from the metropolis, though ironically in doing so the Australians substituted their own version of colonial science for the old. Sadly the experience gained was never reflected back on the study of the Australian reefs themselves.

These continued to be the scene of expeditionary forays from outside, including that by Agassiz (1898) in 1896, and, most important of all, the year-long expedition to Low Isles led by C.M. Yonge in 1928-1929. Even two hundred years after Cook the main initiatives in Australian reef studies were still coming from outside Australia, and though this situation has now changed it does not alter the melancholy picture of initiatives unseized and opportunities lost on the Australian reefs as intellectual horizons broadened through the nineteenth century. In the seventy years before Jukes it is possible to sense the freshness and excitement of coming to terms, using a variety of conceptual tools, with some of the greatest reefs in the world, but it took another ninety years after that before their problems were once more addressed in the manner they deserved (Steers 1929, 1937).

References

Agassiz, A., 1898. A visit to the Great Barrier Reef of Australia in the steamer "Croydon", during April and May, 1896. *Bulletin of the Museum for comparative Zoology, Harvard College*, 28(4), 95-148.

Beechey, F.W., 1831. *Narrative of a voyage to the Pacific and Beering's Strait, to co-operate with the polar expeditions; performed in His Majesty's Ship Blossom*. London: H. Colburn and R. Bentley. 2 volumes.

Beaglehole, J.C., (ed.). 1955. *The Journals of Captain James Cook on his voyages of discovery: The Voyage of the Endeavour 1768-1771*. Cambridge: The Hakluyt Society. 684 pp.

Beaglehole, J.C., (ed.). 1961. *The Journals of Captain James Cook on his voyages of discovery: The Voyage of the Resolution and Adventure 1771-1775*. Cambridge: The Hakluyt Society.

Beaglehole, J.C., (ed.). 1962. *The Endeavour Journal of Joseph Banks 1768-1771*. Sydney: Angus and Robertson. 2 volumes, 476 and 406 pp.

Böttger, L., 1890. Geschichtliche Darstellung unserer Kenntnisse und Meinungen von der Korallenbauten. *Zeitschirft für Naturwissenschaften*, 43, 241-304.

Cameron, H.C., 1952. *Sir Joseph Banks: the autocrat of the philosophers 1744-1820*. London: Batchworth Press.

Chamisso, A. von, 1821. Remarks and opinions of the naturalist of the expedition. *in* O. von Kotzebue: *A voyage of discovery into the South Sea and Beering's Straits, for the purpose of exploring a north-east passage, undertaken in the years 1815-1818*. London: Longman, Hurst, Rees, Orme and Brown, 2, 349-433, 3, 1-318.

Claret de Fleurieu, C.P., 1798-1800. *Voyage autour du monde, pendant les années 1790, 1791 et 1792, par Etienne Marchand*. Paris: Imprimerie de la Republique 6 volumes, 294, 529, 474, 494, 559 and 158 pp.

Dana, J.D., 1853. *On coral reefs and islands*. New York: G.P. Putnam. 143 pp.

Darwin, C.R., 1838. On certain areas of elevation and subsidence in the Pacific and Indian Oceans, as deduced from the study of coral formations. *Proceedings of the Geological Society of London*, 2, 552-554.

Darwin, C.R., 1839. *Journal or researches into the geology and natural history of the various countries visited by H.M.S. Beagle, under the command of Capt. Fitzroy, R.N., from 1832 to 1836*. London; H. Colburn. 629 pp.

Darwin, C.R., 1842. *The structure and distribution of coral reefs*. London: Smith, Elder. 214 pp. *Also*: original text reprinted with an introduction by J.W.Judd (*ed.*) 1890, by Ward Lock & Co, London and New York, page citation is from 1890 edition.

Flinders, M., 1814a. *A voyage to Terra Australis; undertaken for the purpose of completing the discovery of that vast country, and prosecuted in the years 1801, 1802, and 1803, in His Majesty's Ship the Investigator*. London: G. and W. Nicol, 2 volumes, 269 and 613 pp.; atlas.

Flinders, M., 1814b. *General chart of Terra Australis or Australia.*

Forster, J., 1778. The theory of the formation of isles. In his *Observations made during a voyage round the world, on physical geography, natural history, and ethnic philosophy*. London: G. Robinson, pp. 148-159.

Günther, S., 1910. die Korallenbauten als Objekt wissenschaftlicher Forschung in der Zeit vor Darwin. *Sitzungsber. Königlich Bayerischen Akademie der Wissenschaften, Mah.-Phys. Kl*. 1910 (14), 1-42.

Hedley, C., 1896. General account of the atoll of Funafuti. *Memoirs of the Australian Museum*, 3, 1-71.

Hedley, C. and Taylor, T.G., 1908. Coral reefs of the Great Barrier Reef, Queensland. *Reports of the Australasian Association for the Advancement of Science*, 1907, 397-413.

Hill, D., 1982. The Great Barrier Reef Committee, 1922-1982: the first thirty years. *Historical Records of Australian Science*, 6, 1-18.

Hoare, M.E., (ed.). 1982. *The Resolution journal of Johann Reinhold Forster 1772-1775*. London: The Hakluyt Society. 4 volumes.

Ingleton, G.C., 1979. Flinders as cartographer. In R. W. Russell, ed.: *Matthew Flinders: the ifs of history*. Adelaide: Flinders University, pp. 63-80.

Jukes, J.B., 1847. *Narrative of the surveying voyage of H.M.S. Fly, commanded by Captain F.P. Blackwood, R.N., in Torres Strait, New Guinea, and other islands of the Eastern Archipelago, during the years 1842-1846.* London: T. and W. Boone. 2 volumes. 423 and 362 pp.

Lyell, C., 1830-1833, *Principles of geology: being an attempt to explain the former changes of the earth's surface, by reference to causes now in operation.* London: John Murray. 3 volumes.

Macleod, R. & Rehbock, P.E., 1988. Nature in its Greatest Extent: Western Science in the Pacific. University of Hawaii Press, Hawaii.

Peron, F., 1816. Observations zoologiques propres a constater l'ancien séjour de la mer sur le sommet des montagnes de la Terre de Diemen, de la Nouvelle-Hollande et de Timor. *In Voyage de découvertes aux Terres Australes, exécute sur les corvettes la Géographe, le Naturaliste, et la goelette la Casuarina, pendant les années 1800, 1801, 1802, 1803 et 1804. Historique: Tome second.* Paris: Imprimerie royale, pp. 165-183.

Quoy, J.r. and Gaimard, J.P., 1824. Mémoire sur l'accroissement des Polypes lithophytes considére géologiquement. In *Voyage autour du monde entrepris par ordre du Roi par M. Louis de Freycinet: Zoologie.* Paris; Chez Pillet Aine, pp. 658-671.

Spate, O.H.K., 1979. Baudin and Flinders *In* R.W. Russell, (ed.): *Matthew Flinders: the ifs of history.* Adelaide; Flinders University, pp. 87-93.

Steers, J.A., 1929. The Queensland coast and the Great Barrier Reefs. *Geographical Journal,* 124, 232-257, 340-367.

Steers, J.A., 1937. The coral islands and associated features of the Great Barrier Reefs. *Geographical Journal,* 89, 1-28, 119-139.

Stoddart, D.R., (ed.) 1962. *Coral islands,* by Charles Darwin, edited with an introduction, map and remarks. *Atoll Research Bulletin,* 88, 1-20.

Stoddart, D.R., 1977. Darwin, Lyell, and the geological significance of coral reefs. *British Journal for the History of Science,* 9, 199-218.

Stoddart, D.R., 1988. Joseph Beete Jukes, the 'Cambridge connection', and the history of reef development in Australia in the nineteenth century. *Earth Sciences History,* volume 7, Number 2, 99-110..

Stoddart, D.R., 1989. The foundations of atolls: first explorations. *American Association of Petroleum Geologists: Earth Science Monographs,* in press.

Wallace, C., 1984. *The lost Australia of François Peron.* London: Nottingham Court Press. 183 pp.

Williams, J., 1837. *A narrative of missionary enterprises in the South Sea Islands; with remarks upon the natural history of the islands, origin, languages, traditions, and usages of the inhabitants.* London: J. Snow and J.R. Leifchild. 589 pp.

9

European science in High Asia: geomorphology in the Karakoram Himalaya to 1939

Kenneth Hewitt

There is a widely held assumption that human endeavour and thought, especially in science, fall off rapidly in quality or relevance as we move into the past. Historians of science point to the dangers of allowing that to shape their enquiries (Herries-Davis, this volume). It leads to an interest in only those who contributed to matters considered important today, or who seem to echo contemporary language. Such a partial and tendencious view of what is historically relevant, compounds the already severe problems of reconstructing the thought and conditions of the past.

The topic of the present essay involves the risks of a parallel distortion in the *geography* of thought. This is the assumption that knowledge falls off rapidly in quality and relevance as we leave the major centres of learning and, often enough, work in our own language and country. It might be thought this has little relevance here, since European science in the mountain regions of central Asia, is virtually absent from today's histories of geomorphology. The present essay began simply as an effort to redress that. In due course, however, I came to see this as more than accidental neglect. Understandable oversights are involved and the nature of readily available source materials, compounded by language barriers. However, there is also the question of the preoccupations of historians of our field, at least in the English-language. Their enthusiasm has centred upon uncovering the precursors, if not championing the dominant ideas of contemporary Anglo-American geomorphology. It is not an unreasonable motivation given the general state of this subject. Yet, with it has gone a pattern of neglect of alternative frameworks of enquiry. In such a context it seems inevitable that the most substantial work on the Karakoram would be neglected or marginalised, for most of it sits within other intellectual traditions.

The more considerable scholarship prior to 1939 was largely in the understanding of conditions and developments in Inner Asia. Arguably, the problems posed by particular landscapes shaped the major achievements of studies relating to geomorphology almost everywhere, prior to the second half of this century. By contrast, Anglo-American geomorphology of the last several

Figure 1

Karakoram placenames

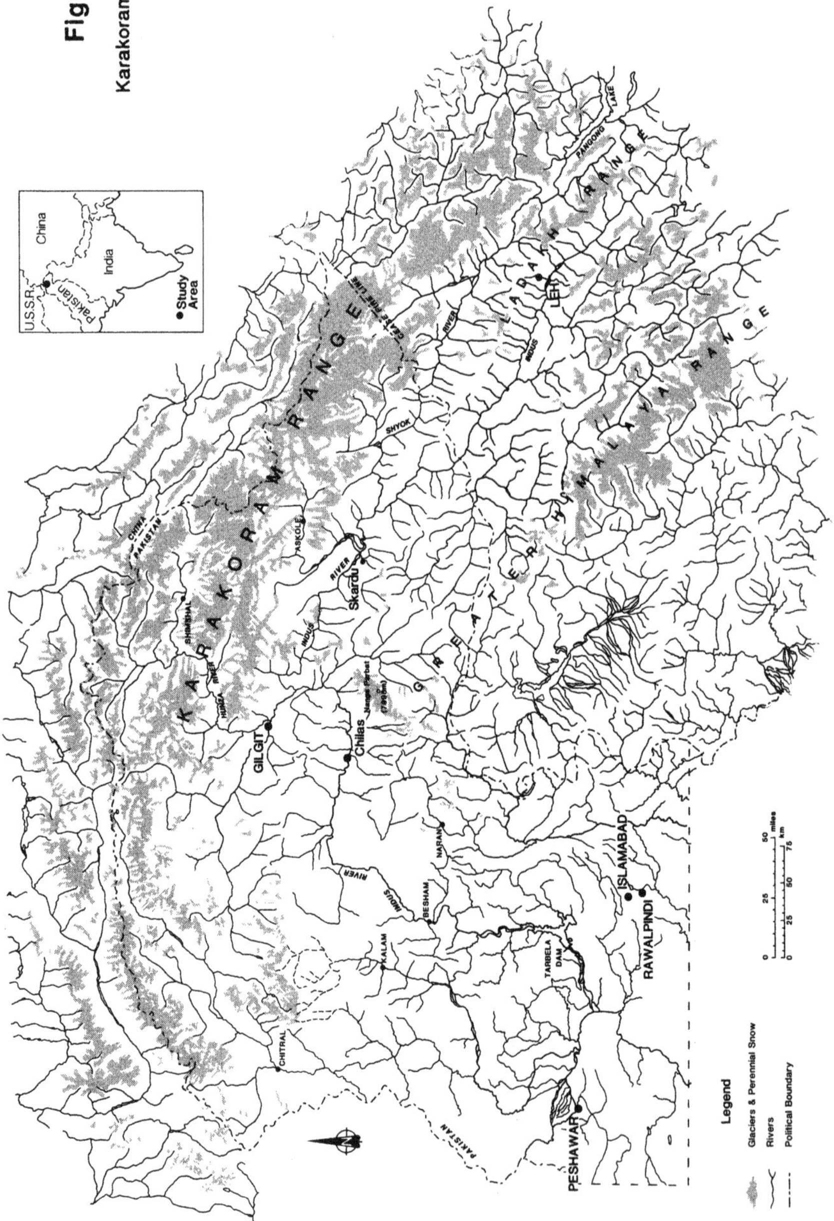

decades has increasingly neglected if not demeaned the empirical, methodological and conceptual questions of such 'geographical' geomorphology. At best, it is countenanced where its examples or practitioners influence the mainstream of abstract, technical generalisation. The interests of historians of the field have tended to follow suit. Their eyes are firmly upon global generalisations about earth time, mechanics of process, properties of materials, or standardised technique.

The style and quality of work in the Karakoram suggests a major distinction between scholars from the continent of Europe, and most British contributions. In part, that reflects the interaction of the distinctive conditions of work, with social and political constructions helping to shape it. Then again, many continental studies might still have gone unrecognised for the linguistic and social reasons identified later. However, it emerges as lost knowledge, even where published in English. Because its major preoccupations and traditions of enquiry were different, even at those points where it has something to say on questions we do debate, or did pioneer work on some of our cherished notions, it is absent. This paper will achieve its purpose if it draws the attention of historians of geomorphology to something of the scope of the historical literature and archives of high Asia, that bear upon their subject. It also leads on to the question of recognising whether and where we mistake a 'socially constructed' consensus, for the logic of enquiry and how it has developed.

Defining the terrain.

The Karakoram Himalaya designates a series of heavily glacierised mountains within a much larger zone of almost equally high ranges in Asia (Fig. 1). They comprise an extensive and complex system of orogeny that includes the Greater Himalaya and its Front Ranges to the south, the Lesser and main Hindu Kush to the West, the Pamirs to the north west and Kun Lun to the north. Eastwards are ranges bounding the high interior basins of Tibet which Sven Hedin called the "Trans-Himalaya." The whole is perhaps best identified as Visser's "South-Central Asian Mountain System" (1935/38 Vol. 1 p. 109). Within it, the Karakoram Range is a series of high ridges, having a north west to south east trend from Batura Mustagh to Saser Muztagh.

There is no intrinsic reason to single out the Karakoram as a *geomorphological* unit. It has, however, been a recognised geographical unit and an objective of investigation for almost two centuries. Its story, as we know it in European languages, is largely the work of explorers and mountaineers, of expeditions, and the missions of individuals sent by one or other agency of the British administration in India. Most of the earliest efforts inch their way into the region from areas to the south and south east. A few observers travelled to it from the north, through Russian or Chinese territories, or from Tibet. As the Karakoram Range itself became known, few who made it their destination would find time or energy for surrounding regions. Most were, in any case, individuals with a main interest and experience in mountains or mountaineering. Thus, knowledge of the area tended to develop in a distinctive way, one factor in the influence or rather lack of it, upon notions of the development of geomorphology.

It is hard to say much about an area whose layout and principal features are hardly known. Geographical discovery, filling in topographical detail and establishing the nomenclature were major concerns in the nineteenth century. They are matters not entirely resolved today. Nor are they purely incidental concerns for us. Good maps and knowledge of the terrain are essential for studies of actual landscapes. They are a major part of the work itself in regions of high relief, organised into a complex series of great ranges, deeply-incised rivers and large glacier systems. Perhaps too much of the Central Asian literature records interminable disagreements about topography and names, some of the least productive reaching into our century (Allen 1982). Yet, a scientist without a map in the Karakoram is confronted with an utterly bewildering topography; as much a "hopeless jumble" as de Saussure found the Alps (Greene 1982 p. 146), and on a much larger scale. No individual high viewpoints will resolve more than a tiny part of it. The great peaks do not stand out so as to define major massifs. The verticality of landscape conditions is nowhere as neatly evident as in von Humboldt's South American volcanoes. Features not linked up to mapped areas were, in a sense, undiscovered, even if known about.

One consequence is that travellers and mountaineers have made a more substantial contribution to scientific developments than is usual in more familiar regions. Their findings loom large in the works of professional earth scientists, while the latter have had to be explorer-mountaineers themselves. Much of the work to be described here was built upon extensive travel and mapping within the region. In addition, the more substantial contributors - the von Schlagintweits, von Richthofen, Oestreich, Dainelli, Hedin, and the Vissers - always filled that out with wide and careful reading of the explorations and discoveries of their predecessors. The willingness and ability to do that effectively, I will argue, is not just a tribute to their personal achievement. It is integral to the intellectual traditions that helped shape the work done here and the way in which we must describe and evaluate its achievements.

At the same time, it is useful to know what European knowledge of the region was like in the period that has been treated as seminal for the emergence of modern earth science. In the nineteenth century, topographical mapping in some of the Himalaya was not so far behind much of that in Europe and well ahead of the exploration of the North American cordilleras. One finds ideas and methods just emerging in work on, say, the European Alps, being considered in relation to the inner Asian Ranges, sometimes before they are accepted or applied in most of Europe.

Geographical knowledge to 1860

Something of the eastern limits of the region, where it borders Tibet and Kashmir, was known to the West before the nineteenth century. Von Humboldt was among the first to realise that ancient Chinese records of "Western Tibet" were superior to anything more modern,- though he gave a rather garbled version of them for our area (Kellner 1963 p. 151). The mediæval Arab geographers were also ahead of European knowledge before the 1830's (Stein 1899, Francke 1907, Hedin 1922, v. VII, p.7 *et seq.*) Little of that was yet available in the West. The Jesuits travelling out of Goa also knew much more, notably Ippolito Desideri, but his great work from the early eighteenth century, was lost until 1875

(Dainelli 1959). François Bernier (1659) wrote of "grand Tibet" as "..a hopeless country full of mountainous deserts..." (Hedin 1922 p. 64). Here is an early hint of the "Semi-Arid Himalaya."

Nicolaus Visscher's map of about 1680 is the first in Europe to show 'Tibet Minor' (Baltistan) and 'Eskerdow' (Skardu). However, they sit in a plain stretching north to 'Kaskar Regnum' (Kashgar). Although the Karakoram Ranges, which occupy this space, have been subject to high rates of uplift, it is fair to assume they were there then! The alternative name for the Karakoram, the Muztagh ("Mus Tag") appeared in a work published in London in 1730. However, Mountstuart Elphinstone's reports on the "Mission to Caubul" first gave the names "Kurra-koorrum" and "Mooz Taugh" currency among English-speaking scholars (Elphinstone 1843, p. 86).

For the first fifty years of the nineteenth century, knowledge of the region was built from the south east, by travellers coming through Kashmir to Ladak (Dainelli 1959, Chapter 2). As a consequence, the arid valleys of Ladak and their features dominated and coloured ideas about the region in general. Discussions of the landscape and geology sought mainly to compare and contrast what was here, with the more humid Vale of Kashmir. Maps appearing between 1790 and 1820 show the Upper Indus in Ladak quite well, but ultimately take it out to the Sutlej, Ganga or Ferghana valleys. William Moorcroft (1865?-1925) was the first Englishman to define the territory north, west and east of Leh. However, before going himself, he had sent Mir Izzet Ullah on reconnaissance northwards. This native of Delhi, pioneered the route over the Karakoram Pass to Yarkand which many would follow or try to follow later. He was the first to describe the Khumdan Glaciers on the Upper Shyok, whose dams and outburst floods would be a major source of interest in the region (Izzet Ullah 1925, 1842-43, p. 283). In 1813, Moorcroft and Izzet Ullah visited the headwaters of the Indus in Lake Manosarovar (Ma'nasalouwochi). Between 1820 and 1822, Moorcroft and his companion George Trebeck visited the upper Shyok and Nubra Rivers, and the enclosed desert lake of Pangong, thus establishing some of the main features at the eastern limits of our region.

Ritter's description in the *Erdkunde von Asien* (Bd. 5, 1837), the best to that time, hints at the existence of the high ranges and glaciers of the Karakoram. However, as the volume was being published, Godfrey Thomas Vigne (1801-1863) was exploring the southern flank of the central Karakoram Range itself (Vigne 1842). Before that, his journeys took him along the east flank of the Nanga Parbat Range with its steeply falling glaciers whose termini lie across the Astor River. He visited the snout of one of the largest Karakoram Glaciers, Chogo Lungma (p. 285). The flavour of his work can be gathered from his description of it:

> I have never seen any spectacle of the same nature so truly grand as the debouchure of the waters from beneath this glacier. The ice is clear and green as an emerald, the archway lofty, gloomy and Avernus-like. The stream that emerges from beneath it is no incipient brook, but a large and ready-formed river, whose colour is that of the soil which it has collected in its course, whose violence and velocity betoken a very long descent, and whose force is best explained by saying that it rolls along enormous masses of ice, that are whirled against the rocks in its bed, with concussion producing a sound resembling that of distant cannon (Vigne 1842, Vol.II p. 286).

Local people told him the glacier was very long, but its full extent - larger than any valley glacier in Europe - was not recognised before the end of the century. With this description alone, however, Vigne defines a major feature of the Karakoram environment; the way most of its large rivers emerge from glacier termini into the semi-arid and arid valleys below them. He also contributed to another recurring theme, the glacier dammed lakes and outburst floods of the Karakoram. He described one that had recently occurred causing great devastation in the Nubra valley (Vigne 1842 Vol. II p. 362). Vigne's work was apparently not known to von Humboldt when he wrote his major work on Central Asia in 1843. Although largely devoted to the mountain ranges and plateaux, there is little sense of the Karakoram or its glaciers. What there is, seems confused with his "Montagnes du Bolor" (von Humboldt 1843 Vol. II pp. 365-412).

Two texts describing the physical geography of the region were published in 1853 and 1854, by Henry Strachey and by Alexander Cunningham (1814-1893) respectively. Both authors had explored parts of Ladak, travelling together in 1847. They provide new but scattered climatic data and estimates of river flow and its variability. Strachey's map gives a good picture of the Indus from Lake Manosarovar to Gilgit, but gives little idea of the Karakoram. Cunningham's map is more accurate but fails to show most of the Karakoram or its glacier cover.

Like many later observers, both were impressed by the abundant superficial deposits along the valleys they visited. Strachey comments:

> the immense extent of existing alluvium, and the uniformity of its maximum elevation, lead me to infer that it must have been deposited under a general sea covering the whole country...and not by lakes, much less rivers...the gigantic ravines in the ancient plateaux are out of all proportion to the contemptible rivulets that now creep along their bottoms, wandering over beds a thousand times too large for them (Strachey 1853 p. 20).

Cunningham agreed that the "vast ravines" of Ladak could not have been carved by "the present streams" (1854 p. 190). These views seem to reflect the 'diluvial' school of thought, favoured in Europe at that time (Chorley *et al.* 1964). Doubts as to the effectiveness of subærial and specifically fluvial erosion, were coupled with a readiness to believe in large scale marine transgressions. In Ladak, it meant accepting they had been preserved through uplift of between 3000 and 5000m. Strachey visited the terminus of the Siachen Glacier, which would turn out to be the largest valley glacier in Asia. He recognised it was large but had no grasp of how large (Strachey 1853 p. 53). Meanwhile, neither he nor Cunningham refers to glacier action as responsible for the 'drift' or valleys they describe. Agassiz's *Études sur les Glaciers* was 14 years old, but glacial action was not widely accepted and substantial glacial erosion was not proposed until the 1860's.

Nevertheless, had it been seen in such terms, the work of these writers gave support to the 'glacial theory,' where they discuss the glacial "debâcles" (ice dam burst floods) from the Shyok drainage. Strachey did not find these "contemptible" processes. He writes of a flood wave that "committed great havoc...throughout 300 miles," rising "50ft at Shayok"[1] and leaving the valley "...full of blocks of ice as big as houses, and so spread with heaps of soft mud that it was impossible to get up the Sasir Pass" (Strachey 1853 pp. 56-7). He documented three such floods and a tradition of many more.

Cunningham also described 'cataclysms' on the Indus due to glacier dam bursts, and from landslide dams on the Indus and Sutlej. He traced their devastations as far as the Plains (Cunningham 1854 p. 99). Moreover, he may have been the first to infer substantial climatic change in Central Asia, believing the region was more humid and 'milder' in the past. His main evidence was that the interior lakes of Western Tibet were once much larger, he reported deposits of once 'vast' lake systems. He described the presence of fresh water molluscs in sediments of now saline lakes. He did not conceive of climatic change having an influence on the glacier cover, but did speculate about a "..gradual elevation..(of) the whole country" (p. 194).

Only two or three years later, 'fluvialist' ideas were advanced to explain the valleys of Ladak. Adolf von Schlagintweit had no doubts concerning "..the great excavation of [these] Thibetan and Himalayan valleys by the action of rivers..." (A. von S. 1856 p. 64). Within six years, Godwin-Austen was attributing much of the extensive superficial deposits to a once far larger glacial cover.

The first scientific explorers of the Karakoram range

The three brothers, Hermann (1826-1882), Adolf (1829-1857), and Robert von Schlagintweit (1833-1885), travelled extensively in the region between 1855 and 1857. They had been recommended to the East India Company by Alexander von Humboldt (Schlagintweits 1855-57 and 1861-66, Kick 1960 and 1982, Polter 1982). They were conducting the Magnetic Survey of India, but investigated a wide range natural and anthropological phenomena (*eds.* Müller and Raunig 1982), providing the first really comprehensive scientific reports of high Asia.

The work of these remarkable individuals has been largely neglected in English-language literatures, whether of natural sciences generally or geomorphology in particular, and even the exploration of inner Asia. Therefore, something of their background should be sketched here. Hermann's 'Habilitation' was in "physical geography" at Berlin in 1851, as was Adolf's at München in 1853. Together they wrote two major texts on the physical geography of the Alps, which were innovative for that time - a comment which applies to the representation of landscape morphology, the grasp of glaciology, and the rôle of glaciers in landform development and topoclimatic conditions (H. & A. von S. 1850 and 1854a). These books and their scientific papers contributed to the integrated geological portrait, and especially glaciology of the Alps, just emerging. They recognise the power of subærial processes to shape landscape, including the action of rivers and glaciers. They pioneered techniques in the representation and analysis of alpine terrain or hypsometry (H. & A. von S. 1855). All of this was the basis for their friendship with von Humboldt, and their work in India and high Asia. Robert, the youngest, was only 21 when they

---

[1] Shayok is an alternative spelling in some literature for Shyok.

left for India in 1854. That year he had completed a dissertation on the physical geography of the Kaisergebirge, south east of München. In India he took some of the earliest photographs - Daguerrotypes - that we have. Later, he played a major rôle in writing up the work and, in 1864, became the first professor of geography at Giessen (Uhlig 1965).

Hermann and Adolf's work explored the south central Asian ranges from Assam to Afghanistan. The three brothers may have been the first Europeans to cross the Karakoram Pass and the Kun Lun, in a journey from Ladak to Chinese Turkestan, where Adolf was killed. Before that, it was Adolf who, in 1856, made major advances upon Vigne's discoveries on Nanga Parbat glaciers, extending them to the 'Diamir Group,' and of the heavily glacierised Central Karakoram ranges. He described the Chogo Lungma, Biafo, Baltoro, Hushe and Bilafond Glaciers. He was the first to give a sense of their great extent, and relatively exact elevations, - with boiling point hypsometer and barometric readings,- of their principal features. He explored the perennial snow zone of glaciers of Upper Hushe, reached the famous Concordia on the Baltoro Glacier, and was the first European to explore the Mustagh ("Muztack") Pass on its watershed. He camped in the firn basin ("firn meer") at 17,990ft, and determined the elevation of the pass as 19,019 ft (1861-67 Vol. II p. 427, H von S. 1869-72 Vol. III, pp. 266-270). As well as fine, accurate watercolours and good sketch maps, his field notebooks contain precise and still useful information on conditions at that time. In particular, they reveal a lively and innovative grasp of erosional questions. He was the one most interested in glaciology and erosion. His tragic death without doubt helped to prevent his many insights reaching a wider audience. Most of these materials may be found in the Schlagintweitiana Collections of the Staatsbibliotek in München.

In high Asia, the brothers dealt with many topics of interest to geomorphology, but couched in the embracing style of 'physical geography' pioneered by von Humboldt (Beck 1961 Vol.II pp. 170-175, Körner 1982). Like him, they attempted to define the interrelated patterns of climate, vegetation and orography. They gathered extensive climatic data. In keeping with von Humboldt's 1843 study of Central Asia, they paid attention to the hypsometry of the mountain ranges (R. von S. 1866). They gave considerable attention to what we would now call "topoclimatic" effects, and how those of this region differed markedly from mountains of lesser relief like the Alps (1861-66 Vol. IV, Pt. II). They accepted that large and relatively recent climatic changes had occurred. Adolph and Hermann reported moraines and other evidence of formerly more extensive glaciation. Of course, they already accepted the emerging "Ice Age" portrait of the Alps, although it was still controversial in, say, Britain. At that time, however, they regarded the glacier fluctuations in Asia as reflecting local rather than global climatic influences, and to have been of lesser extent and more recent origin than those of the Alps. Nine years later, Hermann accepted them as "Vestiges of permanent alteration of our climate" (1861-66 Vol. IV p. 134). He also regarded the erosion of the major river valleys here as themselves factors in the gradual alteration of climate through topoclimatic effects (Vol. IV. p. 539).

The Schlagintweits paid particular attention to the extent of contemporary glaciers and snowcovers. Their determinations and analyses of regional snowlines in the Karakoram are still referred to (1861-66, Vol. IV). The topic exercised many minds, beginning with some observations by R. Strachey (1849) and taken up in his brother's physical geography noted above. However, von Humboldt's authoritative work on the subject in the Andes, and in Central Asia (1820), was the more important influence upon the von Schlagintweits. It was they who first established that the snow limits were higher in the Karakoram than the Greater Himalaya, and higher in the interior ranges than to the north and south. In contrast to the Himalaya, the Karakoram proved to have lower snow limits on the northern side (1861-66 Vol. II pp. 496-99, Vol. IV. pp. 566-75). They also determined the seasonal rise and fall of the lower limit of snow-on-the-ground (Vol. IV p. 571). Later, Hermann completed a comprehensive synthesis of the altitudinal variation of annual isotherms in relation to snowlines for a transect across the Himalaya, Karakoram and Kun Lun (H. von Schlagintweit, 1869-80, Vol. 4, pp. 511-527). The brothers were the first to demonstrate the lower altitudes reached by the Karakoram glaciers compared to the rest of the Himalaya. They suggested that it was due to "...the extent of the[ir] nevé beds...[such that] the streaming together of ice from such large surfaces counterbalances the greater melting away lower down..." (1861-66 Vol. IV pp. 573-4). They recognised the importance of these elements for regional hydrology and hence fluvial conditions. In Ladak Robert drew attention to the "débris flows" consisting mainly of sand, silt and clay, and relating to rapid seasonal snow melt (R. von S. 1866 p. 36).

Hermann also used snow limits to help establish regional and orographic climates. He discussed the interplay of temperature, insolation, and precipitation, their variable quantities and incidence with altitude, and their relations to vegetation zones and altitudinal limits of cultivation. Robert (1866) made a comparative analysis of these conditions to climate in the Himalaya, Alps and Andes. All in all, apart from Dainelli (see below), who relied heavily on them, there would be few significant improvements in knowledge of these aspects of the Karakoram environment until the second half of the present century (Kick 1960).

The question of fluvial erosion is, perhaps, where the Schlagintweits' contribution bears most upon the usual concerns of the history of geomorphology. Adolf's remarks, deserve quoting in full:

> The excavation [of] the valleys [by] the eroding power of streams has been carried to a very great extent, - the considerable fall which rivers like the Indus and Sutlej experience on their course from Thibet down to the low plains of India, has increased their excavating power in a surprising manner.
>
> In the [longitudinal] valley of the Indus near Iscardo, [and] in the valley of Astor near the place where the Indus enters the Himalaya, I had several times to observe gravel and sand beds evidently deposited by these rivers and ancient marks produced by large streams on the rocks at elevations of 3,000 and 4,000 Eng. feet above the present level [to which] the rivers tributary to the Sutlej and Indus have been excavated.... (A. von S. in 1856, publ. 1857 p. 64).

Hermann made a similar statement about southern Tibet, where:

> [amid] elevated basins and slopes gentle enough - only those declivites being steep which are the immediate effect of *the erosion of rivers;* and they

> are gigantic too, for it is not infrequent that the cutting of the rivers can be traced to a depth of 2,000 to 3,000ft... (1861-66, Vol. IV p. 539).

The field notes and sketches of Adolf and Hermann from 1855 and 1856, throughout the Himalaya, point to a clear conviction about deep incision and massive débris transport by rivers at that time.[2] In Western Tibet and the trans-Himalayan Indus Basin, they reversed the view of Cunningham and Strachey, actually explaining the desiccation of the valleys as due to their deep incision by fluvial erosion (Vol. IV p. 539). In other words, they saw the depth of erosion itself leading to the well-known desiccating valley winds.

These observations were made available to, and published in English-language reports of the Magnetic Survey, between 1855 and 1857, and in the *Journal of the Asiatic Society of Bengal* in 1857 and 1858. This bears upon the notion of a 'tropical influence' on fluvial theory, postulated by Chorley et.al. (1964 pp. 385-390), since the observations on gorge-cutting in Himalayan Rivers by Oldham (1859) and Medlicott (1860), come after, and from similar or the same sites as the Schlagintweits'.[3] In the Khasi or Khassia Hills in Assam, Hermann in 1855, postulates 3,000 feet of stream erosion. He notes it is in sandstone, and cites the notoriously heavy rainfall here as a major factor.[4] Oldham's observations were almost identical. Moreover, the Schlagintweits' observations take place 15 years before the surveys in the American West, and two decades before John Wesley Powell's landmark publication on the exploration of the Colorado River, considered a great turning point for a fluvialist geomorphology. The actual scale of erosion envisaged is entirely comparable with that at the Grand Canyon and, for the Tibetan Rivers, is in an equally subdued, surrounding plateau-land. There is a similar 'freshness of approach' and impact of a 'novel environment' (Chorley *et al.* 1964. p. 469). The direct influence upon notions of erosion and Asian landforms on the continent of Europe and, seemingly, upon the officers of the Geological Survey of India has not been noticed.

Meanwhile, this challenges the idea that followers of von Humboldt adopted his "diluvial conservatism" (Tinkler 1985). Again, German geomorphology is often portrayed as much slower than Britain's to accept a 'fluvialist' or subærial view of landscape erosion. Ferdinand von Richthofen's work on China,

---

2 Although, outside our region, mention should be made of the series of surveys of the complex channel reaches of the Brahmaputra, carried out by Hermann in 1855, with the help of a Survey of India officer. In addition to channel morphology, he measured flow rates, some velocity profiles, current patterns, sediment carried and noted variations in bed materials influencing form. The originals of these maps are also in München (Schlagintweitiana IV.5 96-112). The work seems more detailed, though less developed than that for the Mississippi carried out at about the same time by Humphrey and Abbot. Given the conditions and other demands on Hermann's time, it was a remarkable addition to the emerging 'quantitative and morphometric' work on rivers described by Chorley et al. (1964 pp. 436-439), but again making no apparent impact in the literature they study

3 Thomas Oldham (1816-1878) a confirmed 'diluvialist' when he left England to head the Geological Survey of India in 1851, was converted to a fluvial explanation of the gorges of Assam (Oldham 1859). A year later, Medlicott (1860) introduced the view that the gorges of the Ravi, Sutlej and Jumna (Yamuna), across the Himalaya, were cut by the rivers not a marine inundation, and during the rise of the mountains. His evidence overlaps with that of Adolf. This "antecedence" hypothesis of Himalayan rivers was later applied by some to the Upper Indus in our region (R. D. Oldham 1892).

4 Given the importance of this period in histories of geomorphology, we should establish dates, and what the Schlagintweits' were ascribing to subærial, especially fluvial action. Appendix 1 outlines these items with references to their writings.

to which we turn later, is quoted as one of the last major expositions of the 'marine hypothesis' (Chorley *et al.* 1964 pp. 598-601). Yet, von Richthofen largely accepted the von Schlagintweit view of the rivers and "alpine scenery" of our area, although, for his grand perspective on the evolution of the whole of inner and east Asia, it appears as a minor element (von Richthofen 1877-1912 Vol. II, pp. 130-138 and 246-252). The geographer Oestreich (1906) would use a combination of Philippson's and von Richthofen's ideas to argue for the superimposition and antecedence of these parts of the Upper Indus drainage (see below).

Alfred Cloyne (Godwin-) Austen (1808-1884) is also described as a "confirmed anti-fluvialist" in the early 1850's (Chorley *et al.* 1964 p. 331). However, his son, Henry Haversham Godwin-Austen (1834-1923), who arrived in our region in 1860, was soon expressing very different notions on erosion as well as stressing the importance of glaciation. In 1861, he explored the ablation zones of some of the same glaciers Adolf von Schlagintweit had described. He ascended the Panmah, reporting on its moraine-building and ice advance pushing and ploughing up superficial materials. He travelled up the Baltoro Glacier and onto the slopes of Masherbrum (7821m) for a close view and fixing of the watershed situation of K2 (8611m) which, for a long time, atlases named after him.

In an 1862 paper on Karakoram glaciers, presented for him in England by his father, he revealed the great extent of previous glaciation. In a more developed version published two years later he states:

> I have been struck by the indications of considerable amounts of change of temperature within what we may call out own times. The proofs of this ...consist in the enormous terminal moraines which in so many places abut on the larger rivers, down to which point the glaciers must once have descended, and which in some cases must have rivalled in length the present ones of the Mustakh [= Karakoram] Range...it is seen in the long furrows cut out of solid rock as if with a chisel wielded by a gigantic hand... Nowhere are these great striations better seen than in Shigar Loomba [Baltistan, Central Karakoram] (1864 p. 51).

Godwin-Austen's writings leave no doubt as to the power of glacier action. He emphasised the scope and scale of mass movements in the high Karakoram, especially those associated with snowmelt. His is one of the first of many descriptions of a large Karakoram mudflow in action. As he was descending on the path from Skoro La, above Shigar Valley:

> a black mass coming out of a ravine...and moving rapidly over the broad slope of boulders which formed the bed of the valley...(I)t consisted of a mass of stones and thick mud about 30yds in breadth and about 15ft deep... rock(s), some of great size, measuring 10ft x 6ft all travelling together like peas shot out of a bag, rumbling and tumbling one over the other and causing the ground to shake.
>
> ..shortly after (came) another body of stones not so large as the first but travelling much faster.

In due course, these mudflows would become identified as characteristic of the region, and due as Godwin-Austen adds, to:

> the mighty power of transport which the accumulated masses of water and melting snow acquire at these times.
>
> These "shwas" are of frequent occurrence in the ravines, particularly when the sides are of crumbling rock; they originate in landslips which stop the streams for a time (1864 p. 27).

He also concluded that the position and deposits around the Biafo Glacier, one of the largest in the region, confirmed local stories of its having once dammed the Braldu River, causing destructive outburst floods (Godwin-Austen 1864 p. 29, Hewitt 1964). As the first quotation indicates, erosional and depositional evidence convinced him of much more extensive glaciers. He regarded the huge deposits of 'alluvium' in the Skardu Basin as having been washed into an old lake.

The range of Godwin-Austen's observations were, perhaps, the more remarkable since he was constrained in time and concerns by his main mission, survey work for Trigonometric Service of India. Between 1857 and 1863, men like him established a triangulation network into Ladak and as far as Hunza and Nagyr (Walker 1879). However, his major work came later, dealing with the solid geology of Kashmir and Ladak (Godwin-Austen 1883).

In the Karakoram, Godwin-Austen along with the Schlaginweits, epitomised the new "scientific explorer'; interested in almost everything, armed with a plane table and making careful field notes. The importance of skill in drawing needs emphasis here too. It was a necessity for any geologist in this era before the camera (Tinkler 1985 p. 74). In the vast, complex landscapes of the Karakoram, the need was second only to an unusual fortitude. At least, it applies to all who advanced understanding in this first period and even later. The Schlagintweits produced many fine sketches, diagrams and watercolours Godwin-Austen had received a 'certificate of superior qualification' for topographical drawing at Sandhurst, the Military Academy (Keay 1977 p. 194).

In their work, though less well-known, the Karakoram was as well served as areas visited by the more famous traveller-naturalists, such as von Humboldt, Darwin or Hooker. The fame of those men was, however, important for the context and character of work in the Karakoram. Celebrated in the courts of Europe, they were often instrumental in promoting scientific exploration. Interest in and outlets for the resulting reports had grown rapidly in the form of journals, learned societies' meetings, and expanding government 'Surveys' (Tinkler 1985 p. 75). The strategic, trading and administrative interests of imperial powers were, of course, the underlying engine of the interest in nature. What is puzzling, in view of later developments, was the cosmopolitan range and extent of knowledge sharing across Europe at the time.

'Extreme' events

In 1841, there had occurred the most devastating modern geomorphic event known in our region. This was a dam burst flood which came from an earthquake-triggered landslide in the Indus Gorge, under Nanga Parbat. The barrier, over 350m high, lasted some six months, then drained catastrophically in 24 hours. The flood, the largest recorded on the Indus, rose more than 24m. at Attock some 550 km downstream. Peak discharge there I estimate to have been close to 2 million cubic feet per second. Two decades later, considerable scientific interest developed in this flood (Becher 1859 p. 226, Montgomerie 1860, Drew 1875 p. 417, Hewitt 1968).

In 1858 a slightly smaller landslide dammed the Hunza River. Its bursting generated the second largest flood wave known from the Attock gauge records, and reports of massive erosion along the way (Henderson 1859, Montgomerie 1860, Todd 1930). This flood inspired some of the earliest theoretical studies of flood wave behaviour (Obbard 1860, Pratt 1860). The glacier dam burst floods of the region have already been mentioned.

Later, we will encounter work on the scale and recurrence of large landslides, including those triggered by earthquakes. The large mudflows have been introduced, and so must the reports of large-scale avalanche action. H. C. B. Tanner (1891) described the "havoc" he saw committed by snow avalanches in 1879 in many valleys between Kashmir and Gilgit. This included widespread destruction of forest, landslips and blocking of streams. Being impressed by the power displayed, he made a strong plea for the importance of snow avalanches "..in the conformation of the topography... (which) has not received the attention it deserves at the hands of geologists." . His suggestions seem not to have been taken up, but many other travellers would add further evidence of large scale or recurrent erosion by avalanches. However, along with the other seemingly extreme processes, they helped to maintain a certain 'catastrophism' in reports of the Karakoram landscape. This contrasts with the ultra-uniformitarian notions coming to hold sway in European geology. We will return to this question in discussion later. However, in the last three decades of the nineteenth century most of the more substantial work relating to Karakoram geomorphology, adopts another perspective.

The legacy of "The Ice Age"

Frederick Drew worked and travelled in the region in the 1860's and 1870's. He reviewed the evidence of outburst floods from glacial 'barrages' and 'landslips,' reporting on outburst floods from the Shimshall and Ishkoman valleys (1875 pp. 418-420). He gave a detailed description of the 1850 Tarshing Glacier dam across the Astor River and disastrous outburst flood (pp. 400-1). He traversed the lower 30km of the Chogo Lungma Glacier, where he described the filling and sudden drainage of ice-margin lakes (p. 369). However, his primary interest was in the the past development of the landscape, evidenced by the same superficial deposits in the valleys of the Upper Indus we have already met. From them, he elaborated notions of much more extensive glaciers and lakes, and the downcutting of the rivers (Drew 1873). He was ready to conceive of ice dammed lakes in the more distant past. The moraines of Katzura, at the mouth of the Skardu Basin, he explained as remains of a dam that had ponded the Indus there, thus explaining the extensive lacustrine deposits noted by Godwin-Austen (pp. 372-373). It was a bold conjecture. There is no glacier within 40 km of that place, and only little cirque glaciers at the head of the Shigarthang Valley, whence he believed the great glacier damming the Indus had come. Drew had no training in sedimentology or geological mapping, his being the work rather of a keen amateur. His work, however, and his explanations of the Karakoram valleys in terms of Pleistocene climatic change and glaciation became more widely known than that of many trained scientists working there.

Drew signals a strong shift in the emphasis of work bearing upon the landscape. Previously, the discovery and description of the contemporary scene took precedence, and evidence of changes in the past were incidental. From now on, and to the end of our period, the more substantial contributors would all tend to treat the landscape as the product and expression of more or less long-term evolutionary developments. This had two consequences which may help explain the absence of the work from recent concerns of geomorphology. While present-day processes were brought to light by explorers and mountaineers and some on-going government surveys, the major scientific thrust now, had to do with historical geology and denudational history. Technically, that led to an over-riding concern with many and particular local features and sediments, that aided in the detective work of reconstructing those chronologies.

Another important but shadowy figure in the early development of ideas about the evolution of the region and climatic change was the geologist Ferdinand Stolicza (1838-1874). He was a student of Suess at Vienna, who worked on the geology of Kashmir, Ladak and Yarkand between 1862 and 1874, as part of the "Second Yarkand (Forsyth) Mission." He is best known as one of the pioneer palæontologists (Zittel 1901 p. 398). In India, he published some of his unique discoveries concerning the palæontology of Yarkand, Ladak, and Kashmir (Stolicza 1866), but died before writing up most of it. This was done from his notes by W. T. Blandford (Government of India 1878), including observations concerning erosion on the eastern borders of the Karakoram, where:

> debris is brought down in large quantities by the melting of snow in the valleys...[and]... at some remote, - say, diluvial, - period this state of things operated on a far greater scale. Not only were [land-locked, salt] lakes like Pankong much more extensive, but valleys like Chang-chenmo, or the Tankse...became temporarily blocked up by glaciers or great landslips...Near Aktágh...[resulting] deposits of stratified clay exist of about 160 feet in thickness, and extend over an area of 100 square miles...when these large sheets of water were in existence the climate of these... cold and arid regions was both milder and moister (p. 180).

Stolicza made a major contribution to geological knowledge of this region and it is clear his work was relied upon by the Geological Survey of India for the rest of the century. Like Adolf Schlagintweit, his achievements were cut short by untimely death. Unlike Blandford, others who owed much to him may have failed to acknowledge the fact.

Between 1877 and 1881, the geologist R. Lydekker travelled in the region. In "upper" Baltistan he reported abundant evidence of formerly more extensive glaciation, and the linking up of the Baltoro, Panmah and Biafo Glaciers (1881 p. 46). He wrote:

> Polished rock surfaces, groovings and perched blocks, occur abundantly in the Braldu valley...and from their position prove that the ice was in places...at least 1,500ft. thick.

> At the mouth of [Panmah] valley... on the summit ridge...at least 2,000ft above the Braldu [River]...[are] gigantic blocks of porphyritic gneiss which must have been brought from far up the valley...(1883 p. 35).

He also believed the ancient lacustrine deposits in Skardu Basin were due to an ice-, or moraine-dammed lake there (1881 p. 71). In these matters he amplified and confirmed what Godwin-Austen and Drew had said. Dainelli felt, however, that his work was largely derivative, the geological parts coming, often unacknowledged, from previous work, notably that of Stolicza (Dainelli 1934 Vol. 1 pp. 167-174).

Medlicott and Blandford, in the Second Edition of their *Manual of the Geology of India* (1893), probably the most widely read work involving our region at that time, accepted a great expansion of the Himalayan glaciers, similar to that well-established in Europe. C. Diener reviewing the problem of the Ice Age in the Himalaya in 1896, concluded that the evidence showed a maximum descent of the glaciers to 2100m. Although he was not explicit about it, that implied, as Dainelli points out (1924-1935 Vol. 3 pp. 6-7), the possibility of a glacier filling the main Indus Valley at Skardu and below. It was, however, Oestreich (1906) who gave the first detailed reconstruction that established so extensive a glacier cover (see below).

The era of major expeditions: 1892-1939.

In the last decades of the nineteenth century there were many more explorers, officials and military men on hunting treks, whose reports helped to build a portrait of the region (see Dainelli 1924-35, Vol. 1, for a detailed survey). In most cases, observations relevant to geomorphology were of events or features encountered by chance, and mixed in with many other concerns. That has continued down to our own time. However, from the 1890's onwards, the major efforts in exploration and scientific work occurred in the context of more or less major expeditions (Dainelli 1959). It was partly a matter of fashion, but the remaining unexplored, unmapped areas if not the major scientific questions, involved high altitude areas or those furthest removed from the established valley routes and populated areas.[5]

The tone of the new era is set by the Conway Expedition of 1892. William Martin Conway (1856-1937), led an extended trek to the higher peaks of the Central Karakoram. With the fine water-colourist, A. D. McCormick, and several experienced mountaineers, he broke new ground, combined scientific collecting with improvements in the mapping of the glaciers.

Conway, an art historian as well as explorer, wrote in a compelling way. His book was more widely read and generated more interest in the region than anything before. It contains many descriptions of earth surface features and processes repeatedly used as evidence of geomorphic conditions, and the glaciers in the highest parts of the Karakoram. His companion, the soldier-mountaineer, Charles Granville Bruce - leader of the 1922 assault on Everest - explored the Hopar and Barpu Glaciers of Nagyr. The main party traversed the length of the

---

[5] Most of what had passed for 'discovery and exploration' followed routes that local peoples and merchants used constantly, had known for centuries, and guided the foreign 'explorers' along!

Hispar and Biafo Glaciers. Conway brought back the first description of the "Snow Lake" (Lukpe Lawo), the vast accumulation basin that feeds the latter - though he exaggerated its dimensions (Conway 1894 pp. 334-412). He traversed the Baltoro Glacier, climbed "Golden Throne" (6888m) and provided a beautifully drawn if somewhat simplified map of its basin.

With respect to processes, he was the first to stress the exceptional scale and frequency of avalanches at high altitudes, as compared to any other mountains he knew (1894 p. 321). This must be added to Tanner's remarks, noted earlier. As more visitors actually reached the high glacial areas they reinforced Conway's observations (Curzon 1896 p. 20, Workmans 1910 p.117, Guilliarmod 1904 p. 23).

Conway refers to "mud avalanches" eight times in a few weeks of travel through the lower valleys of Hunza and Baltistan (pages 127, 170, 323, 411, 427, 618, 625 and 628). His description of one near Hispar village, if owing more to Milton than geological language, adds something to Godwin-Austen's. The ".. vast black wave, advancing down the *nala* [ravine]

> was a horrid sight. The weight of the mud rolled masses of rock down the gully, turning them over and over like so many pebbles, and they dammed back the muddy torrent and kept it moving slowly but with accumulating volume. Each of the rocks forming the vanguard of this avalanche (sic.) weighed many tons; the largest were about 10ft cubes. The stuff that followed them filled the nala to a width of about forty and a depth of about fifteen feet. The thing moved down at a rate of perhaps 7 mph. Looking up the nala we saw the sides constantly falling in and their ruin carried down.
>
> Three times did the nala yield a frightful offspring of this kind and each time it found a new exit into the main river below and entirely changed the shape of its fan" (p. 323).

The geologist Thomas George Bonney (1833-1923) who, with C.A. Raisin, analysed the rock samples returned by Conway (Bonney and Raisin 1894), drew some broader conclusions from the "mud streams" (Bonney 1902). He thought that here and in the Alps they had created many deposits identified as 'moraines,' but were less frequent in the Alps because these "have passed through a phase of denudation which is still in process in the Himalayas" (p. 15). Bonney attributed the earth pillar formations on valley sides in the Alps and widely reported in the Karakoram, to mudstream deposits (Hewitt 1968 Note 9.1 p. 302) and noted the important fan deposits of mudflow material.[6] However, if seeming to anticipate the idea of a "paraglacial phase" following ice recession, it reflects, rather, Bonney's glacier "protectionist" ideas (Church and Ryder 1972, Tinkler 1985 p. 132). By contrast, Guilliarmod thought the 'torrents de boue' in the Braldu valley, were due to springtime saturation of otherwise dry, 'old glacier' deposits (1902 pp. 127-8).

More or less dramatic descriptions of mudflows continued to pour from the pens of travellers and scientists. (Tanner 1891 p. 407, Dunmore 1893 Vol.

---

[6] A little later, the widespread existence of mudflow-generated fans was reported throughout the Central Asian mountains by Rickmers (1913). He called them "lion's paw fans" (p. 205) - deposits of mudflow débris whose surface is scored by the changing channels and levées of successive events, a typical feature in the Karakoram valleys below about 3000m (*see* Durand 1900 pp. 33-4).

1. p. 190, Curzon 1896 p. 20, Tanner 1897 p. 407, Durand 1910, de Filippi 1912 pp. 147-9, Hedin 1917-22 Vol. IV p. 177 and Vol. VII p. 421). A widely quoted review by the French geographer C. Rabot (1904), presented the mudflow as a dominant feature of the Karakoram environment.[7]

Shortly after Conway, and with the assistance of his alpine guide Mattia Zurbriggen, an American husband and wife team began a unique series of major Karakoram expeditions. Fanny Bullock Workman (1859-1927) ardent Wagnerite, bicyclist and suffragette, played an equal part in all aspects of these expeditions with William Hunter Workman (1847-1937). They led seven expeditions between 1899 and 1912, visiting more of the high Karakoram glaciers than anyone else to that time.[8] Over the years their descriptions and mapping improved, although the more refined results came from the scientists Karl Oestreich (see below), Cesare Calciati and Matthias Koncza, who accompanied them, and the surveyor B. H. Hewlett (Calciati 1910, 1923).

The huge Workman literature contains, within much wordy, tedious daily expedition materials, a unique set of descriptions of glacier conditions at that time. For instance, they described and photographed many of the unusually well-developed lateral morainic ridges and depositional troughs between them and the valley walls; the so-called 'ablation valleys' (Oestreich 1911/12). With the almost universal glacier recession over most of this century, the ablation zone ice has thinned to well below the lateral moraines nearly everywhere. In 1908, however:

> The Hispar...and also the Biafo at certain points are, at present, actively engaged in building lateral moraines. At the edges of these glaciers great ragged, perpendicular ice-walls rise high above the glacier bed [sic.], their summit as well as their substance heavily loaded with debris which is constantly showered down...upon the moraines at their bases...as these [ice] walls move down they...apparently exert no lateral pressure on their moraines, so that there is no question of their forming...by pushing or ploughing up of ground moraine material. The moraines appeared to be built up by the deposition of the debris borne on the glacial surface as well as in the crevasses and substance of the ice walls...(1910 p. 123).

William Hunter Workman was a student of 'nieves pénitentes,' describing ice melt features of many kinds (Workman 1908 and 1913/14). His observations and ideas have been ignored in the literature, perhaps justifiably, but they provide many useful photographs and descriptions from the Karakoram glaciers.

---

[7] Mudflows are largely confined to a zone between about 2000 and 3500m - a little higher in the eastern fringes of the region. As Robert Schlagintweit and, later, Guilliarmod realised the abundant, normally dry deposits become occasionally saturated in spring and summer by snowmelt. But this zone represents about one fifth of the Karakoram surface area. The many reports of mudflows relate not just to their erosional importance, and dramatic impression of an encounter with them, but the predominance of visits confined to, or expedition time spent moving through, the lower, dry valley areas in spring and summer.

[8] The Workmans have not had a good press in the mountain literature. Even as late as 1984 they could receive an ungenerous and patronising dismissal as "...introducing a slight note of comedy into the awe-inspiring world of the high peaks" (Middleton 1984). Perhaps, they appeared 'amateurish' and prone to exaggeration, but so do many who criticised them for it. Then there was Fanny, an outspoken suffragette in what so many of her day considered a 'man's world'. She not only questioned Sir Martin Conway's description of the "Snow Lake," but planted "Votes for Women" signs above it!

The geomorphic significance of earthquakes is also brought out by these and other travellers. On the Kondus Glacier in August 1912, after an earthquake "...the thunder of avalanches and rockfalls continued for forty minutes..." (Workman 1917 pp. 215-216). The effects were even more severe in the lower Kondus and Saltoro river valleys, where landslides continued over several days, an individual case lasting nearly ten hours. Near Tirich Mir on the Hindu Kush flank of the Karakoram, Cockerill described an earthquake of 1892, where "...suddenly, steep mountain sides on both sides of the [Chitral River] gorge grew 'alive' and huge masses of rock fell crashing into its valley.." (1922). There are many other examples of the same sort (Hedin Vol. VII p. 297, Mason 1914, Coulson 1937, Hewitt 1975).

Karl Oestreich's study of northwest Himalayan valleys (1906), is the first strictly geomorphological text on the region, based on his wide-ranging observations with the Workmans. His analysis of the past action and extent of glaciers was a major, systematic improvement with careful, imaginative use of topographical detail, superficial deposits, morphology, and assumptions about erosive action. Along with striations and moraines, he employs the notion that U-shaped valleys are due to glacial action. From that he concluded that the whole of the Shigar Valley, for example, was carved by former glaciers (p. 78). He interpreted stream thalwegs of the south flank tributaries of the Upper Indus, in terms of base level control (p. 46) and other ideas from Philippson (1886a, 1886b). He rejected the idea that the Ladak Indus could be "superimposed" on morphological grounds, and because it does not cut across geological structures. However, with the aid of von Richthofen's ideas, he argued that the transverse Himalayan courses of the Indus and other rivers are "antecedent and epigenetic" (pp. 101-10, *c.f.* R. D. Oldham 1892, 1907). Meanwhile, he introduces Davisian ideas to elaborate upon the work of Adolf von Schlagintweit on the high, subdued Deosai Plains, calling it an old, uplifted "Peneplain."

In a description of Chogo Lungma Glacier, he introduced the 'ablation valley' into the literature, as a significant feature of Karakoram glacial geomorphology (Oestreich,1911/12). He defined the "Muztagh-" or "Firnkessel" glacier which, along with the "Turkestan-" or "Avalanche-" glacier are recognised as distinctive morphogenetic types of the Karakoram and surrounding ranges of Central Asia (von Klebelsberg 1925/26, Visser 1933-34).

Imperial initiatives

The British administration of India had already played a large role in regulating visits to the north, in mapping and geological surveys. This activity was stepped up in the early years of our century. Official surveys reached further into the Karakoram, while officers were seconded to private expeditions to gather further intelligence. From the turn of the century, systematic statistical and demographic surveys were under way (Clarke 1901). In 1905, a scheme for regular observations and reports on Karakoram glaciers was initiated by the Geological Survey of India, partly in reponse to a request by Douglas W. Freshfield for the "International Commission on Glaciers" (*Records of the Geological Survey of India*, Vol. 5 pp. 127-157). The first major report was on the Hunza glaciers in 1906 by H. H. Hayden (1907).

The direction and preoccupations of this work are evident in an official publication, *A sketch of the Geography and Geology of the Himalaya Mountains and Tibet* (1907), by Colonel S. G. Burrard, (Superintendent, Trigonometrical Surveys), and H. H. Hayden (Superintendent, Geological Survey of India), whose Preface states:

> " Our subject has fallen naturally into four parts, as follows :-
> PART I. - The high peaks of Asia.
> PART II.- The principal mountain ranges of Asia.
> PART III. - The rivers of the Himalaya and Tibet.
> PART IV. - The geology of the Himalaya." (p.i)

The first three parts are a gazetteer of peaks, ranges and rivers described in terms of elevation, direction, areas and other strictly topographical features. (Incidentally, the first volume of the Schlaginweits' 1861, official English-language report was the same, reflecting what was asked, or suited the form of such 'government science') Here and there, and in the last part, Hayden provides information on rock types and, very briefly, some geological notions of how the Ranges were formed. Essentially, the work provides an extreme kind of 'surveyors' view' of the physical geography. Geomorphological ideas, or the discussion of processes shaping the topography are absent or relegated to a minor position and brief notes. This remains, however, the only comprehensive 'physical geography' of the region in English in our century, by individuals actively engaged in working it out.

In 1913 a project to join the geodetic survey network of India to that of Russia was finally negotiated. The link was to go through the Gilgit-Hunza region, northwards to the Pamir. Kenneth Mason (1887-1970), who had joined the Topographic Survey of India in 1909, was in charge of the surveying (ed. Burrard 1914, Mason 1914a, Crompton 1984). It began Mason's lifelong association with the Karakoram. His major interest became to the glaciers of the region (Mason 1914b). He led an expedition to the Upper Shaksgam River from th Shyok-Yarkand River side (Mason 1928). Thereafter, he became especially involved in the well-established problem of those glaciers that interfered with rivers and created outburst floods (Rabot 1905, Hedin 1910). He co-ordinated glacier information for the Geological Survey of India and wrote about it for the *Himalayan Journal* (Mason 1926, 1929a, 1930, 1931). His rôle in the monitoring of the Khumdan Glacier dams on the Upper Shyok, and of large outburst floods in 1926, 1929 and 1931, is of enduring interest (Mason 1929a, 1932, Lyall-Grant and Mason 1939). He brought together the dam-burst evidence, with that of sudden, exceptional rates of advance of some Karakoram glaciers, now recognised as glacier surges (Hewitt 1969). From this came his hypothesis that ice dams only formed through such sudden or, as he termed them, "accidental" advances (Mason 1935). He did accept that enlarged glaciers of the past had created even bigger ice-dammed lakes (1926), along similar lines to Dainelli's (1922) picture of the Upper Indus Basin. Mason made a significant contribution to the nomenclature and mapping of the region (1929b, 1938).

Perhaps because he was a surveyor, the mapped topography or mappable behaviour of, say, glaciers or floods, were the frame of reference for his work. His discussions of the monitoring and prediction of glacier behaviour, relate wholly to the positions of their termini. He tried to establish rhythms or their absence from this evidence, but paid scant attention to the then emerging science of of glacier system dynamic and metabolism, or particular problems of Central Asian glaciology (von Klebelsberg 1927, Visser 1933-34, Washburn 1939). In arguing that only "accidental" glacier advances would dam rivers - hence fluctuations that were unrelated to climatic variations and peculiar to each glacier - he provided no explanation as to how this could happen (1930, 1935). His case depends entirely upon reports of terminus positions, but he plays down or ignores that which does not fit his preconceptions (Hewitt 1969 p. 1015).

In general, Mason's approach to processes and features in the environment was to try and generalise from the local and particular, but also scattered and isolated 'facts,' to an overall pattern, but without an embracing explanation, or physical theory. Such were his general rules in the case of glacier fluctuations. It recalls Burrard's efforts to deduce general principles from the height and distribution of 'great peaks' in the Himalaya (Burrard and Hayden 1907 p.31).

Mason's main importance is, perhaps, in introducing the role of 'applied geomorphology' into our history, especially concerning what he called 'threatening glaciers' (1935). In the 1930's, he became Professor of Geography at Oxford. His *Abode of Snow* (1955), a popular account of exploration and climbing in the Himalaya, was well-received and widely read. I am not aware that, at Oxford, he made any further contributions to either Karakoram or geomorphological understanding.

Getting high

The rise of mountaineering as a major activity, had a direct impact upon science in the region. More and more of those who explored new areas or shaped popular and sometimes scientific interest, were mountaineers. After Conway came Collie, Longstaff, Eckenstein, the Duke of Abruzzi, Dyhrenfurth, Tilman, Shipton and Hunt to name just a few (Dainelli 1959). Such men, with their eyes on the highest places, if not pushing the limits of endurance in climbing to them, seized the imagination of the public, the geographical societies and governments. As a result, not only *their* activity, but much of the scientific work, was attached to attempts to scale some high peak, or explore the more inaccessible areas. It is a development that continues into our own time. In the conclusion, we will examine how this coloured the work and what we are and are not aware of, in it. However, within the essentially mountaineering and "expedition science" framework, a few projects and individuals stand out in the scope of their contributions to earth science.

Major scientific studies by continental Europeans.

Between 1899 and 1907, the eastern portions of the Karakoram were visited by the Swedish explorer-geographer Sven Hedin (1865-1952). It was but a small part of his journeys and research in the interior of Tibet and Chinese Turkestan (Hedin 1903, 1907). Yet he provides much information and some of the most scholarly commentary relating to Karakoram geomorphology. The bulk of his systematic statements in this regard occur in various parts of the 13-volume study *Southern Tibet; etc.* (1917-1922). It includes the most detailed examination of the history of geographical knowledge and discovery in the region to that time (Vol. VII). References to landforms and earth surface processes are too numerous and scattered to summarise. Unfortunately, no landforms of the Karakoram engaged his attention as, say, the distal region of the Tarim river did. His maps and interpretations of the changing river patterns, dunes and other desert conditions there, are outstanding geomorphology for their time (1904 pp. 227-309).

In our region, he referred to Penck's idea of "an upper limit of denudation" as an erosional factor influencing the mountain ranges (Vol. IV p. 262, Vol. VII p. 459, 506). He argued that the morphology and development of the landscape in the eastern part of the region and Tibet, reflected progressive desiccation and climatic oscillations. He was particularly interested in the evidence for and impact of, climatic change. His reconstructions of lake level variations and the fluctuations of the Khumdan Glaciers in relation precipitation in Northern India, were a major new contribution (Vol. II). He also remarked, that attempts to correlate precipitation, lake levels and glaciers, were suspect unless they took account of the very different rates of response to climate (p. 199). A world-wide glacier recession was then being reported, but Hedin noted that certain Karakoram Glaciers were advancing (Vol. II p. 199, Holland 1908). While it does not support Mason's "accidental" ideas, his more detailed evidence on historical changes of the Khumdan Glaciers and dams formed by them, showed that Longstaff's (1910) 35-year cycles "nearly disappear" (p. 198). We hope that was not because, in a bitter dispute, Longstaff had helped to make his "Trans-Himalaya" disappear! (Allen 1982 Chapter 10).[9]

With respect to landscape evolution Hedin relied heavily upon the ideas of Ferdinand P.W. Freiherr von Richthofen (1833-1905). The Karakoram lies at the margins of von Richthofen's concerns in his great work on China (1877-1912), and he never visited the region. Nevertheless, as Hedin shows, this founding figure of modern physical geography and geomorphology, propounded one of the major hypotheses of landform development involving our region (volume 7 p. 322). This was his notion of expanding river systems in Central Asia, related to climatic change and orogeny, with glaciation and fluctuating lake levels playing important roles. Thus, he considered the Upper Indus, like the Upper Hwang-ho, to have evolved from a series of basins of internal drainage. Dozens of these lie east and north of its headwaters, while two are still incorporated within it (Hewitt 1968 Vol. I, Note 3.4 p. 299). Rather than 'base level' control, he invoked episodes of high humidity as promoting phases of stream cutting and the integration of basins into larger drainage networks. In the interior plateaux areas, dry epochs, by reducing lake levels, would tend to cut them off, as was becoming well-established for Lake Pangong immediately east of the Karakoram. However, as the trunk streams cut further into and 'captured' the high sub-basins, they were more likely to stay open in dry periods too. Here

---

[9] Hedin became embroiled in bitter debates over nomenclature and priority in discoveries at the Royal Geographical Society. This along with his later relation with National Socialist Germany, overshadowed the enormous achievements of his exploration and interpretations of Inner Asia. However, even at the time when he received medals from the R. G. S., his work recieved scant attention in the Anglo-American geographical or earth science literatures (Allen 1982).

von Richthofen relied heavily on others we have considered, especially Hermann von Schlagintweit's work on the salt lakes. (H. von S. 1871, and 1872 Vol. III, iv). Hedin also showed that the great majority of existing basins of interior drainage, south of the Kun Lun, are less than 10 square miles in area, indicating that relatively small increases of precipitation might cause them to overflow (Vol. VII pp. 493-494).

These ideas have received little notice in the literature in English. As noted, von Richthofen is usually regarded as one of the last great exponents of the marine denudation of land surfaces (Chorley *et al.*, 1964 pp. 598-601). His views on the development of river basins in Central Asia seem an important departure from that. Of course, to the extent that the more widely quoted Anglo-American literature involves our region, it concerns rivers cutting across the main Himalaya and its Front Ranges, and the competing hypotheses of antecedence, superimposition and headward capture (Oldham 1888, Pascoe 1919, Wager 1937, Davies 1940, Oberlander 1965). Von Richthofen, like Oestreich and Hedin, recognised the relevance of these questions to the south of our region and contributed to them (Vol. IV p. 450). He approached denudation in the high, climatically distinct and vast interior of Asia, as a different problem.

Giotto Dainelli

No one else's work quite equals Dainelli's in our period, where the Karakoram and its environs are concerned. In detail, depth, consistency of development and attempted synthesis it stands alone. Only Sven Hedin showed an equal capacity for monumental and scholarly - and prolix! - exposition, relating to an even more encompassing curiosity and endurance in field work. But his work rarely deals with more than the eastern margins of our region.

Dainelli enters the Karakoram scene as a scientist attached to the de Filippi Expedition of 1913-14 (de Filippi 1932). The foundation of his achievement is seen in the twelve volumes of "Geographical and Geological Results" which he edited and in large part wrote. It is impossible to do justice to his contribution here - accepting that a good editor could cut the number of his words in half to good effect! None of the volumes was ever translated into English and one doubts few have ever read more than a fraction of them in the original. We can most usefully summarise the volumes where Dainelli's contributions to scientific understanding of the Karakoram are most evident.

Volume I *La esplorazione della regione fra l'Himalaja Occidentale e il Caracorum*. (1934) 430p. (History of Exploration).

It is appropriate to draw attention to this massive work, sufficient to give Dainelli a high place in the study of geographical discovery alone However, what he introduces here, is also integral to the rest of his work. He refers to the same literature in each phase of his scientific exposition. It is a set of source materials, almost as important to the development of his enquiries and conclusions, as his own field work, and the scientific ideas that he applies.

Volume II *La Serie dei Terreni* (1933-34). Part I. 458p. Part II. 1099p. (Historical Geology).

An extraordinary 1557 pages, and more than 800,000 words(!) describe and reconstruct the regional geology since the Palæozoic. Evidence is drawn from hundreds of field locations to define rock types and provinces, fossil sequences, tectonic development and orogeny. The reconstructions of Asian palæogeography in successive eras are monumental (Pt II pp. 834-970). However, it is the discussion of tectonics and rapid uplift in the Himalaya and Karakoram that are central to any consideration of contemporary geomorphology and to his Pleistocene reconstructions in the following volume (pp. 972-1050). He draws upon the geological work of Godwin-Austen, Wyss (1931) and de Terra (1932), and concepts developed by Suess (1883-1904)[10] and Argand (1922). The concepts already have a late twentieth century quality, accepting Argand's essentially "Plate Tectonic" picture of Himalayan tectonics (see Fig.112 p. 1007). He leaves no doubt of his debt to other geologists, but the overall synthesis is his. It is not flawless but it is unsurpassed as yet.

Volume III *Studi sul Glaciale* (n.d.) Pt. I 651 p. Pt.II. Illustrations ("Ice Age Studies").

This is his most sustained and influential study relating to geomorphology. The reconstruction of Pleistocene glaciations are the basis for most of what he has to say about the development of the present landscape. Others had proposed a Pleistocene glacier expansion as in Europe. His are the first substantial reconstructions of a series of major glaciations and interglacials. His notion of nearly total glacier cover of the whole region in the First Glacial at least, and of an Upper Indus Basin dramatically interfered with by glaciers entering and blocking the main stream in the final phases of the Pleistocene, represent one of the more painstaking, yet daring reconstructions in modern geology. They are still broadly accepted. He also made a major study of the Ice Age in the Vale of Kashmir (pp. 459-588). Some aspects, such as his postulated "Indus Glacier" all the way to the Plains (pp. 619-628), have been found wrong. Since he worked mainly from deposits and morphology without any direct dating techniques, recent work is superseding his detailed stratigraphic notions. None have demonstrated an overall grasp sufficient to challenge his major theses.

Volume IV *Le condizione fisiche attuali* (1928) 479 p., with O. Marinelli (The Contemporary Physical Environment)

A large part is devoted to describing the Karakoram glaciers, which he integrates with discussions of snowlines, fluctuations of termini, moraine-building, glacial lakes and outburst floods. A revised picture of snowlines between the Himalayan Front Ranges and Kun Lun, is shown in Plate XII.

Dainelli also wrote the huge volume on human settlement and life, and contributed to others. However, his place in relation to Himalayan geoscience, rests upon the volumes cited. He produced a comprehensive overview of most of what was known to the 1930's, with any bearing upon geomorphology. Like many physical geographers of his day, however, he had only a limited interest in contemporary processes and forms. His notions of synthesis, the major questions that preoccupied him, concerned the development of the present landscape, and, in turn, its use to reconstruct that development on geological time scales.

---

[10] Dainelli actually quotes from the French translation (1902-18) published by Armand Colin, Paris

He returned to the Karakoram with his own expedition in 1930, and carried out a major exploration and mapping programme on the Siachen and Rimo Glaciers (Dainelli 1933). Much later he wrote a more popular work on Karakoram exploration, published in 1958. I am not aware of any further contributions relevant to our concerns.

Ardito Desio (1897-?), the Italian geologist who led the first successful assault on K2 in 1954, represents the links between the two thrusts of major mountaineering expeditions and science. However, while an important figure in Karakoram geoscience since the 1930's, his more significant studies come well after our period. Before 1939, his contributions have a 'reconnais-sance' quality, comparable to Godwin-Austen's or Longstaff's reports on their expeditions. Much derives from rapid surveys, fitted in between efforts to reach new areas or support the climbing, and written up with little analysis or conceptual development Desio (1930, 1940). Since the 1950's Desio has made or shared in major geoscience studies in the Karakoram, mostly concerning the solid geology and tectonics (Desio 1954, 1960, 1964.

Between 1922 and 1935, another husband and wife team, this time from the Netherlands, led a series of scientific expeditions to the Karakoram. They were Philips Christiaanson Visser (1882 - ?) and Jeanette Visser-Hooft (1888 - ?). In their main three-volume study (1935-38), the higher valleys, mountain ranges and glaciers of the Karakoram receive most attention. However, it was never translated from the German, and we have only minor papers in English describing their work (Visser 1934). They also provide a detailed review of previous exploration and science in the region (1935-38, Vol. 1), and an overview of the orography and nomenclature (Vol. 1. pp. 87-114). P.C. Visser determined the distribution of some 15,000km^2 of glacier cover, and the relative proportions in the Hindu Kush, Sarikol, Aghil and Karakoram Ranges (Vol. 2 Table 1 p. 7). He summarises the patterns of snowlines throughout these South Central Asian mountains. The morphological features of the glaciers and their typologies are discussed. Specific processes such as glacier 'surges,' ice dams and the 'ablation valleys' receive their most complete treatment to that time.

Much of the physical geography (Vol. 1), and the glaciology (Vol. 2), is developed in terms of features encountered in their three expeditions, or by valley and mountain range. It is hardly possible to summarise this, and a large part of it is 'mere description.' It is valuable in order to fill out details of terrain but, except for some development of glaciological ideas, their approach to landscape and environment is 'topographical' and encyclopædic. There is not, as in Dainelli, any effort to orchestrate the observations and specific processes into a coherent view of the development of the landscape, or of overall systems of environmental control or morphology. Nevertheless, their's is the last, and most comprehensive survey of the physical geography of the Karakoram to 1939 (Washburn 1939).

Stemming from another set of important research expeditions, the work of Helmut de Terra (1900-1981) and E. Trinkler (1896-1931) and their colleagues in the 1930's, mainly addressed denudation chronology, or Pleistocene and Recent climatic change. The expeditions were to the eastern fringes of the region (Trinkler 1930, 1931, de Terra 1933, 1936). They added to, but largely seemed to confirm Dainelli on the 'Ice Age' (de Terra 1934, de Terra and Patterson 1939 p. 220). De Terra agrees with Dainelli's estimate of "8000ft" of uplift in the Pir

Panjal Range, to the south, "since the second half of the Pleistocene," but adds that such "young uplifts" affected all of the Himalaya and Karakoram and have been decisive for the development of the modern landscape (1934 p. 12). He interpreted the main valley sections of the Ladak Indus, Shyok and Chang-Chenmo Rivers as following the strike of the mountain ranges along major faults parallel to them (1934 p. 38). He echoed the von Schlagintweits in placing the "pre-glacial" floor of the Upper Indus "3000-4000ft" above its present level (p. 40). For the longer term, de Terra also postulated series of ancient erosion surfaces and a high level "widely extended relief of maturity or old age," indicating low elevation until late Tertiary times, with three successive, increasingly large phases of uplift since (1934 pp. 31-38).

During the 1930's there were great improvements in knowledge of topography and conditions in the remoter, higher parts of the region. They include work in the upper Baltoro by Dyrenfurth (1935, 1939), and the expeditions of Shipton and Tilman to Shaksgam (1935) and Hispar-Biafo (1939). These provided much better maps, especially of the upper reaches large glacier systems, the Shipton maps remaining the best available for the Central Karakoram (Shipton *et al.* 1938, 1940, Mott 1950). The 1934 German Expedition to Nanga Parbat resulted in even more magnificent maps, and the most detailed and accurate morphological and movement studies of glaciers in this region to that time (Finsterwalder *et al.* 1935, 1937).

From the same field work came some revolutionary notions on granites and orogeny, but that takes us beyond my period (Misch 1949, 1950). This is also the place to note the path-breaking work of the great Indian geologist Wadia, concerning Himalayan structure and tectonics. His work would deal with the exceptional crustal thickness, tectonic activity and rates of uplift in what he called the "North-West Himalayan Syntaxis," (Wadia 1930, 1931, 1933, 1939). That has had an important bearing upon the interpretation of the morphology and development of the region, but this development was also cut short, to be resumed after the Second World War (Gansser 1964).

Before leaving this story, we should note Wadia was not alone among contributors from the region itself, notably in surveying. In the class-conscious style of the times, Europeans tended to wield the theodolites, but local surveyors did much of the detailed mapping. I have not seen the originals of the work of Afraz Ghul Khan Sahib and Mohammed Akram, surveyors with the Visser expeditions. However, the plane table work of Fazal Ellahi with the 1939 Shipton expedition, is superb in draughtsmanship and accuracy for such terrain. Minutely faithful to the morphology, the originals of his maps, now in the Royal Geographical Society's Library, remain an invaluable tool.

One last person deserves a place as an important observer of the Karakoram landscape: the photographer Vittorio Sella. His plates adorn many Italian publications from de Filippi (1912) to Dainelli (1958), and also the Dhyrenfurth books. They are a systematic and enduring visual archive of these landscapes. Magnificent in their very unpretentiousness, their æsthetic appeal does not diminish their value for the scientist, unlike many more showy, 'coffee table' pictures since (Sella Foundation 1987).

Problems of interpretation

The Karakoram is, arguably, an extraordinary tribute to the action and power of those orogenic and erosional forces central to the concerns and history of geomorphology. Yet it is simply absent from our literature. Its features and those who investigated them fail to appear in the major discussions. That is surely not just a question of them being overlooked. I submit that it is not explained by an absence of substantial, sometimes outstanding and innovative work. Rather, it is a question of the practical and social conditions shaping the development of science in high Asia on the one hand, and the dominant concerns in the histories of geomorphology, on the other.

With respect to the development of physical geology and physical geography, there seem to be three main factors at work :

i) the practical and environmental circumstances of enquiry in the Karakoram;
ii) the influence of certain social 'constructions,' especially national, political conditions, upon the styles of work, but also the acceptance and dissemination of results;
iii) the way the preoccupations of historians of geomorphology tend to exclude or ignore the intellectual concerns, as well as substantive achievements, of the major figures who worked in the region.

"Expedition science"

The portrait of the Karakoram that we have found in the literature, is work by visitors from distant places. Most came in more or less large expeditions or government missions, with specific geographical objectives. Even where trained earth scientists or physical geographers were involved, exploration and mapping, scaling a peak or hunting, subduing 'the natives' or defining a political boundary, were more often the primary concerns (Keay 1977, 1979).

The remoteness, great scale, extremes of climate, altitude and ruggedness of the region must not be forgotten. They have absorbed much more effort than scientific tasks. In the period that concerns us, there were months of travel from Europe to the borders of the region, and then weeks of walking to reach the higher valleys and ranges. So much of the work cited, derived from only a few weeks or even days, out of the many months a whole expedition took. This helps to explain why much of the observation of geomorphic conditions has an *ad hoc* or chance character. A great deal of what we have is 'reconnaissance' rather than sustained and systematic observation.

The geographical and environmental coverage of this 'expedition science' were highly constrained and partial. Within the Karakoram, most observations are actually focussed upon particular small areas or limited environments within the whole. The lower valley floors are the most accessible. Expeditions spent most of their time in them, even if heading for the higher areas. Scientists have said far more about processes, forms and superficial deposits in the semi-arid river valleys, than anything else, more about the termini of glaciers that intrude into these valleys, than about the glacier basins as a whole. Conversely, most expeditions after the 1890's, often the best equipped and most widely reported, were headed for a few giant peaks and their immediate environs. The Baltoro, surrounded by most of the highest mountains, has been photographed, described and mapped more often than all other glaciers combined, - yet we know less about its geomorphology, glacial and otherwise, than several others. Most large and small glacier basins, lacking some major peak, are known from just one or two flying visits. Few expeditions had any opportunity to work in the *much larger* fraction of the Karakoram landscapes between the dry valley floors and high altitude extremes.

The expedition context explains the lingering 'catastrophism' in the literature. It depended upon encounters with extreme or unusual events. Processes not unusual in this environment seem very large in relation to other regions. There is a greater likelihood of encountering them, even in short, summertime expeditions. Only the largest and most destructive events, the great landslides or glacier outburst floods, would be heard of from areas and at times where there was no expedition. That is why they prevail in the reports. With the exception of dam burst floods, there was little appreciation of their frequency and role in shaping the Karakoram landscape. This, then, is not 'catastrophism' as either a general idea, nor specific explanation, but a reflection of the information system of 'expedition science.'

While the problems of work in the Karakoram were important, they do not support an 'environmentalist' interpretation: namely, that ideas about landscape reflect what is distinctive about the particular environment (Ellenberger 1980). On the contrary, the expedition context made it *harder* to envisage the region, so to speak, on its own terms. Rather, the tendency was to make a story out of the exotic, extreme or chance phenomena, without an ability to determine their overall significance. The superlatives abound. But they did not generally provide for a coherent portrait. Instead - in curious way that may have much to do with 'environmentalism' in all its forms - the decisive influences were actually operating from a distance, and in terms of distant rather than local priorities.

The practical constraints did not preclude major scientific achievements, but they greatly influenced what approaches would, or would not be intellectually robust. It required unusual commitment and talent but also, I suggest, an intellectual approach adequate for the task. Nevertheless, whether and where such an approach developed, also depended upon the pre-selection of the actors involved, their social origins and intellectual horizons. Through that, political and national factors could dominate what was sought, accepted and disseminated. In other words, the familiar themes in "the social construction of knowledge" were present, albeit refracted by the prism of Central Asian expedition science.

Politico-military versus academic priorities

It is not an accident that, wherever non-British scientists worked in our region, elite influence was at work. Von Humboldt's stature in the courts of Europe made his recommendation of the von Schlagintweits successful. The Duke of Abruzzi was behind de Filippi's and Dainelli's visits; the Duke of Aosta in Desio's and the King of Sweden Hedin's. Even so, borders were often closed and permission refused, as happened to von Humboldt himself and Hedin.

Another indication of the socio-political context, is the professional division along national lines. Almost every British figure of note was either a soldier or a government servant. Cunningham, Godwin-Austen, Mason, Burrard, were professional soldiers working for such colonial services as the Topographical or Geological Surveys of India. Drew, Younghusband, Longstaff, Hayden, were mostly travelling in official capacities or on 'missions' Soldiers and civil servants took part in other expeditions - C.G. Bruce with Conway, Wood with de Filippi, Mott with Shipton, for example.

Contrast this with all the important Continental figures: the Schlagintweits, Stolicza, von Richthofen, Calciati, Oestreich, Hedin, de Filippi, Dainelli, Desio, Trinkler, the Vissers. It is true that some of them also were employed by colonial agencies, and parts of their work reflect that, for example most of Schlagintweit, 1861-66. However, they are all academics, trained in physical geography, geology or other physical sciences. That comes out in the work equally strongly.

There is, then, throughout this period a marked distinction between most work of British and continental contributors, and their professional backgrounds. Again, however, before pursuing this we must resist the temptation to see some general 'determinism' at work. "National character," or the overall qualities of science in the particular nations may not be irrelevant, but they are not the issue. It is rather a question of what influenced the deployment of scientific work in Central Asia, and dissemination of its results.

Initially I thought we had here a continuing legacy of that "British, amateur, decentralized naturalist tradition" - albeit now turned professional - that Porter finds in the eighteenth century (1977 p. 170). He contrasts it with the "range and distinction" of the continental tradition exemplified by Buffon, von Humboldt and Lamarck. Certainly, the work of the Schlagintweits', von Richthofen, Hedin and Dainelli, belongs with that. By comparison, from Vigne and Drew, to Longstaff and Mason, most British contributors, however remarkable in other ways, seem 'amateur' scientists: 'amateurish' in the geology, glaciology, climatology, plant geography of their day, and casual about the scholarly homework their notions seem to require. Indeed, what *is* surprising about them is the range and quality of their interest in these subjects, when they are not trained nor current in ideas about them. That may reflect the admirable 'decentralised naturalist tradition' in the education and culture of the English middle class. Yet, it is simply not true that British *geology* and Anglo-American *geomorphology*, failed to achieve 'range and distinction' in this period. The best of it was not indifferent to, nor unwilling to praise and borrow from *some* continental scholars - Agassiz, Suess or Penck. And if there were scandalous misrepresentations such as Greene (1982) identifies in the Hutton-Werner controversy or in efforts to demean Wegener's qualifications, one could document similar narrow or xenophobic constructions in any one of the continental traditions. Indeed, the Karakoram work is a remarkable contrast to the "range and distinction" of other phases of British scientific endeavour, notably in the biological and physical sciences and including Darwin's work, that did depend upon 'expedition science' and evidence from remote regions. One could also note the extraordinary breadth and depth of Sir Aurel Stein's researches in Central Asia which, unfortunately only touched the fringes of our subject.

It would be my contention, therefore, that we are looking at a particular set of national influences with a particular bearing upon our region. The practical difficulties did mean that, usually, those few expeditions which remained for a long time; those individuals who returned several times; and those with the time and patience to carefully read and evaluate the work of their predecessors, provide the most substantial studies. The primary concern with a scholarly and scientific audience is at work too. The Schlagintweits were sending their materials directly to von Humboldt. Oestreich or Dainelli returned to university posts and addressed their work to learned societies. It is not the same as the Military unit or the Government bureaucracy to which so many of the British observers were answerable. Clearly, there is no neat division in terms of ability, or along national lines. Godwin-Austen's substantial work elsewhere in India has been noted. In the early phase, Cunningham and to some extent Drew, display a scholarly approach. For the most part, however, what was done, or valued, was "physical geography and geology" as epitomised by Burrard and Hayden (1907). Rather than being a peculiarly dry and stiff bit of writing, their text crystallises the central, *professional* concerns of the British effort. It is geography in the service of empire, and at its narrowest focus. Imperial interests prevail in this 'surveyor's' science, or cartography shaped by geo-strategic goals.

The mission of most expeditions, from Godwin-Austen to Shipton, was to establish territory, natural resources, the routes and river courses, the watersheds, passes and relation of great peaks to them. Nomenclature, and its official sanctioning, was an integral part of this process. Good maps and knowledge of terrain, rather than providing the background to understanding, became its main solution. Processes, features, historical geology were, at best embroidered upon the establishment of the terrain. The discussions following paper after paper at the *Royal Geographical Society*, and those who took part in them, exemplify these preoccupations throughout this period. The issues that they stress, over and over again - and that a visiting lecturer would ignore at his peril - were topographical accuracy, priority of claims to discovery, and nomenclature. The men of experience and distinction who spoke to this in the meetings, were those for whom the burdens of military and government duty, the social setting and preoccupations of Imperial India, shaped what was looked for and emphasised; who was to be championed or criticised. It led to a pragmatic, soldier-statesman approach. It was sympathetic to mountaineering and feats of exploration by almost anyone, but scholarly and intellectual endeavours, or extravagant claims by foreign nationals, received short and not always courteous attention. And that tended to decide who was valued and what was passed on. With that went a remarkable neglect of other concerns and especially the work of other European nationals in the region, even though nearly all of them addressed the Society. Its influence in promoting and disseminating geographical knowledge in this period was a decisive factor.

That the Karakoram also coincided with the poorly known, insecure limits of British control on its borders with one jealous and another expansionist imperial power, China and Russia respectively, became an obsession over the period that concerns us (Keay 1979). All the important British travellers were thoroughly briefed in, or for a part of these machinations. It was heady stuff, "the great game" as some called it. As far as the individual was concerned, it meant that reputations turned upon who was first in an area, established its geographical names or, at least, put it on the map.

I suspect the same applied in many other remote and insecure colonial territories, and in the *imperial* geography of other European states. There were not only the problems of security and the influence of a colonial administration upon what work could be done and by whom. There was also the 'public,' the media-conscious home environment that developed in this period. It was reflected in the pressures to fashion myths about colonial territories, performance and national heroes in them. One should add that a sympathetic hearing here between different European contributions was made vastly more difficult by the animosities generated by the two world wars with which our period ends. However, this is an indicator of conditions. It does not explain what distinguishes the character and quality of the continental work, nor its absence from or misrepresentation in the history of geomorphology.

Against the grain: Karakoram geomorphology and historians of the science

The more strictly geomorphological issues here, turn upon the history and ideas of "physical geography," rather than of geology. By contrast, the history of geomorphology has definitely leaned towards the latter as its inspiration. What is striking about British physical geography in our period is, firstly the dearth of general ideas and major figures contributing to earth science - as compared with such impressive figures as Mackinder in human or general geography. Secondly, the work done is almost wholly of a gazetteering, regional kind. By contrast, many important contributors to the general development of earth science on the continent were geographers. One of the first claims upon the scientific work of their earth science was what von Richthofen called "chorological representation" (Zittel 1901 p. 82). This required a regionally or globally oriented geographical approach that, as he said of Ritter, "gathered information from the most varied sources and kneaded it into an organic and intellectual whole, united by the principle of causality." He was speaking here of 'general geography' but his essential inspiration was the natural science tradition of geography exemplified by von Humboldt (von Richthofen 1883). It is from the same root that the work of the von Schlagintweits, Oestreich, Hedin, Dainelli and, to some extent, the Vissers belongs. It is not only language, nationality or the work of academics versus soldier-bureaucrats, that sets it apart from British work: but its goals and methods.

However, the direction in which Anglo-American geomorphology has gone since the Second World War is even less sympathetic to regional or geographically-based notions. It sits more comfortably within the "physical geology" of, say, Arthur Holmes, or even civil engineering, than the geographical tradition of von Humboldt. Indeed it can hardly fail to reinforce the view of von Engeln (1942 p. 6) among others, that the continental geomorphologists had been wilfully racing off in other directions, too conservative or too nationalistic to learn from the more 'advanced' Anglo-American work.

The ideas in question in the latter are almost always of an abstract and general kind, supposed to apply globally, or to represent fundamental conditions in nature: the supposedly universal abstract languages of "structure, process and stage," morphometry, process mechanics and rates, mass and energy systems, experiments and modelling. Like so much of our physical and social science, the preferred ideas and methods are those that seek to go behind and ultimately dispense with the distinctive historical and geographical conditions of particular times or places.

The history of geomorphology has, in the main, been written as an interplay of these sorts of dominant ideas and the individuals who promoted or opposed them. The historiographic detail and focus of debate concerns the sayings, doings and personalities of individuals - Hutton, Lyell, Gilbert, Davis, - who contributed to the development of presently accepted, universal principles. Unless features and observations in particular areas have been used to develop or demonstrate these principles, or the approaches lying behind them, they do not appear in the story. Only here do scientific expeditions and reports from remote places come into play, if they do at all.

By contrast, the most substantial efforts on Karakoram geomorphology were not only by continental figures, but 'chorological' in intellectual approach. As we have seen, in the Karakoram much of the best work strives to reconstruct the present landscape in terms of historical geology and Pleistocene change. But it does so in the style, and with the methods and massive commitment to local detail of the 'chorological' tradition. I suggest that this is one, though not the only style of work, which allows one to overcome the limitations of expedition science. That does not mean 'pure' and unique regional description. Rather that is something which defines the 'topographical' or gazetteering approach. The continental tradition certainly worked towards comprehensive, causal syntheses that could then be used in global reconstructions.

I would argue, then, that the interpretive questions here require us to pay attention to the different emphases of physical geography, as against physical geology, if not physical sciences generally. I offer it firstly as a thesis as to why high Asian geomorphology is largely absent from our work. Secondly, there seems to be a clear weakness in historiography which ignores the substantial 'chorological' base of past geomorphological studies. Thirdly, it is interesting to reflect upon this, given that Anglo-American *physical geographers* have made such an impressive contribution to the history of geomorphology.

Of course, it would be wrong to ignore the considerable, regionally based work of, say, Murchison, Lyell, G. K. Gilbert or W. M. Davis. Not only work in the Karakoram, but so much of the history of landform science has concerned reasoned understanding of given places and regions, or their unique relations to the globe. It has been fundamentally 'geographical,' even when used as the basis for the ageographical formulations of "structure, process and stage" geomorphology.

Finally, I do not make these points either to defend the former or to condemn the latter. Recognising them does seem crucial to a balanced approach to the past, and the varieties of geomorphology. If a personal bias is involved, it is the conviction that the best work reflects a commitment to, and cross-fertilizing between these approaches; their relative importance depending upon the questions at issue, the location of research and state of knowledge relevant to it, rather than an ideological commitment to one or the other, the 'chorological' or 'abstract, theoretical,' as more 'scientific.' The more substantial work on the Karakoram mostly reflects that sense of problem.

Appendix 1

The Schlagintweitiana Collection of their field notes and sketches in München, include the following:

Schlag. II. 1. Geologische Beobachtungen.

Bd. 29. 1a. pp. 276-281 "**erosionsthaler**" of the Diamir, Tarshing and Astor valleys around Nanga Parbat, identifying "3,000 - 4,000 ft " river erosion (Adolf, 1856).
Bd. 30. 6. (p. 96) Khassia Hills, Assam, "3,000ft" erosion in sandstone ascribed to strong rainfall (Hermann, Adolf, 1855).
5-9. erosional valleys and slopes in Tibet, Nepal, Ladak, Simla, and Assam (Hermann, Adolf). Hushe Valley and Masherbrum south flank (Adolf).

Schlag. III. "Erläuterungskizzen und Pausen...Hermann und Adolph.."

Bd. 14. Bl. (= Field Sketch) **#28**, 'Bhagratti Thal', notes ascribe deep gorge and steps in valley profile to fluvial erosion (Adolf 1855).
Bl. 35 "Erosion of lateral valley of Sutlej opposite Babe in Spiti" (Adolf July 8th, 1856).
Bl. 40 The erosional gorge (..**Die Erosions Schlucht**...) at the junction of the Tirlen and Sains with the Bias [rivers].." (Adolf June 3, 1856) [with declinations of slopes 50-60°].
Bl. 42. (Deep) " Erosions Schlucht des Bias..." with geology identified "all crystalline".
Bl. 55 & 56. Jhelum valley erosion in lateral streams and transport of alluvium and large blocks by stream (Adolf November 11, 1856)
III. 15.
Bl. 16,17 & 18. Mountains and Upper Indus near Gartok, describing material in valleys as "Alt..Diluvialschutt", and gorge-like "Nord Chako La Thal" as a "....Tiefer (deep) Erosions Schlucht..." (Adolf July 27, 1855)
III. Bd. 16.
Bl.1 Spiti area erosional valleys and slopes (Hermann 1856).
Bl. 6-14. River erosion of gorges with steep cliffs in Spiti, Zanskar, Khango Thal etc., (Adolf June, 1856).
Bl. 37. Mountain ranges around Shigar, Baltistan attributing deep lateral valleys of Shigar and Indus Rivers, to erosional "..valleys of excavation..", and also gullies on slopes as "**... erosions rinnen..**". (Adolf August 30, 1856).
III. 18-19.
Bl. 28. Lahaul, Boru Nag Glacier, with comments on erosional features.
Bl. 36. "Mustack Glacier" (=Baltoro) terminus, "...covered entirely with debris.." and notes "...old alluvium of river cut through by new excavations..." (Adolf.August 16, 1856)
IV..
Bd. 39 "Banks of detritus and its erosion...Spiti (Hermann June 13, 1856)

Bibliography

Allen, C., 1982. *A Mountain in Tibet: The search for Mount Kailas and the sources of the great rivers of India.* Andre Deutsch: London,.254p.

Argand, E., 1922. La tectonique de l'Asie. *Comptes Rendus 8th Congrés Internationale de Géologie*, Bruxelles, 1922: Liége, 171-372.

Becher, J., 1859. Letter addressed to R.H. Davies Esq., Secretary to the Government of Ounjab and its Dependencies, 1st July 1859, (On the 1858 Indus flodd). Journal of the Royal Asiatic Society of Bengal, 28, 219-228.

Beck, H., 1961. *Alexander von Humboldt.* Franz Steiner: Weisbaden, 2 volumes.

Bernier,F., 1699. *Voyages, contenant la description des etats de Grand Mongol de l'Hindoustan .. etc.*, 2 volumes.

Bonney, T.G., 1894. Notes on Mr. Conway's Collection of Rock Specimens for the Karakoram Himalaya, *in* Conway (1894 *see below*) Volume 2, 41-73.

Bonney, T.G., 1902. Moraines and mud-streams in the Alps, *Geological Magazine*, n.s., Vol. 9. 8-16.

Botting, D., 1973. *Humboldt and the Cosmos.* Michael Joseph: London, 295p.

Burrard S. G. (ed.), 1914. Completion of the link connecting the triangulations of India and Russia, *Records of the Survey of India*, Volume 6, p. 116.

Burrard, S. G. (ed.), 1914. Completion of the Link connecting the triangualtion of India and Russia, 1913, *Records of the Survey of India*, 6, 116.

Burrard, S. G. and Hayden, H. H., 1907. *A Sketch of the Geography and Geology of the Himalaya Mountains and Tibet.* Government of India: Calcutta, 230p.

Calciati, C., 1910. Les fronts des glaciers de Jengutsa et d'Hispar, *La Géograhie*, 22, 241-247.

Calciati, C., 1923. Li piu grande ghiacciaio della Terra. Il Siascen nel Caracroum. *Emporium* 55.

Chorley, R.J., Dunn. A.J. and Beckinsale, R.P., 1964 *A History of the Study of Landforms, or the Development of Geomorphology*, Volume I : *Geomorphology before Davis.* Methuen: London 678p.

Church, M. and Ryder, J.M., 1972. Paraglacial sedimentation; a consideration of fluvial processes conditioned by glaciation. *Geological Society of America, Bulletin*, 83, 3059-3072.

Clarke, R.T., 1901. *Assessment Report of the Kargil Tahsil of Balistan, 1901*, Lahore: Gulab Singh.

Cockerill, G.K., 1902. Byways of Hunza and Nagar. *Geographical Journal*, 60, 98-112.

Conway, W.M., 1894. *Climbing and Exploration in the Karakoram Himalaya*, London, 3 volumes.

Coulson, A.L., 1938, The Hindu Kush earthquake of 14th November, 1937. *Records of the Geological Survey of India*, 73, 135-144.

Crompton, T.O., 1984. The Pakistan to Russia triangulation connection : past and projected error analysis, *in* **Miller, K.J.** (ed.), 1984. *The International Karakoram Project*, 2 volumes. Cambridge University Press. Cambridge, Vol. 1, 171-185.

Cunningham, A., 1854. *Ladak, physical, statistical and historical, with notices of surrounding countries.* London

Curzon, G.N., 1896. The Pamirs and the source of the Oxus. *Geographical Journal* 8, 15-54 & 97-119 & 239-263.

Dainelli, G., 1924-1935. *Relazione Scientifiche della Spedizione Italiana de Filippi, nell'Himalaia, Caracorum e Turchestan Cinese (1913-14)*. Serei II. Resultati geologici e geografici. 10 volumes, Bologna.

Dainelli, G., 1933. *Buddhists and Glaciers of Western Tibet*, London

Dainelli, G., 1959. *Esploratori e Alpinisti nel caracorum*. Torino

Davies, L.M., 1940. Note on three Himalayan rivers. *Geological Magazine*, 77, 410-412

Davis W.M., Huntingdon, E and Pumpelly,R., 1905. *A journey across Turkestan.etc. 1903*. Carnegie Institute, Publication. 26, Washington, 21-119.

de Filippi, F. and Dainelli, G, 1922-34. *Spedizioni Italiana de Filippi nell'Himalaja, Caracorum e Turchestan Cinese (1913-14*. Series I. 3 volumes. Nicholla Zanichelli: Bologna.

de Filippi, F., 1912. *Karakoram and Western Tibet, 1909*. Constable: London.

de Filippi, F., 1932. *The Italian Expedition to the Himalaya, Karakoram and Eastern Turkestan, (1913-14)*. Arnold, London.

de Terra, H. & Patterson, T.T., 1939. *Studies in the Ice Age in India and the associated Human Cultures*. Carnegie Institute of Washington D.C., Publication # 493, 354p..

de Terra, H., 1933. A scientific exploration of the Eastern Karakorma and the Zaskar Himalaya. *Himalayan Journal*, 5, 33-45.

de Terra, H., 1934. Physiographic results of a recent survey in Little Tibet.*Geographical Review*, 24, 12-41.

de Terra, H., 1936. Joint geological and prehistorical studies of the late Cenozoic in India. *Science*, 83 (2149), 233-236.

Desio, A., 1940. Il Baltoro. *Le Alpi*, 59, Numbers 10-11, 977-978.

Desio, A., 1930. Geological work of the Italian expedition to the Karakoram, *Geographical Journal*, 75, 402-411.

Desio, A., 1954. An exceptional glacier advance in the Karakoram-Ladakh region. *Journal of Glaciology*, Volume 2, no.16, 383-385.

Desio, A., 1960. Sull'estensione dei plutoni granitici nel Karakoram e nell'Hindu Kush (Asie Centrale). *Rend. Accad. Naz. Lincei.*, ser. 8, volume 28/6, 783-786.

Desio, A., 1964. *Geological tentative map of the Western Karakoram - 1:500 000*, Instituto di Geologia, Universita di Milano.

Diener, C., 1896. Die Eiszeitim Himalaya. *Mitt. d. k.k. Goegr. Gesell. Wein*, Bd.39, S, 1-35.

Drew, F., 1873, Alluvial and lacustrine deposits and glacial records of the Upper Indus Basin, *Quarterly Journal of the Geological Society of London*, 29, 441-471.

Drew, F., 1875, *The Jammoo and Kashmir Territories: a geographical account*. Stanford: London.

Dunmore, Earl of, 1893. *The Pamirs*. 2 volumes. Murray: London.

Durand, A.G.A., 1910. *The making of a Frontier: five years' experiences and adventures in Gilgit, Hunza, Nagar, Chitral and the Eastern Hindu Kush*. Second edition, London, 383p..

Dyhrenfurth, G., 1939. *Baltoro; ein Himalaya-Buch*, Benno Schwabe: Basel.

Dyhrenfurth, G.O., 1935. *Damon Himalaya; Bericht der Internationalen Karakorum Expedition*, 1934.

Ellenberger, F., 1980. De l'influence de l'environment sur les concepts: l'exemple des théories géodynamiques au XVIIIe siècle en France. *Revue d'Histoire des Sciences*. XXXIII (1), 33-68.

Elphinstone, M, 1843. *An Account of the Kingdom of Caubul and its Dependencies...etc*, 2volumes, 3rd Edition, Longmans: London.

Engeln, von A., 1942. *Geomorphology,*

Finsterwalder, R, Raechl, W, Misch P. and Bechtold. F., 1935. *Forschung am Nanga Parbat: Deutsche Himalaya-Expedition 1934*. Geograph. Gesell. Hannover: Hellwingische Verlags Buchhandlung: Hannover, 143p.

Finsterwalder, R., 1937. Die Gletscher des Nanga Parbat. *Zeitschrift für Gletscherkunde*, 25, 57-108.

Franke, A.H., 1907. *A History of Western Tibet: one of the unknown empires.* Partridge: London.

Gannser, A., 1964. *Geology of the Himalayas.* Wiley-Interscience: New York), 289p.

Godwin-Austen, H.H., 1862. On the glacier phenomena of the valley of the Upper Indus (Abstract). *Report of the 32nd Annual Meeting of the British Association for the Advancement of Science.* John Murray: London.

Godwin-Austen, H.H., 1864. On the glaciers of the Mustagh Range (Trans-Indus), *Proceedings of the Royal Geographical Society*, 34, 19-56.

Godwin-Austen, H.H., 1883. Geology of the Western portion of the Himalaya. *Proceedings of the British Association for the Advancement of Science.* New Series, 5, 610-625.

Gordon, T.E., 1876. *The Roof of the World.* Edmonston and Douglas: Edinburgh.

Government of India, 1878-1891. *Scientific Results of the Second Yarkand Mission*, 15 Parts, H.M.S.O. - Eyre and Spottiswode: London, p. 67.

Greene, M.T., 1982, *Geology in the Nineteenth Century* Cornell: Ithaca, 324p.

Guillarmod, J.J., 1904. *Six Mois dans l'Himalaya, le Karakorum et l'Hindu Kush.* W. Sando: (Neuchâtel.

Hayden, H.H., 1907. Notes on certain glaciers in Northwest Kashmir. *Records of the Geological Survey of India*, 35, 127-137.

Hedin, S., 1904-07. *Scientific Results of a Journey to Central Asia, 1899-1902*, 8 Volumes. Lithographic Institute, Swedish Army: Stockholm)

Hedin, S., 1909-1913. *Transhimalaya. Discoveries and Adventures in Tibet.* 3 volumes, MacMillan: London.

Hedin, S., 1910. The Kumdan Glaciers in 1902, *Geographical Journal*, 36, 184-194.

Hedin, S., 1917-1922, *Southern Tibet: Discoveries in former times compared with my own researches in 1906-1908*, 13 volumes, Lithographic Inst. Swedish Army: Stockholm.

Henderson, W., 1859. Memorandum on the nature and effects of the flooding of the Indus on 10th August 1858 as ascertained at Attock and its neighbourhood. *Journal of the Asiatic Society of Bengal*, 28, 199-219.

Hewitt, K., 1964. A Karakoram ice dam. *Indus (Lahore)*, 5, 18-34.

Hewitt, K., 1968a. *Studies of the geomorphology of the Mountains Regions of the Upper Indus Basin.* Unpublished Doctoral Thesis, University of London. 2 volumes.

Hewitt, K., 1968b. The freeze-thaw environment of the Karakoram Himalaya. Canadian Geographer, 12, 85-98.

Hewitt, K., 1969. Glacier surges in the Karakoram Himalaya (Central Asia), *Canadian Journal of Earth Sciences*, 6, 1009-1018.

Hewitt, K., 1975. Earthquake hazards in the mountains. *Natural History*, LXXXV (5), 30-37.

Hewitt, K., 1982. Natural dams and outburst floods of the Karakoram Himalaya *in* **Glen, J., (ed.),** *Hydrological Aspects of Alpine and High Mountain Areas.* International. Hydrological. Association, (I.A.H.S.) Publ. 138, 259-269.

Holland, T.H., 1908. Observations of glaciers movements in the Himalaya, *Geographical Journal*, 31, 315-317.

Humboldt, Freiherr H.A. von, 1820, Sur la limite inférieure des neiges perpétuelles dans les montagnes de l'Himalaya et des régions équatoriales, *Annales de Chimie et de Physique*, Tome 14, p.5.

Humboldt, Freiherr H.A. von, 1843. *Asie Centrale. Recherches sur les chaines de montagnes et de climatologie comparé*. 3 volumes, Paris.

Huntingdon, E., 1907. *The Pulse of Asia. A journey in Central Asia illustrating the geographic basis of history*. Constable: London.

Izzet-Ullah, (Mir), 1825. Travels beyond the Himalaya, *Calcutta Oriental Magazine and Review*, p. 103.

Izzet-Ullah, (Mir), 1842-43, Travels beyond the Himalaya. *Journal of the Asiatic Society of Bengal*, 7, 297.

Keay, J., 1977. *When Men and Mountains Meet: the explorers of the Western Himalayas, 1820-75*. John Murray: London.

Keay, J., 1979. *The Gilgit Game: the explorers of the Western Himalayas, 1865-95*, John Murray: London.

Kellner, L., 1963. *Alexander von Humboldt*, Oxford University Press: Oxford.

Kick, W., 1960. The first glaciologists in Central Asia, *Journal of Glaciology*, 3, 687-696.

Kick, W., 1982. Alexander von Humboldt und die Brüder Schlagintweit, *in* **Müller, C.C. and Raunig, W. (eds.)**, 1982. *Der Weg zum Dach der Welt.*, Pinguin-Verlag, Innsbruck, 75-77.

Klebelsberg, R. von, 1925-26. Der Turkestanische Gletschertypus. *Zeitschrift für Gletscherkunde*, Bd. 14, 193-209.

Körner, H., 1982. Die Brüder Schlagintweit etc., *in* **Müller, C.C. and Raunig, W. (eds.)**, 1982. *Der Weg zum Dach der Welt.*, Pinguin-Verlag, Innsbruck, 62-75.

Longstaff, T.G., 1910. Glacier exploration in the Eastern Karakoram, *Geographical Journal*, 35, 622-658.

Lyall-Grant, I.H. and Mason, K., 1940. The Upper Shyok Glaciers, 1939. *Himalayan Journal*, 1, 52-63.

Lydekker, R., 1881. Geology of part of Dardistan, Baltistan and Neighbouring Districts. *Records of the Geological Survey of India*, Vol. 14.

Mason, K., 1914a. The Indo-Russian Triangulation connection, 1911-1913, *Geographical Journal*, 43, 664-672.

Mason, K., 1914b. Examination of certain glacier snouts of Hunza and Nagar. *Records, Trigonometric Survey of India*, 6, 49-51.

Mason, K., 1926. Movements of Indian Glaciers, *Geographical Journal*, 68, 57-62.

Mason, K., 1927. The Shaksgam Valley and Aghil Range, *Geographical Journal*, 69, 289-332.

Mason, K., 1929a. Indus floods and Shyok glaciers, *Himalayan Journal*, 1, 10-29.

Mason, K., 1929b. The representation of glaciated regions on maps of the Survey of India, *Professional Paper, No. 25*, Survey of India, Two Parts. (Dehra Dun).

Mason, K., 1930. The glaciers of the Karakoram and neighbourhood, *Records of the Geological Survey of India*, 63, 214-278.

Mason, K., 1935. The study of threatening glaciers, *Geographical Journal*, 85, 24-35.

Medlicott, H.B. and Blandford W.T., 1879. *Manual of the Geology of India*. Government of India Press: Calcutta, 445-817.

Medlicott, H.B., 1860. On the geological structure and relations of the southern portion of the Himalayan Range between the Rivers Ganges and Ravee. *Memoirs of the Geological Survey of India*, 3, 206p.

Middleton, D., 1984. Karakoram history; early exploration, in **Miller, K.J.(ed.)**, 1984. *The International Karakoram Project*, 2 volumes. Cambridge University Press. Cambridge, Volume 2, 17-31.

Miller, K.J. (ed.), 1984. *The International Karakoram Project*, 2 volumes. Cambridge University Press: Cambridge.

Misch, P., 1949. Metasomatic granitization of batholithic dimensions, I: synkinematic granitization in the Naga Parbat area. *America Journal of Science*, 247, 209-245, 372-406 and 673-705.

Misch, P., 1950. Diastrophism in the Himalayan Mountains, *Earth Science Digest*, 5, 19-22.

Montgomerie, T. G., 1860. Memorandum on the great flood of the River Indus which reached Attock on 10th August, 1858. *Journal of the Asiatic Society of Bengal*, 29, 128-135.

Mott, P., 1950. Karakoram Survey, 1950. *Geographical Journal*, 116, 89-95.

Müller, C.C. and Raunig, W. (eds.), 1982. *Der Weg zum Dach der Welt*. Pinguin-Verlag, Innsbruck. 407p.

Oberlander, T., 1965. *The Zagros Streams: a new interpretation of transverse drainage in an orogenic zone*. Syracuse Geographical Series, 1, Syracuse University Press, 168p.

Oestreich, K., 1906. Die Täler des nordwestlichen Himalaya. *Petermanns Mitt.* Hft. 155 Bd.33.

Oestreich, K., 1911-12. Der Tschotschogletscher in Baltistan, *Zeitschrift für Gletscherkunde*, 6, 1-30.

Oestreich, K., 1914. Himalaya-Studien, *Zeitschrift Gesellenschaft für Erdkunde*, Berlin, #6, 417-451.

Oldham, R. D., 1888. Some notes on the geology of the north-west Himalayas. *Records of the Geological Survey of India*, 21, 149-157.

Oldham, R. D., 1893. The river valleys of the Himalaya. *Journal of the Manchester Geographical Society*, IX, 112-125.

Oldham. T., 1859. On the geological structure of a portion of the Khasi Hills, Bengal. *Memoirs of the Geological Survey of India*, 1, 99-210.

Pascoe, E.H., 1919. Early history of the Indus, Brahmaputra and Ganges. *Quarterly Journal of the Geological Soicety of London*, 75, 138-159.

Philippson, A., 1886a. *Studien über Wasserscheiden*, Ver. Erdkunde, Leipzig, 163p.

Philippson, A., 1886b.Ein Beitrag zur Erosionstheorie, *Petermann's geogr. Mitteilungen*, 32, 67-79.

Polter, S.B., 1982. Nadelschau in Hochasien : Enlische Magnetforschung und die Brüder Schlagintweit. *in* **Müller, C.C. and Raunig, W. (eds.)**, 1982. *Der Weg zum Dach der Welt*. Pinguin-Verlag, Innsbruck. 407p., 78-98.

Porter, R., 1977. *The Making of Geology : earth science in Britain 1660-1815*, Cambridge University Press: Cambridge.

Pratt, J.H., 1860. On the physical difference between a rush of water like a torrent down a channel and the transmission of a wave down a river, - with reference to the indunation of the Indus as observed at Attock in August 1858. *Journal of the Royal Asiatic Society of Bengal*, 29, 274-282.

Price, L.W., 1981. *Mountains and Man: A study of process and environment*, University of California Press: Berkeley 506p.

Rabot, C., 1904. Explorations des glaciers du Karakorum, *La Géographie*, 9, 374.

Rabot, C., 1905. Glacial reservoirs and their outbursts. *Geographical Journal*, 25, 545-54.

Richthofen, F. Fr. von, 1877-1912. *China: Ergebnisse eigener Reisen darauf Gegründeter Studien*, 5 volumes. Berlin.

Richthofen, F. Fr. von, 1883. *Aufgaben und Methoden de heutigen Geographie.* Akad. Antrittsrede, Leipzig.
Rickmers, W.R., 1913. *The Duab of Turkestan,* Cambridge University Press: Cambridge.
Schlagintweit, A. von, 1857. Report on the progress of the magnetic survey of India and researches connected with it from May to November 1856, in, *Himalaya Mountains and Western Tibet,* Survey of India Report No. 9 (Dehra Dun.)
Schlagintweit, H and A., 1855. *Stereoscopisches Bilder nach den Schlagintweit'schen Reliefen...etc.* J.A.Barth: Leipzig.
Schlagintweit, H. von, (-Sakünlünski), 1869-80 *Reisen in Indien und Hochasien,* 4 volumes. Constable: Jena.
Schlagintweit, H. von, 1866. Die thermischen Verhältnisse der tiefsten Gletscherenden im Himalaya und Tibet. *Ebenda,* 1, 290-93.
Schlagintweit, H. von, 1871. *Unterschungen über die Salzseen im westlichen Tibet und in Turkestan.* Abh. der k. bayer. Akad. der Wissenschaft. Math-Physikal. Klasse II, 1. München, 74p.
Schlagintweit, H. von, 1877. *Vergleichende Angaben über die Scneegrenzen in Hochasien und in den Alpen, in.* Beilage zur Allgemeinen Zeitung, Augsburg, #20, 287p.
Schlagintweit, H., A. and R, von, 1861-1866. *Results of a Scientific Mission to India and High Asia, undertaken between the years 1854 and 1858...etc.,* 4 volumes and Atlas, Trübner: London.
Schlagintweit, R., 1857 Über Erosionsformen der indischen Flüsse. *Ebend,* 3, 428-431.
Schlagintweit, R., 1866. Comparative hypsometrical and physical tableau of High Asia, the Andes and the Alps, *Journal of the Asiatic Society of Bengal,* 35, Part 2.
Schlagintweit, S., 1982. Die Brüder Schlagintweit - ein Abriss ihres Lebens. *in* **Müller, C.C. and Raunig, W. (eds.),** 1982. *Der Weg zum Dach der Welt.* Pinguin-Verlag, Innsbruck. 407p. , 11-13
Sella Foundation, 1987. *Vittorio Sella with the Italian Expedition to the Karakorum in 1909,* (Exhibition Catalogue, April 1987) Fiatagri, 64p.
Shipton, E.E., 1940. Karakoram 1939. *Geographical Journal,* 95, 409-427.
Shipton, E.E., Spender, M. and Auden, J.B., 1938. The Shaksgam expedition 1937. *Geographical Journal,* 91, 313-339.
Stein, M.A., 1899. *Memoir of Maps Illustrating the Ancient Geography of Cashmir,* Calcutta.
Stolicza, F., 1866. *Geological sections across the Himalayan Mountains, from Wangtu-bridge on the River Sutlej to Sungdo on the Indus, etc.* Memoir of the Geological Survey of India, 5, Calcutta.
Strachey, R., 1849. On the snowline of the Himalaya. Jouranl of the Royal Asiatic Society of Bengal, 29.
Strachey, R., 1853. A physical geography of Western Tibet. *Journal of the Royal Geographical Society,* 23, 1-69.
Suess, F.E., 1883-1904. *Das Antlitz der Erde,* 4 volumes, Vienna, The edition cited by Dainelli is **Suess, E.,** 1897-1918. *La face de la terre* (Translated by E. de Margerie): Paris, 3 volumes (1897, Vol. 1; 1900; Vol.2, 1902-1918, Vol. 3).
Tanner, H.C.B., 1891. Our present knowledge of the Himalayas. *Proceedings of the Royal Geographical Society,* N.S.,13, 403-423.
Tinkler, K.J., 1985. A Short History of Geomorphology, Barnes and Noble: New York, 334p.
Todd, H.J., 1930. Gilgit and Hunza River Floods (Correspondence). *Himalayan Journal,* 2, 173-175.

Trinkler, E., 1930. The Ice-Age on the Tibetan Plateau and in the adjacent regions. *Geographical Journal*, 75, 225-232.

Trinkler, E., 1931. Morphologische studien aus den Hochregion Zentralasiens, *Zeitschrift für Geomorphologie*, 6.

Uhlig, H., 1965. Die Giessemer Geographen Robert Scgalgintweit und Wilhelm Sievers. Festkolloquium 100 Jahre Geographie in Giessen, 94-103. *Giessener Geogr. Schriften*, 6, Wilhelm Schmitz Verlag, Giessen.

Vigne, G.T., 1842. *Travels in Kashmir, Iskardo and Ladak*. Colburn: London, 2 volumes.

Visser Ph.C., and Visser-Hooft, J., (*eds*.) 1935-38. *Karakoram: Wissenschaftliche Ergebnisse der Neiderlandischen Expeditionen..etc* 3 volumes, Leiden.

Wadia, D.N., 1930. Hazara-Kashmir syntaxis. *Records of the Geological Survey of India*, 63, 129-138.

Wadia, D.N., 1931. The Syntaxis of the North-West Himalaya: its rocks, tectonics and orogeny. *Records of the Geological Survey of India*, 65, 189-220.

Wadia, D.N., 1933. Notes on the geology of Nanga Parbat (Mt. Diamir) and the adjoining parts of Chilas, Gilgit District, Kashmir. *Records of the Geological Survey of India*, 66, 212-234.

Wadia, D.N., 1939. The structure of the Himalayas and the Northern Indian Foreland. Proceedings of the 25th Indian Science Congress. (Presidential Address; Geology Section), *Royal Asiatic Society of Bengal*, 25(2).

Wager, L.R., 1937. The Arun River drainage pattern and the rise of the Himalaya, *Geographical Journal*, 89, 239-249.

Walker, J.T., 1879. *Synopsis of the results of the operations of the great Trigonometrical Survey of India, Vol. VII*, Trigonometrical Survey of India, Dehra Dun.

Washburn, A.L., 1939. Karakoram glaciology, *American Journal of Science*, 235, 138-146.

Workman F.B. and W.H., 1910. The Hispar Glacier, *Geographical Journal*,.35, 105-125.

Workman F.B., 1899. Ascent of the Biafo and Hispar Glaciers: two pioneeer ascents in the Karakoram, *Scottish Geographical Magazine*, 25, 523-526.

Wyss, R., 1931. Vom zentralasiatischen Hochgebirge zwischen Vorindien und Ost-Turkestan. *Die Alpen*, 7, 281-301.

Zittel, K.A., 1901. *History of Geology and Palæontology, to the end of the nineteenth century*. Walter Scott: London.

10

Eustasy to plate tectonics: unifying ideas on the evolution of the major features of the earth's surface

R.P. Beckinsale and R.D. Beckinsale

Introduction

During the last 200 years geomorphology has emerged and developed into a respectable science. One of its greatest problems was the inheritance of a diversity of models, mainly catastrophic, put forward to explain the general evolution of continents and oceans. The search for a rational model of this evolution inevitably became entangled with the search for the cause of major landforms such as mountains, plains and plateaux. Such theorising on the origins of the chief features of the earth's surface has never lost its popular appeal.

Since about 1880 the most significant models concerning major global earth-features have been based successively on:

1. Changing sea-level (*eustasy*) combined with stable continents.
2. Vertical instability of landmasses (*diastrophism*).
3. Horizontal displacements of continents.
4. Ocean-floor spreading.
5. Plate tectonics.

The tenet of the whole theme has been gradual acceptance of the mobility, both vertical and horizontal, of continental blocks and ocean basins.

Changing sea-level and stable continents

The prime advocate of eustatism, or world-wide changes in sea-level, was Eduard Suess who was born in London in 1831 and died in Austria in 1914. Early on he produced a volume *Die Enstehung der Alpen* (1875), concerned largely with the origin and structure of mountain chains. In this he discussed the festooned pattern of mountain ranges and distinguished between orogenic forelands and hinterlands. In the zone between foreland and hinterland he considered that earth movements probably were tangential and unilateral.

In 1883, Suess began to publish his *Das Antlitz der Erde*, the first large-scale presentation of his geophysical ideas. Two years later he went twice to the vicinity of Tromsö where the terrace levels seemed to him so flat that they confirmed his earlier conviction that such terraces must have been caused by changes of sea-level rather than by uplift of the landmass. Therefore in the second volume of *Das Antlitz der Erde*, Suess proposed the principle of eustatic levels (II, 680 - 700). Ocean basins were formed by major collapses between stable continental blocks. Changes of world sea-level occurred because of deformation of the ocean floor. Sedimentation caused sea-level to rise. The geological evidence available appeared to indicate that marine transgressions were lengthy and were occasionally interrupted by short-lived regressions. Vertical elevation of landmasses was largely restricted to localized belts which were experiencing crustal compression. But there was no doubt that "the history of the continents results from that of the seas." (II, 700).

When completed (1883 - 1908), Suess' monumental work dominated geological thought for half a century, partly because Emmanuel de Margerie translated it into French (3 volumes 1897 - 1918: *La face de la terre*) and richly annotated it. Its opinions were supported by the findings of the *Challenger* deep-sea expedition of 1873-6, which drew attention to land-derived deposits on the ocean floor and to the existence of great deeps. Suessian notions were bolstered also by important geographers and geologists such as Albrecht Penck (particularly in *Morphologie der Erdoberfläche*, 1894) and Emile Haug who in 1900 and 1907 suggested that oceanic trenches ("foredeeps") were geosynclinal downbuckles associated with mountain-building, especially in mobile belts marginal to continental shields.

Not surprisingly the eustatic theory seemed to be strongly supported also by the abundance of marine terraces which today are known to be the direct result of ocean-level changes associated with ice-sheet advances and retreats during the Pleistocene. This glacio-eustatism helped to create a strong following for Suess' concepts and in the 1930s drops of sea-level of 500m and even of 1000m were being postulated.

The meetings of the International Geographical Union illustrate vividly the dilemma of trying to distinguish between old and new terraces. In April, 1925 a commission was set up for the study of Pliocene and Pleistocene terraces. By 1952 it had issued 7 Reports, including Henri Baulig's *Problèmes des terrasses* (1948). But in the meanwhile, in 1931, another commission was established to study the cartographical representation of surfaces of Tertiary flattenings. This subject attracted many discussants, including Baulig who spoke on the Central Massif of France on which he had just published a book dedicated to Eduard Suess and Paul Vidal de La Blache (Baulig 1928). At the same meeting there was a discussion about mid-Tertiary erosional flattenings in southeastern England (Linton 1931). Shortly afterwards Baulig lectured in London and strongly influenced many leading British geomorphologists (Baulig 1935). Nearly twenty years later S. W. Wooldridge wrote of "a great unifying generalization in geomorphology."

> The latter-day record of the continents seems to have been one of 'uplift' starting long before Pleistocene time — or, more probably, in view of the uniformity of the record, of successive negative eustatic shifts of sea-level (Wooldridge 1951, 173).

However, by this time field investigations were beginning to throw more doubt on the adequacy of any theory based on eustatism and continental stability. In 1949, the two I.G.U. commissions on terraces and Tertiary flattenings were combined and in August 1952 the theme ceased to function. Today it seems surprising that the simplistic eustatic theory—as distinct from its glacial aspects—should have survived so tenaciously. In 1968 Jean Tricart could safely write, "Si l'eustatisme proprement dit est à reléguer au rang des théories inutiles, il n'en pas de même du glacio-eustatisme."

Vertical instability of landmasses

The Suessian idea of vertical landmass changes being restricted to narrow belts is less complementary to the concept of landmass vertical instability than appears at first sight. Both Suess (1875) and Albert Heim (1878) had suggested that mountain ranges were produced by tangential compression directed away from ocean basins and so involving great thicknesses of marginal marine deposits. Alpine orogenic movements were elaborated further, under Suess' inspiration, by Marcel-Alexandre Bertrand (1884, 1897) who developed the theory of *grandes nappes* (overthrusts) and demonstrated that Europe had probably been built up from north to south by successive periods of intense orogeny, called Caledonian, Hercynian and Alpine. The overthrust theme was carried a stage further by Maurice Lugeon (1901) who described the general structure of the whole Alpine chain and presented, probably for the first time, a synthesis of its complicated series of successive recumbent folded thrust-sheets (*nappes de charriage* or *Decken*). This method was continued by Émile Argand (1916) who was well versed in Suess' writings. A gifted artist with a three dimensional outlook, he introduced the idea of kinematic analysis and synthesis (embryotectonics) in which the structure was depicted right back to the original sedimentary terrain.

The complicated tectonic themes put forward by Argand and other European geologists did not necessarily refute Suess' concept of eustatism imposed upon stable continental blocks. Ideas of vertical landmass stability were weakened much more seriously by increasing knowledge of isostasy, of earth heat, and of the visible relationship between surface slopes and land uplift. All these three topics underwent either a revolution or a popularization between 1880 and 1910 during the time when Suess' influence was climbing towards its zenith.

Firstly, in 1889 C. E. Dutton popularized for geographers and geologists the concept of isostasy which had originated with the work of J.H. Pratt and G.B. Airy in 1855. His theory was adapted by other geomorphologists to explain vertical crustal movements, (e.g. Gilbert 1890) and ever since it has been used with increasing regularity and subtlety as a component of crustal vertical motion and often, indirectly, of lateral movements also.

Secondly, there was the topic of earth heat. Because isostasy was based on measurements of earth gravity or density, it inevitably became associated with ideas on the nature of the earth's interior. For example, in 1897 Emil Wiechert furthered the idea that the existence of an iron core (5000 km radius) covered by a rock mantle (1400 km thick) would explain earth-density characteristics. This suggestion was clarified in 1906 by Richard D. Oldham from the evidence of earthquake waves. Four years later, Andrija Mohorovíčič also from seismic wave velocities, deduced the presence of a distinct outer layer in the earth. Today this boundary between the "crust" and "the mantle" is called the Moho. At this boundary, the velocity of Primary seismic waves accelerates almost regularly from 6 to 8 km/sec and the Secondary waves from 4 to 5 km/sec. Thereafter geophysical progress was such that by 1914 Beno Gutenberg had outlined the earth's density layering (outside the inner core) in terms that are recognizable today. More recent improved methods of measuring terrestrial gravity have further enhanced the idea of a vertically mobile crust in which the Moho discontinuity lies at an average depth of about 32 km beneath the continents and nearer 8 km beneath the oceans.

Ideas on "floating continents" fitted in well with the concept of a more molten or more plastic layer at the top of the "mantle." This "floating" or buoyant suggestion was encouraged by abundant observations of increasing heat with increasing depth below the land surface. We may quote, for example, T. Mellard Reade's *The Origin of Mountain Ranges* (1886).

> Accepting as an inference of high probability that the increase of heat ... progresses downward until the temperature of molten matter—or matter that would be molten if freed from the super-incumbent pressure—is reached, so enormous a reservoir of heat lies below, that it would seem puerile to doubt that the extremely thin cooled crust must respond to internal changes of temperature that in all probability take place in the enormous mass of heated interior rock (p. 3)... So many and so great have been the difficulties connected with the contractional hypothesis [of the Earth], that several eminent geologists and physicists have assumed a layer of molten matter lies between the solid outer shell and a solid inner spheroid (p. 6).

The author goes on to notice that Professor Prestwich favours a collapsing solid shell floating on an intercalated zone of molten rock while Osmond Fisher in his *Physics of the Earth's Crust* (1873, 146-7) supports his mountain ranges by flotation in a similar zone of heavier matter.

The suggestions of downward-increasing earth heat and of a possible plastic layer were greatly strengthened when Henri Becquerel in 1896 inferred that luminescence was accompanied by X-ray emission. Soon this emission was recognized as a production of heat or radioactive disintegration. As early as 1904 Dutton in his last major geological work, *Earthquakes in the Light of the New Seismology*, linked volcanicity to radioactivity. Five years later, John Joly drew full attention to radioactivity as a source of terrestrial heat, with a reasonably constant rate of decomposition. His *Radioactivity and Geology* was followed in 1925 by *The Surface History of the Earth* which cemented his great geological influence.

Thirdly there is the matter of the correlation between surface slopes and land uplift. In the late nineteenth century an increasing number of physical geographers and geologists began to associate the shape of landforms and of landscapes with changes in land level (diastrophism) rather than with world-wide changes of sea-level (eustatism). W. M. Davis published the first complete picture of his "cycle of erosion" in 1889 and with other American geologists established a "denudation chronology" in which base-level changes were best ascribed to regional diastrophy (Chamberlin 1909). Davis told of "peneplains" that obviously had been tilted during their uplift and of "complications of the geographical cycle" based mainly on vertical movements of the landmass. He deliberately oversimplified his popular model with sudden uplift but he believed also in slow or gradual or any other rate of uplift. His compatriot, Douglas Johnson, based a denudation chronology of the Appalachians on up-arching and discontinuous uplift (1931).

In Europe, Emmanuel de Martonne strongly followed Davisian methods when studying the Transylvanian Alps (1907) and much later, in 1929, keenly attacked Henri Baulig's excessive use of eustatism. Some Europeans considered that a wide variety of uplift patterns was reflected in the gradients of existing slope forms. For example, Albrecht Penck envisaged a wide range of uplift-slope patterns while his son Walther devised a grand slope-development generalisation and provided details of great folds and regional arching (W. Penck 1920, 1924, 1925). However, the point we wish to make is that numerous denudation studies assumed changes in the level of landmasses.

Diastrophic ideas were also helped by findings on the nature of sediments. Thus Joseph Barrell in 1914 showed that many sedimentary strata were not marine. He thought that geological processes were halting rather than continuous and that the earth's crust consisted of a stronger lithosphere of varying density above a zone of flowage (Barrell,1914, 1915, 1917).

One other broad aspect of diastrophism must be mentioned as it still plagues geomorphologists. In 1890 G.K. Gilbert in his monograph *Lake Bonneville* (p. 340) described broad "swelling" movements of the land platforms which, he thought, created the continents. These he named *epeirogenic* to distinguish them from narrower orogenic or mountain-building movements. The concept at first attracted few allies although by the 1920s some geologists, such as L.W. Stephenson (1928), were ascribing major marine transgressions and regressions to differential warping of the continents. However, the general epeirogenic concept was broadened by Hans Stille (1936, 1955) in his syntheses of global tectonics. He envisaged a crust separated into mobile belts, with marginal oceanic low cratons and continental high cratons with some subsiding regions (parasynclines). Ocean basins were foundered continental "high cratons," except the Pacific which was permanent. The stability of the craton was linked to that of the geosyncline and marginal orogenies which could be correlated with major marine regressions.

At about the same time O. Jessen was discussing epeirogenic themes in his *Die Randschwellen der Kontinente* (1943). Nearly twenty years later Lester C. King in his *Morphology of the Earth* (1962) found it necessary to introduce a concept called "cymatogeny" which he considered was distinct from epeirogeny. In cymatogenic activity on a broad scale

> the crust accommodates itself to vertical movements which are regional ... at depth by flow in the solid state, expressed in steeply-dipping gneisses... attributed to... profound, and broad, vertical laminar movement of the shields, with recrystallisation.. (p. 65).

Such vertical deformations create in the landscape broad flexures or gigantic undulations (cymatogeny) in the form of arches of uplift and basins of depression.

These broad concepts concerning vertical motions on shields have at times been supplemented or supported by ideas on large-scale pediplanation and concomitant isostatic uplift. There appears to be a continuing demand for some form of epeirogeny or cymatogeny associated perhaps with rock re-crystallisation at depth.

Horizontal displacements of the continents

The lateral migration of continents is an idea held in a crude form in Europe since the early seventeenth century and based largely on the apparent outline-fit of opposite sides of the Atlantic. Its notable modern exponents include Antonio Snider-Pelligrini who in his *La Création et ses mystères dévoilés* (1858) used as evidence the coastal fit of the New and Old Worlds which had, he thought, been fractured and severed catastrophically. In 1873, Osmond Fisher suggested that if, as had been postulated, the moon had been separated from the Earth, one likely result would be lateral displacement and fragmentation of the cooled granitic crust. With extraordinary insight, he also suggested that the relatively fluid interior of the Earth experienced convection currents which rose beneath the oceans and descended beneath the continents.

However, probably the first coherent hypothesis of continental displacement was put forward in 1910 by an American, Frank Bursley Taylor, who based his scheme on the patterns of Tertiary mountain chains as outlined by Suess. But whereas Suess assumed that the northern hemisphere pattern of chains was attributable to oceanic subsidence and tangential thrusts southward from northern horsts, caused by contractional cooling of the Earth, Taylor replaced global contraction by some nondescript force that generated "a mighty creeping movement" also equatorward from the poles. His regional details included the notion that the separation of Africa and South America began along the line of a mid-oceanic ridge.

At the same time Alfred Wegener (1880-1930) was developing independently his views on the lateral displacement of continents. His first published account (1912) was expanded within three years into *Die Enstehung der Kontinente und Ozeane* which was revised three times before its author disappeared on the Greenland ice-cap in 1930. One posthumous revision and translations into a least five main languages also appeared.

Wegener's general theme involved the splitting up of a super-landmass, Pangæa, into continental blocks which floated in an isostatic state upon a pliable medium. Mountains were created where the advancing front of a moving continent encountered the resistance of the ocean floor. The motive force was attributed tentatively to tidal friction for westward movement and to "Pohlflucht" or flight from the flattened poles under gravity impulses, for continental movement toward the equator. The author admitted that lateral motion could also be generated by any other deviations in the Earth's spheroidal shape but he preferred to emphasize polar wandering and the dilemma of whether it involved displacement of the whole crust upon its substratum or the internal shift of the axis of rotation. He believed that the Earth's equatorial bulge could move.

The thesis involved a large amount of geophysical, geological, palæoclimatic and bio-geographical data, especially in dealing with the possible former connection of opposite sides of the Atlantic. The steadily increasing evidence that material under the ocean floors was denser than that under continental blocks was shown to militate against the popular belief until recent geological times, in the existence of transoceanic land-bridges. Continental displacement, or "drift" as it came to be known in English, was suggested as a rational alternative to foundered land-bridges.

These various arguments aroused much scepticism and generally were not accepted at the time. Among the more potent criticisms were the crudity of some of the continental outline fits, the probability of inter-continental land connections, the lack of geodetic measurement, and the scientific inadequacy of the suggested mechanisms. Many famous geophysicists, such as Harold Jeffreys of Cambridge, disliked the idea of polar wandering and of continental displacement on a rather rigid Earth.

However, Wegener attracted some notable followers. For example, at Harvard University, a Canadian, R.A. Daly (1871-1957) accepted the concept of drift but was sceptical of the proposed mechanism. In Europe, tectonists such as É. Argand and R. Staub, who associated the lateral movements of continents with mountain-building, were willing adherents. Some leading Dutch geologists, such as Vening Meinesz (1940) and Kuenen (1950, 175-208), found parts of the concept helpful in interpreting East Indian tectonics. But undoubtedly Wegener's two most influential supporters were Arthur Holmes (1890-1965) in Britain and Alexander L. Du Toit (1879-1948) in South Africa.

In 1929 Holmes in an account of *Radioactivity and earth movements*, proposed a model of the Earth's outer layers, in which the uppermost layer consisted mainly of granites, the intermediate layer mainly of diorites, and the lower layer mainly of peridotites beneath the oceans (with some eclogite beneath the continents). The Earth's crust consisted of the upper and intermediate layers together with the top of the lower layer. The basal part of the lower layers was "glassy" or "thermally fluid". Holmes went on to devise a system of convection currents in this fluid substratum (or mantle) beneath the crust. These currents, he considered, would tend to rise under the continents, where radioactivity was greatest, and sink towards the edges of the continents. They caused the lateral displacement of the continental blocks. This hypothesis received far wider advertisement when incorporated into a popular *Principles of Physical Geology* (1944) which was re-issued in much enlarged editions in 1964 and, revised by others, in 1978.

Wegener's other great supporter, Du Toit, studied and lectured for some time in Glasgow and London before returning to field geology in South Africa. He was the first to suggest that Pangæa (Wegener's world landmass) had been subdivided at least since late Palæozoic times into a southern supercontinent (Gondwanaland) and a northern supercontinent (Laurasia). The division was occupied by the "Tethys Sea." In his *Our Wandering Continents* (1937) Du Toit reconstructed Gondwanaland in greater and more accurate detail than hitherto, especially with regard to probable connections between Africa and South America. The book was a trumpet blast played with invigorating virtuosity. It eliminated some of Wegener's errors and made an important positive contribution to the hypothesis of continental lateral shift.

In retrospect the support for Wegener outlined above is seen to be inadequate to upset the general balance of contemporary geophysical beliefs. The judgement by Ph. H. Kuenen in 1950 (p. 129) seems impartial and sound:

> If the conclusions of adherents to drift in some form or other, such as Du Toit, Wegmann, Gutenberg, and Kirsch, are confronted with the opinion of opponents, for instance Bucher, Umbgrove, Stille, and Cloos, it becomes obvious that neither of the two camps can claim a decisive victory. But the evidence favorable to drift often proves illusive, or at least open to serious doubt, on closer inspection. For the time being most geologists appear to have lost faith in continental drift as a sound working hypothesis.

From about the mid-twentieth century, the bias against the theory of continental displacement began to be steadily modified because of new knowledge of ocean floors and the application of improved palæomagnetic techniques and radioactive age-measurements in geophysical problems. The permanent remanent magnetisation, especially of basaltic lavas, provided a key to the contemporary latitude of the ambient rock and so to apparent changes in the geographical position of the rock and of the poles. Polar wandering is, for geophysicists, quite an old concept and in 1906 W. M. Davis considered that further evidence might force "this daring, gratuitous, and discredited hypothesis to be taken seriously into account." Modern palæomagnetic research, in spite of difficulties in the interpretation of results, has "stimulated great interest, and if it failed immediately to make large numbers of converts to continental drift, at least many Earth scientists became tolerant towards the hypothesis" (Hallam 1973 p. 41).

The concept of ocean-floor spreading

Between 1946 and 1966, increasing knowledge of the oceans gave rise to the idea that ocean floors were expanding contemporaneously with the splitting up of the two supercontinents. Many new exploration and measurement techniques and long oceanic traverses by the *Vema* (1954-60), *Glomar Challenger* (1968 onward) and many other vessels transformed the detailed picture of ocean floors. The major discoveries and revisions or confirmations included at least five facets.

(a) Numerous seamounts (guyots) were found at depths of between 1000m to 1700m. These were for the most part volcanic in origin and flat-topped. They seemed to indicate a widespread lowering of the ocean floor (Hess 1946, 772-91).

(b) A strong negative gravity was associated with the trenches of island arcs. This had been discovered in the early 1930s by F.A. Vening Meinesz, who attributed it to downward buckling of the crust which formed a deep wedge of less dense sial in the heavier substratum of sima. His interpretation was generally confirmed (Meinesz 1940, Kuenen 1950).

(c) The presence of a mid-oceanic ridge in the Atlantic (recognized in the 1870s) was confirmed, defined and now linked up with similar features throughout the worlds' oceans. The "ridge" was not always central and locally widened into a "rise" but it could be traced for a total distance of 60 000 km into all the main oceans.

(d) This mid-oceanic ridge was found to develop, in parts, topographical features that closely resembled major rift valleys on continents. As early as 1937 it had been recognised that submarine ridges in the Arabian Sea consisted of a trough between two parallel ridges, a pattern similar to that of the East African rift valley (Wiseman and Sewell 219). But the authors considered that the submarine rifts had been formed on land previous to the submergence beneath the ocean. In 1953 a steep-sided median valley was found in the mid-Atlantic oceanic ridge and later this rift-like feature was traced, with breaks, into the Gulf of Aden and so to the East African rift system. This median valley became increasingly recognised as a prime seat of seismic and volcanic activities.

(e) The physico-chemical nature of the basement beneath the sediment cover of the ocean floor was better defined. Except for small areas, such as the Seychelles Bank, the ocean basement seemed to consist of a basaltic layer, averaging 1.0 to 1.5 km thick, and a high density layer of ultrabasic material averaging about 5 km thick. At about 100 to 200 km below the surface is a world-wide, low velocity layer, the existence of which had been predicted much earlier by B. Gutenberg. This general layering differs slightly below mid-oceanic ridges where a relatively narrow zone commonly has seismic velocities of 7 to 8 km/sec—as opposed to over 8 km/sec elsewhere—and tends to show high heat-flow. It is interesting that Vening Meinesz considered that the general positive gravity over the oceans and deficiency over the continents could only be satisfactorily explained by an hypothesis of convective currents operating in the substratum.

The various findings and ideas outlined above gradually encouraged the concept of ocean-floor spreading. The idea was circulated in typescript by Harry Hess in 1960 and expanded in print in 1962. During the interval, R. S. Dietz proposed the term "spreading of the sea floor" in describing the same process. Each suggested that mid-oceanic ridges are situated above convective currents rising from the mantle. The currents move outward from the ridge and carry the continents with them like parcels on a conveyer belt. At the margins of an ocean the currents sink, their descent being marked by marginal trenches with strongly negative gravity anomalies. The ocean floor remains young geologically because it is constantly being renewed at the mid-oceanic ridges and re-absorbed into the mantle at the oceanic marginal trenches.

Readers will recall that Holmes had envisaged a somewhat similar mechanism thirty years earlier but his ideas were cruder and in some instances wrong as when, for example, he attributed the East African rift system to compression rather than tension. Hess and Dietz faced a much more receptive audience and, in addition, were supported immediately by a mass of new scientific details.

Recordings of palæomagnetism appeared to indicate frequent short-term reversals in the general direction of the magnetic field. These reversals were represented by peculiar striped patterns of magnetic anomalies near an oceanic ridge. The patterns showed a succession of changes in polarity that could be rationally attributed to repetitive splitting and outward movement, from a central ridge, of a succession of lava outflows. In places the succession of polarity changes could be matched on each side of a medial ridge. This lateral succession allowed a chronology, by potassium-argon (^{40}K: ^{40}A) dating, to be obtained for the lavas and for changes in polarity. It was reasonably certain that the rocks became older away from the axis of the medial ridge. By 1966 the sea-floor spreading hypothesis had gained respectability and with its emergence the concept of continental drift was received with less skepticism and in some academies with new and growing enthusiasm. Within two years, a time-scale had been drawn up for magnetic events in the Tertiary and, assuming constant lateral expansion, the spreading rate of different ocean floors was assessed at 1 cm up to 10 cm a year (Heirtzler *et al.* 1968).

The year 1968 also saw the start of the extensive oceanic traverses of the *Glomar Challenger* with its highly efficient drilling equipment. Analysis of numerous ocean-floor cores confirmed that the age and thickness of the sedimentary cover increases outward from mid-oceanic ridges; that the oldest sediments overlying the oceanic basaltic basement are Middle Jurassic; and that the greater part of the ocean floor covering is post-Cretaceous or Cainozoic.

Seismologists also continued to contribute significantly to the status of the sea-floor spreading hypothesis. The foci of earthquakes seem to indicate that major faults, or some form of deep-seated collapse or phase-change, occur when lithosphere is thrust downward in marginal trenches. Here is the source of deep-seated earthquakes (down to 700 km) whereas other parts of the ocean floor are characterized by shallow-seated earthquakes caused by shallow fractures.

The concept of plate tectonics

The hypotheses of sea-floor spreading and continental drift evolved directly into the theory of plate tectonics. The merger was helped by fresh ideas on earth-fractures. It had become widely accepted that some great faults, such as the Great Glen in Scotland and San Andreas fault in California, had experienced a dominant and considerable horizontal movement. For example, the horizontal motion at the so-called transcurrent fault of the Great Glen exceeded 100 km, (Kennedy 1946).

In the mid-1960s ocean-floor surveys began to reveal that segments of most oceanic ridges were offset laterally by what appeared to be long, linear fracture belts. These lateral features, which regularly interrupted the general continuity of a mid-ocean ridge, had rift-like characteristics such as relatively narrow valleys flanked by tall, steep scarps. In the northeast Pacific some of the east-west faults extended about 3700 km from near the west coast of California to the Hawaiian Ridge. It seemed highly probable that these rifts were associated with the outward motion of an expanding sea-floor.

These submarine details, together with observations on the global pattern of volcanos and earthquakes to which they often contributed, drew ever increasing attention to mid-oceanic ridges and certain ocean margins. It was soon realised that the Earth's crust consisted of narrow active belts which outlined blocks or large spaces where volcanic and seismic activity was largely or entirely absent. This concept of "plate tectonics" may be said to have flourished after 1965 when J. Tuzo Wilson stressed that crustal movements or activities tend to be concentrated in belts characterized by either mountain ranges, including island arcs, or mid-oceanic ridges, or major faults with a long horizontal displacement. These mobile belts interconnect to form a continuous network which divides the Earth's crust into large rigid "plates." Although the term plate had been in use since at least 1904 when H. B. C. Sollas used it in his translation of Suess[1] (I, 600), the concept of plate tectonics had now emerged.

Wilson went on to try to explain why linear features of the mobile belts so often were *transformed*, or changed, at their extremities and junctions. Faults that ended abruptly or changed direction sharply, he called "transform faults." These differed from transcurrent faults in that their crustal blocks moved relatively in the same direction. For example, the San Andreas fault, hitherto regarded as transcurrent, may be assumed to be a transform fault linking up the axes of the East Pacific rise from near the Gulf of California to the Juan de Fuca oceanic "ridge" southwest of Vancouver Island. Along this fault, the Pacific plate has a northward motion. Suggestions such as these began to allay the qualms of some geophysicists, and the theoretical side of plate tectonics developed rapidly in the late 1960s.

The concept was applied to a sphere with a crustal mosaic of 20 rigid blocks divided from, or joined to, each other by either oceanic rises (forming new crust) or marginal trenches (destroying crust) or transform faults where crust was neither made nor destroyed. Later, the postulated mosaic was modified to 6 major plates and several minor plates, and their relative direction and rate of movement or expansion was defined. Rejecting the idea of a rapid increase in the Earth's radius during the last 200 million years, it has been shown that the rotation of a plate about its own axis could explain much plate movement and the shape of some existing oceans and bays. For example rotatory action might have opened up the Bay of Biscay, a notion envisaged decades earlier by Argand.

In addition, attempts were made to improve upon existing sub-crustal convective models, particularly by the involvement of phase-change, or re-crystallisation to higher density minerals in the upper mantle. At the same time theories on the evolution of mountain chains turned increasingly to subduction and the collision of plate margins. Tuzo Wilson considered the motion, relative to the deeper mantle, of two converging plates. Where an advancing continental plate or margin overrides an oceanic plate that is stationary relative to the mantle, the subduction zone comprises coastal mountains (as in Chile) with a marginal trench pushed ahead of the continent. Where an advancing oceanic plate passes beneath a relatively stationary continental margin, island arcs and deep-sea trenches form offshore, as in eastern Asia.

Today the evolution of mountains and ocean-floor trenches is commonly expressed by reference to four models, bearing in mind that plates can be oceanic, or continental, or oceanic with continental blocks.

---

[1] *Editor's note:* Russell (1936) used the word *plate* in his discussion of crustal structure in the Mississippi Delta without any sense of special pleading.

(a) At the convergence of two oceanic plate margins, the leading edge of one plate is forced down into the asthenosphere, so creating a deep trench, and a volcanic-dotted island arc above the subduction zone.
(b) At the collision of an aggressive oceanic plate and a relatively stationary continental-block margin, the oceanic plate pushes itself beneath the continental block which remains relatively little disturbed and island-arcs and trenches are formed offshore.
(c) At the collision of a more aggressive continental block margin with a less aggressive oceanic plate, the latter is subducted beneath the continental margin which is buckled into a lofty cordillera topped by volcanos.
(d) At the collision of two continental plates, as in the Himalaya, the upthrust creates a lofty cordillera markedly lacking in volcanoes.

To these alternatives must be added the splitting up of continental blocks as in eastern Africa where the mobile belt is still part of a landmass.

The theory of plate tectonics has continued to improve in scientific quality. It has stimulated most branches of physical geography, and probably a large majority of earth scientists would agree with A. Hallam that it

> has proved highly successful in integrating diverse geological phenomena and contributed towards a more coherent and intelligible picture of the Earth's evolution....It must rank as the biggest advance in Earth Science since...the paradigms of uniformitarianism and stratigraphical correlation of fossils established geology as a true science (1973, p. 114).

Geomorphologists and plate tectonics

The concept of plate tectonics has absorbed and enlarged the implications of continental drift for geomorphologists for whom in spite of its germinal state, it offers many attractions.

It has a truly global application. But geographers and geomorphologists must ensure that it is represented only on equal-area projections. Existing world maps of plates that are not on an equal area grid showing, for example, an Antarctic plate covering nearly 25 per cent of the earth's surface, are ridiculous and harmful to the plate concept.

It draws attention mainly to geographical conditions during the last 200 million years and emphasizes the changes in landmass and ocean positions during that time. Such an approach stimulates an appreciation of changing environments and the reasons for them, and creates an awareness that many landsurfaces by virtue of plate motions over the earth's surface may be only superficially related to physical conditions on their present locations.

It incorporates radioactivity and so draws attention to geothermal heat and its possible exploitation for human uses, including mineral-content expectation. It is concerned with the nature of ocean-floor topography and sea-floor spreading both of which are highly relevant to geomorphological studies and simultaneously, it emphasises the nature of mobile belts and their association with earthquakes, volcanoes and fractures.

It stresses the complexity of mountain systems. For example, it shows clearly that long mobile belts become active at widely different places at widely different times; that in some areas downthrusts are more important than upthrusts; and that most ancient surfaces (pre-Triassic) are excessively complicated, being apparently partly the product of earlier ocean-spreading and ocean-closing. The possibility of assessing the rates of some of these mechanisms, oceanic and orogenic, and of recording the time and place of the occurrence of violent earth-phenomena will eventually acquire some value for prediction. Geophysical science is like geomorphology, still in its infancy.

Inevitably geologists will find some of the above concepts grossly over-simplified. They are faced with many further problems, concerning, for example, accreted terranes; the role of oceanic plateaux; the possibility that a continental margin, such as westernmost North America, may change from being passive to being active; that orogenesis may not always occur when an ocean oceanic plate is subducted beneath a continental plate; and that the effect of plate collisions may extend across vast distances upon the surface of a continental plate.

Geomorphologists, however, will be grateful for the unifying approach of plate tectonics. Some may recall the work of Walther Penck who, for more than a decade before his premature death in 1925, tried to correlate crustal mobility with surface topography. His ideas on uprising domes and on great folds bring to notice many areas—such as Basin and Range regions and mountain ranges (Pamir; Tien Shan etc.)—that still fascinate modern tectonists (Chorley and Beckinsale 1989). Perhaps we should turn anew to crustal mobility and assume the universal presence of sub-crustal heat, whose upward escape is controlled partly at least by the insulating strength of the superincumbent rock.

References

Airy, G.B., 1855. On the computation of the effect of the attraction of the mountain masses disturbing the apparent astronomical latitude of stations in geodetic surveys; *Philosophical Transactions of the Royal Society of London*, 145, 101-104.

Argand, É., 1916. Sur l'arc des Alpes occidentales. *Eclogae Géologique Helvetiæ*, XIV, 145-191.

Bailey, E.B., 1935. *Tectonic Essays : Mainly Alpine.* Oxford.

Barrell, J., 1914. The Strength of the Earth's Crust. *Journal of Geology*, 22, 425-33.

Barrell, J., 1915. The Strength of the Earth's Crust. *Journal of Geology*, 23, 27-44; 425-44.

Barrell, J., 1917. Rhythms and the measurement of geologic time. *Geological Society of America Bulletin*, 28, 745-904.

Barrell, J., 1919. The Status of the Theory of Isostasy. *American Journal of Science*, 4th series, 48 291-338.

Baulig, H., 1928. Le Plateau Central de la France et sa bordure Méditerranienne : *Étude morphologique;* Paris.

Baulig, H., 1928. Les hauts niveaux d'érosion eustatique dans le Bassin de Paris; *Annales de Géographie*, 37, 385-406.

Baulig, H., 1935. *The changing sea level.* Institute of British Geographers, Publication No. 3.

Baulig, H., 1948. *Problèmes des Terrasses*, Paris.

Beckinsale, R.P., 1972. The I.G.U. and the development of geomorphology. In *Geography Through a Century of International Congresses*, I.G.U., 114-130.

Becquerel, H., 1896. Sur les radiations émises par phosphorescence.., *Comptes Rendus*, 122, 420, 501, 689.

Ben-Avraham, Z. *et al.*, 1981. Continental accretion : from oceanic plateaus to allocthonous terranes. *Science*, 213, 47-54.

Bertrand, M., 1884. Rapports de structure des Alpes de Glaris et du bassin houiller du Nord. *Géologie Société de France, Bulletin*, 3d ser., 12, 318-30.

Bertrand, M., 1897. Structures des Alpes françaises et récurrence de certains facies sédimentaires: *16th International Geological Congress.* (1894), Zürich, Rept., 163-77.

Boillot, G., 1978 (trans. 1981). *Geology of the Continental Margins.* London: Longman.

Chamberlin, T.C., 1909. Diastrophism as the ultimate basis of correlation. *Journal of Geology*, 17, 685-93.

Chorley, R.J., 1963. Diastrophic background to twentieth-century geomorphological thought. *Geological Society of America Bulletin*, 74, 953-70.

Chorley, R.J., & Beckinsale, R.P., 1980. G.K. Gilbert's Geomorphology, in Yochelson, E.L. (ed.) 1980, (see below) 129-42.

Chorley, R.J., and Beckinsale, R.P., 1989. (in press) *The History of the Study of Landforms.* 3. The Suess - Davis - Penck Era. London: Methuen.

Chorley, R.J., Beckinsale, R.P., & Dunn, A.J., 1973. *The History of the Study of Landforms*, 2. The Life and Work of W.M. Davis. London: Methuen .

Chorley, R.J., Dunn, A.J., & Beckinsale, R.P., 1964. *The History of the Study of Landforms*, 1. Geomorphology before Davis. London: Methuen.

Cox, A., & Hart, R.B., 1986. *Plate Tectonics & How it Works*, Oxford: Blackwell.

Daly, R.A., 1926. *Our Mobile Earth.* New York.

Daly, R.A., 1940. *Strength and Structure of the Earth.* New York.

Davies, G.L., 1969. *The Earth in Decay.* London: MacMillan.

Davis, W.M., 1889. The rivers and valleys of Pennsylvania: *National Geographical Magazine*, 1, 183-253.

Davies, W.M., 1906. Observations in South Africa. *Geological Society of America Bulletin*, 17, 377-450.

Dewey, J.F., 1973. Plate tectonics and the evolution of the Alpine system. *Geology of Society America Bulletin*, 84, 3137-80.

Dewey, J.F. & Bird, J.M., 1970. Mountain belts and the new global tectonics. *Journal of Geophysical Research*, 75, 2625-47.

Dictionary of Scientific Biography, **G. C. Gillispie, (ed.)** 1970-80; 16 New York: Scribner's.

Dietz, R.S., 1961. Continent and ocean basin evolution by spreading of the sea-floor. *Nature*, 190, 854-57.

Dietz, R.S., 1964. Collapsing Continental Rises. *Journal of Geology*, 71, 314-33.

Du Toit, A.L., 1937. *Our Wandering Continents.* Edinburgh: Oliver and Boyd.

Dutton, C.E., 1889. On Some to the Greater Problems of Physical Geology. *Philosophical Society of Washington Bulletin*, No. 11, 51-64.

Fisher, O., 1873. *Physics of the Earth's Crust*, London, (2nd. edition 1889).

Gilbert, G.K., 1890. Lake Bonneville, *United States Geological Survey, Monograph* 1.

Green, H., 1977. The evolution of the Earth's crust In Heacock, J.G. (ed.) *The Earth's Crust.* American Geophysical. Union. Monograph 20.

Griggs, D., 1939. A Theory of Mountain Building. *American Journal of Science*, 237, 611-65.

Gutenberg, B., 1914. Über Erdbebenwellen VIIA. Beboachtungen an Registrierungen von Fernbeben in Göttingen and Folgerungen über die Konstitution des Erdkörpers: *Nachrichten der Akademie der Wissenchaften in Göttingen. Math e-physikal.* Klasse, 166-218.

Hallam, A., 1973. *A Revolution in the Earth Sciences: From Continental Drift to Plate Tectonics.* Oxford: Clarendon Press.

Haug, E., 1900. Les geosynclinaux et les aires continentales: contribution à l'étude des transgressions et des regressions marines. *Géologie Société de France, Bulletin*, 3rd ser., 28, 617-711.

Haug, E., 1907. *Traité de géologie; Les phénomènes géologiques.* 1. Paris.

Heim, A., 1879. *Untersuchungen über den mechanismus der gebirgsbildung.* Basel, Benno Schwabe, 2 volumes.

Heim, A., & de Margerie, E., 1888. *Les dislocations de l'écorce terrestre.* Zürich.

Heirtzler, J.R., *et al.*, 1968. Marine magnetic anomalies. *Journal of Geophysical Research*, 73, 2119-36.

Hess, H.H., 1946. Drowned ancient islands of the Pacific Basin. *American Journal Science*, 244, 772-91.

Hess, H.H., 1962. History of Ocean Basins. In *Petrologic Studies* (ed. A.E. Engel *et al.*) Geological Society of America, 599-620.

Holmes, A., 1931. Radioactivity and Earth Movements, *Transactions Geological Society of Glasgow*, 18, 559-606.

Holmes, A., 1944. *Principles of Physical Geology* London: Nelson (and later editions in 1964 and 1978).

Hurley, P.M., 1968. The confirmation of continental drift. *Scientific American*, 208, 53-64.

International Geographical Union, 1928-1952, *Reports of the Commission on Pliocene and Pleistocene Terraces*: no. 1, 1928; no. 2, 1930; no. 3, 1931; no. 4. 1934; no. 5, 1938; no. 6 1948; no. 7, 1952.

International Geographical Union, 1956. *Premier rapport de la Commission pour l'étude et la corrélation des niveaux d'érosion et des surfaces d'aplanissement autour de l'Atlantique*: 9th General Assembly, Rio de Janeiro, 4 Volumes.

International Geographical Union, 1972. *Geography Through a Century of International Congresses*

Jeffreys, H., 1923. *The Earth, its Origin, History, and Physical Constitution.* Cambridge (and later editions).

Jessen, O., 1936. *Reisen and Forschungee in Angola.* Berlin.

Jessen, O., 1943. *Die Randschwellen der Kontinente.* Gotha.

Johnson, D.W., 1931. *Stream sculpture on the Atlantic slope.* New York,.

Johnson, D.W., 1932. Principles of marine correlation. *Geographical Review*, 22, 294-98.

Joly, J., 1909. *Radioactivity and Geology.* London.

Joly, J., 1925. *The Surface History of the Earth.* Oxford.

Kennedy, G.C., 1959. The origin of Continents, Mountain Ranges and Ocean Basins. *American Scientist*, 47, 491-504

Kennedy, W.Q., 1946. The Great Glen Fault. *Quarterly Journal of the Geological Society*, 102, 41.

King, L.C., 1962. *The Morphology of the Earth* Edinburgh: Oliver & Boyd.

Kuenen, Ph. H., 1950. *Marine Geology.* New York: Wiley.

Linton, D., 1931. On the relations between the early and mid-Tertiary planation surfaces of south-east England. *Comptes Rendus* (IGU), 2, 498-503.

Lugeon, M., 1901. Les grandes nappes de recouvrement des Alpes du Chablais et de la Suisse. *Géologie Société de France Bulletin*, 4, 723-825.

Margerie, E. De, 1897-1918. *La Face De La Terre*, (See. Suess, E. 1883-1908).

Martonne, E. De, 1907. Recherches sur l'évolution morphologique des Alpes de Transylvanie, *Annales de Géographie*, 1, 1-247.

Martonne, E. De., 1929. La morphologie du Plateau Central de la France et l'hypothèse eustatique. *Annales de Géographie*, 38, 113-32.

Meinesz, F.A., Vening, Umbgrove, J.H.F. & Kuenen, Ph. H., 1934. *Gravity Expeditions at Sea, 1923-1932*, 2. Delft.

Meinesz, F.A. Vening., 1940. The Earth's Crust deformation in the East Indies *Proc. Koninklijke Nederlandse Akademie van Wetenschappen, Amsterdam*, 43, 278-93.

Mohorovíčić, A., 1910. Das Beben von 8.X. 1909. *Jahrbuch Meteorol Observatory, Zagreb*, no. 9, 4, A. 1.

Oldham, R.D., 1906. The constitution of the earth as revealed by earthquakes. *Quarterly Journal of the Geological Society of London*, 62, 456-575.

Ollier, C.D., 1981. *Tectonics and Landforms*. London: Longman.

Paton, T.R., 1986. *Perspectives of a Dynamic Earth*. London: Allen & Unwin.

Penck, A., 1894. *Morphologie der erdoberfläche*. Stuttgart, 2 volumes.

Penck, A., 1934. Theorie der bewegung der strandlinie. *Sitzungsberichte d. preuss Akademie der Wissenschaften, Mathe-Physik. Klasse*, 19, 321-48.

Penck, W., 1920. Wesen und Grundlagen der morphologischen Analyse. *Bericht Sachsischen Akademie der Wissenschaften, Mathe-Physik Klasse*, 72, 65-102.

Penck, W., 1921. Morphologische Analyse. *Verhandlungen Deutscher Geographentages*, (Berlin), 122-28.

Penck, W., 1924. *Die morphologische Analyse: Ein Kapitel der physikalischen Geologie*; Geographische Abhandlungen, 2nd Series, Vol 2, 1-283. (Published separately in the same year by J. Engelhorn of Stuttgart. Translated into English by H. Czech and K.C. Boswell as *Morphological Analysis of Landforms*, London: MacMillan, 1953).

Penck, W., 1925. Die Piedmontflächen des südlichen Schwarzwaldes. *Zeitschrift der Gesellschaft für Erdkunde zu Berlin*, 83-108.

Pratt, J.H., 1855. On the attraction of the Himalayan mountains and of the elevated regions beyond on the plumb line in India. *Royal Society, Philosophical Transactions*, 145, 53-100.

Pratt, J.H., 1859. On the deflection of the plumb line in India caused by the attraction of the Himalayan mountains and of the elevated region beyond. *Philosophical Transactions of the Royal Society*, 149, 745-96.

Prestwich, J., 1886-1888. *Geology*. 2 volumes.

Reade, T.M., 1886. *The Origin of Mountain Ranges*. London.

Reade, T.M., 1903. *The Evolution of Earth Structure with a Theory of Geomorphic Changes*. London.

Runcorn, S.K. (ed.), 1962. *Continental Drift*. New York: Academic Press.

Russell, R.J., 1936. Physiography of the lower Mississippi River Delta. *Reports on the geology of Plaquemines and St Bernard Parishes*, Geological Bulletin No 8, Baton Rouge: Lousiana Department of Conservation, 3-193.

Shepard, F.P., 1923. To question the theory of periodic diastrophism. *Journal of Geology*, 31, 599-613.

Shepard, F.P., 1933. Submarine Valleys. *Geographical Review*, 23, 77-89.

Shepard, F.P., 1948. *Submarine Geology*. New York: Harper, 3rd. edition 1973.

Snider-Pellegrini, A., 1858. *La Création et ses mystères dévoilés*. Paris.

Stacey, F.D., 1977. *Physics of the Earth*. New York: Wiley.

Staub, R., 1924. *Der Bau der Erde.* Beitr. Geol. Karte der Scheiz. Bern.

Stephenson. L.W., 1928. Major marine transgressions and regressions and structural features of the Gulf Coastal Plain. *American Journal of Science.* 5th ser., 14, 281-98.

Stille, H., 1936a. Die Entwicklung des amerikanischen Kordillerensystems in Zeit und Raum. *Preussisches Akad. Wiss. Sitzungsber.* 15, math-physikal, kl., 134-55.

Stille, H., 1936b. The present tectonic state of the earth: *American Association of Petroleum Geologists Bulletin,* 20, 849-80.

Stille, H., 1955. Recent deformations of the earth's crust in the light of those of earlier epochs, 171-192 in Poldervaart, Arie, Editor, *Crust of the earth.* Geological Society of America, Special paper 62.

Suess, E., 1875. *Die Entstehung der Alpen.* Wien.

Suess, E., 1883-1908. *Das antlitz der erde.* Wien, 3 volumes.

Suess, E., 1897-1918. *La face de la terre* (Translated by E. de Margerie): Paris, 3 v. (1897, v. 1; 1900, v.2; 1902-1918, v. 3).

Suess, E., 1904, (Translated by C. Schuchert) Farewell lecture on resigning his Professorship. *Journal of Geology,* 12, 264-75.

Suess, E., 1904-1924. *The face of the earth* (Translated by H.B.C. and W.J. Sollas). Oxford, 5 v. (1904, v. 1. 1906, v. 2, 1908, v.3, 1909, v.4, p.1924, v. 5).

Tapponier, P. & Molnar, P., 1976. Slip-line field theory and large-scale continental tectonics. *Nature,* 264, 319-24.

Taylor, F.B., 1910. Bearing of the Tertiary mountain belt on the origin of the earth's plan. *Geological Society of America Bulletin,* 21, 179-226.

Tricart, J., 1968. *Géomorphologie Structurale.* (Société d'Edition d'Enseignement Supérieur, Paris). Translated into English by Beaver, S.H. and Derbyshire, E., 1974, as *Structural Geomorphology.* London: Longman.

U.S. Geodynamics Committee, 1980. *Geodynamics in the 1980's.* National Academy of Sciences, Washington, D.C.

Wegener, A., 1912. Die enstehung der kontinents. *Petermanns Mitteilungen,* 185-195, 253-256 and 305-309.

Wegener, A., 1915. *Die Enstehung der Kontinente und Ozeane.* Brunswick. French translation (3rd edition 1923, 5th edition 1937, English translation, 3rd edition 1924).

Wiechert, E., 1897. Ueber die massenverteilung im Innern der Erde. *Nachrichten der Akademie der Wissenchaften in Göttingen. Math e-physikal. Klasse.,* 221-43.

Wilson, J.T., 1963. Hypotheses of Earth's Behaviour, *Nature,* 198, 925-99.

Wilson, J.T., 1965. A new class of faults and their bearing on continental drift, *Nature,* 207, 343-47.

Wilson, J.T., 1969. Aspects of the Different Mechanics of Ocean Floor and Continents. *Tectonophysics,* 8, 281-83.

Wilson, J.T., (ed.) 1977. *Continents Adrift and Continents Aground.* San Francisco: Freeman.

Wiseman, J.D.H. & Sewell, R.B.S., 1937. The Floor of the Arabian Sea. *Geological Magazine,* 74, 219-30.

Wooldridge, S.W., 1951. The progress of geomorphology. In Taylor, G., (Ed.) *Geography in the Twentieth Century.* London: Methuen, 165-77.

Wright, A.E., 1976. Alternating subduction direction and the evolution of the Atlantic Caledonides. *Nature,* 264, 156-60.

Wyllie, P.J., 1976. *The Way the Earth Works.* New York: Wiley.

Yochelson, E.L. (ed.), 1980. *The Scientific Ideas of G.K. Gilbert.* Geological Society of America, Special Paper 183.

11

Reconstructing the chronology of Lake Bonneville: an historical review

Dorothy Sack

This place, which we named Llano Salado, because we found some thin white shells there, seems to have once had a much larger lake than the present one. We observed the latitude and found it to be 39° 34' 36" (in the diary of Silvestre Veléz de Escalante, under the date October 2, 1776).

Lake Bonneville was the largest of the 94 Pleistocene lakes that formed in the Basin and Range physiographic province of the western United States (Williams and Bedinger 1984). At its greatest extent Lake Bonneville occupied approximately 19,800 sq. mi (51,282 sq. km) in the northeastern portion of the Great Basin (Figures 1 and 2), and attained a maximum depth of about 1220 ft (372 m) (Currey *et al.* 1984b p. 230). If the lake had not achieved external threshold control to the Snake River-Columbia River system at this level, paleoclimatic factors would probably have caused it to expand further. Lake Bonneville has been the object of scientific interest since the western geological and geographical surveys of the 1870s (e.g. Gilbert 1874, 1875), and it is still being studied intensively.

Lake Bonneville defined

In 1875 Gilbert (p. 88-90) (Figure 3) named Lake Bonneville and its two major shorelines, the highest, or "Bonneville beach," and the most prominent lower shoreline, the "Provo beach" (Figure 2). He also designated the body of sediments deposited by Lake Bonneville as the "Bonneville group" (Gilbert, 1875, p. 89). Within the Bonneville group Gilbert (1882 p. 174-175) included the distinctive yellow clay and white marl units, which he considered to represent two separate epochs of high water. Confusion concerning the definition of Lake Bonneville arose about mid-20th century when workers began to consider

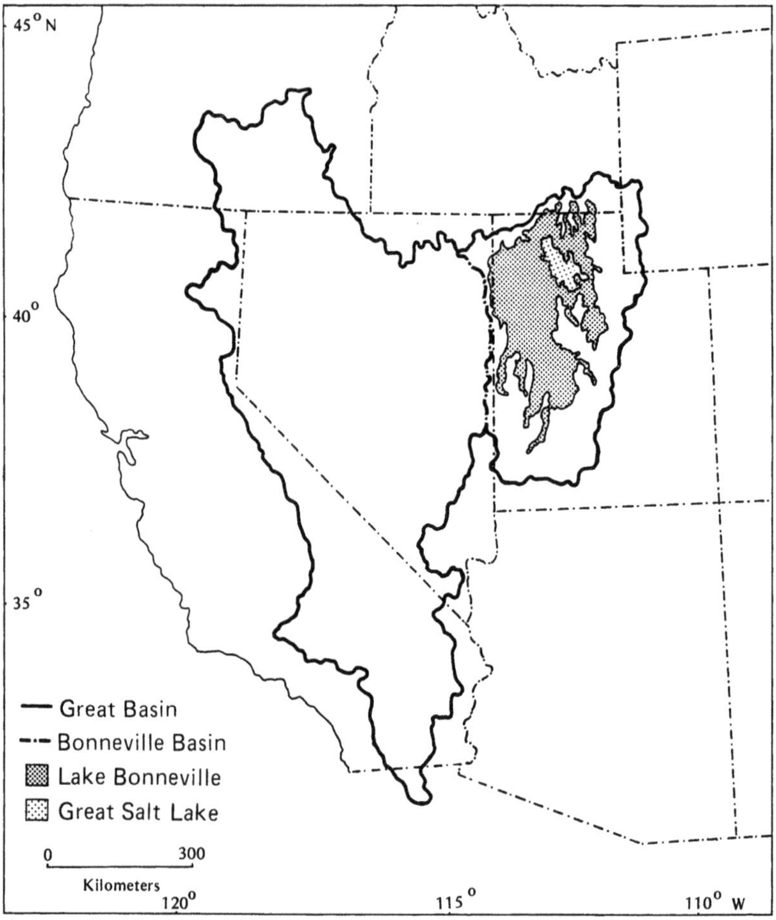

Figure 1

Map showing the location of Lake Bonneville within the Great Basin.

evidence for preceding lakes in the basin (Gvosdetsky and Hawkes 1953, Marsell and Jones 1955). Some investigators (Morrison and Frye 1965, Morrison 1966; Scott 1980, Scott *et al.* 1982, 1983a, 1983b) applied the name Lake Bonneville to all Pleistocene lakes that occupied the Bonneville basin; other investigators (Pac 1939, Gvosdetsky and Hawkes 1953, Hunt *et al.* 1953, Marsell and Jones 1955, Crittenden 1963b, Eardley 1964, Currey 1980e, Currey *et al.* 1983b; Oviatt 1987, Oviatt and Currey 1987, Oviatt *et al.* 1987) reserved the name only for the latest of the several Pleistocene lakes. Because it is now known that Gilbert's yellow clay and white marl were both deposited during the last Pleistocene lake cycle in the basin (Oviatt 1987), using Lake Bonneville for only the latest Pleistocene lake upholds Gilbert's (1875, 1882) original use of that term. With one recent exception (Van Horn 1979), researchers place the end of Lake Bonneville, the transition from Lake Bonneville to Great Salt Lake, at approximately the Pleistocene-Holocene boundary (e.g., Rudy 1973, Currey and Madsen 1974, Currey 1980a, Scott *et al.* 1983, Currey *et al.* 1984a).

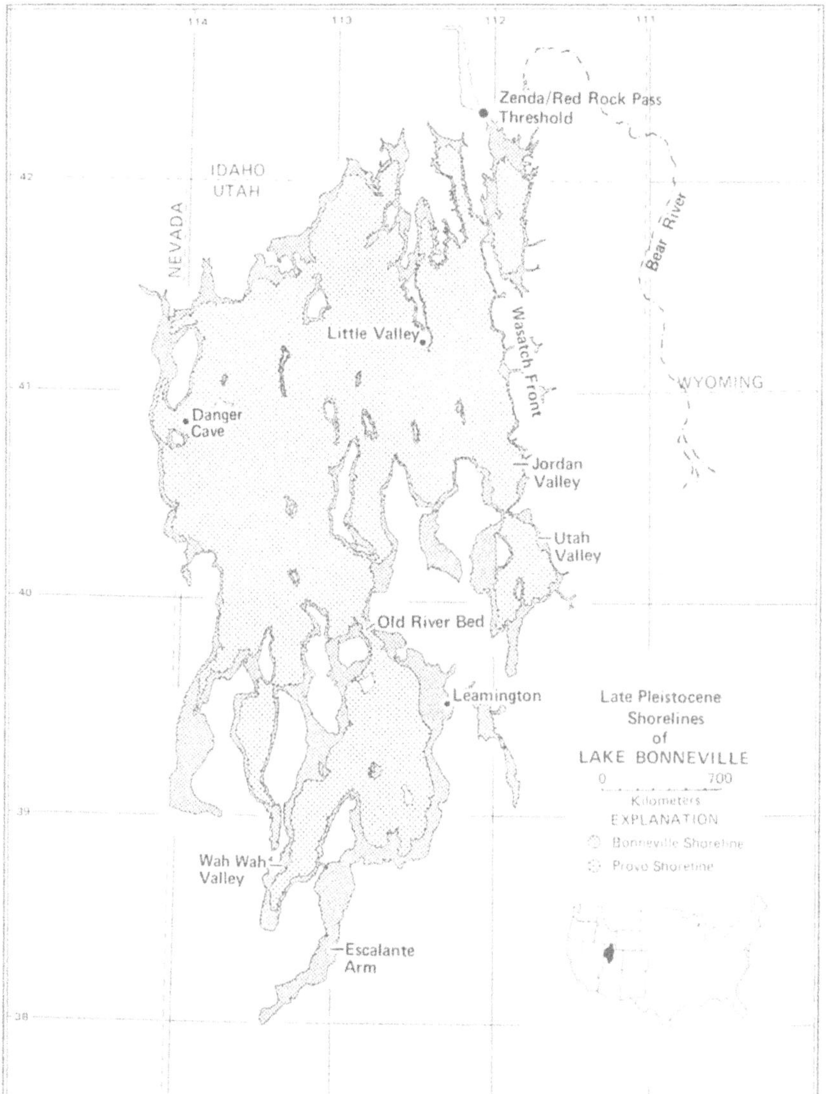

Figure 2

Map showing the extent of Lake Bonneville at the Bonneville and Provo Shorelines and the location of sites discussed in the text. Base map adapted from Currey 1982. (After Burr, in preparation.)

Figure 3

Grove Karl Gilbert (1843-1918). Photograph taken in 1898. U.S. Geological Survey Photographic Library, Portrait 111.

Nature of Lake Bonneville research

Lake Bonneville research may be categorized as either (1) studying the lake phenomena directly for its own sake, or (2) using dated lacustrine features as applied tools to help solve paleoenvironmental, geophysical, or geomorphic problems. Investigations in the first group analyze lake-related geomorphology, sedimentology, and stratigraphy in order to understand the lake's characteristics and history. Besides contributing to such fields as process geomorphology (e.g. Gilbert 1885) and depositional environments (e.g. Eardley 1938), these studies are necessary for reconstructing the paleohydrologic history of Lake Bonneville (Gilbert 1882 p. 71, 1890 p. 23), and they are often undertaken specifically for that purpose.

The second class of Lake Bonneville research involves using age-controlled shorelines and lake deposits--information resulting from the first category of investigation--to aid reconstructions of paleoenvironmental variables (especially paleoclimate) and to examine the nature, magnitude, and timing of geophysical and geomorphic events. Lake Bonneville is especially useful in this applied sense because it (1) responded in a sensitive fashion to environmental change, (2) recorded these changes geomorphically in originally horizontal shorelines and stratigraphically in distinctive lithostratigraphic units, such as Gilbert's (1882, 1890) white marl, and (3) left its evidence over a large area of the continental interior. In addition, its shorelines and sediments contain datable material, interrelate with a variety of geomorphic regimes, and occur in association with various seismotectonic and other environmental hazards in the Wasatch Front urban corridor.

Besides net paleohydrologic balance, Lake Bonneville provides evidence for reconstructing other aspects of the paleoenvironment, including climatic (e.g. Gilbert 1890, McCoy 1981), botanical (e.g. Bright 1966, Madsen and Kay 1982), zoological (Hubbs and Miller 1948), and archaeological (e.g. Antevs 1945, Madsen 1980) components. Paleoclimatic reconstructions are currently the most critical type of reconstruction, because developing an explanatory climate model for the present and future requires construction from and testing against the climate of the past (Flint 1974 p. 6). Although major uncertainties exist in estimating specific paleoclimatic parameters from paleolake evidence (McCoy 1981, Smith and Street-Perrott 1983, Street-Perrott and Harrison 1985, Benson and Thompson 1987), amino-acid racemization and stable-isotope fractionation may prove helpful for deducing paleotemperature.

Gilbert (1875 p. 93) first recognized the nonhorizontality of Lake Bonneville shorelines and their utility in geophysical and geomorphic problems (Gilbert 1875, 1880b, 1882, 1886, 1890). Gilbert (1890) analyzed the amount of shoreline deflection around the Bonneville basin (Figure 4), and concluded that the warping resulted from application and removal of the water load. Geochronometric control of the major shorelines enabled subsequent workers to model the isostatic response rate, degree of isostatic compensation, upper mantle viscosity, and regional lithospheric thickness (Crittenden 1963a, 1963b, 1967, Walcott, 1970, Passey 1981, Nakiboglu and Lambeck 1982, 1983, Bills and May 1987). Faulted lake sediments and shorelines provide temporal control for Bonneville basin seismic events (Gilbert 1880b, 1890, McCalpin 1986).

Conclusions drawn from applying the Lake Bonneville chronology to geomorphic, geophysical, and paleoenvironmental problems can only be as valid as the chronology on which they are based. As our knowledge of the chronology of Lake Bonneville has changed, so have the interpretations that are made as the result of using it as an applied tool.

Discovery of the ancient lake features, 1776-1873

Father Francisco Silvestre Velez de Escalante traversed the southern part of the Bonneville basin in 1776 with a small group of Spaniards and Indian guides

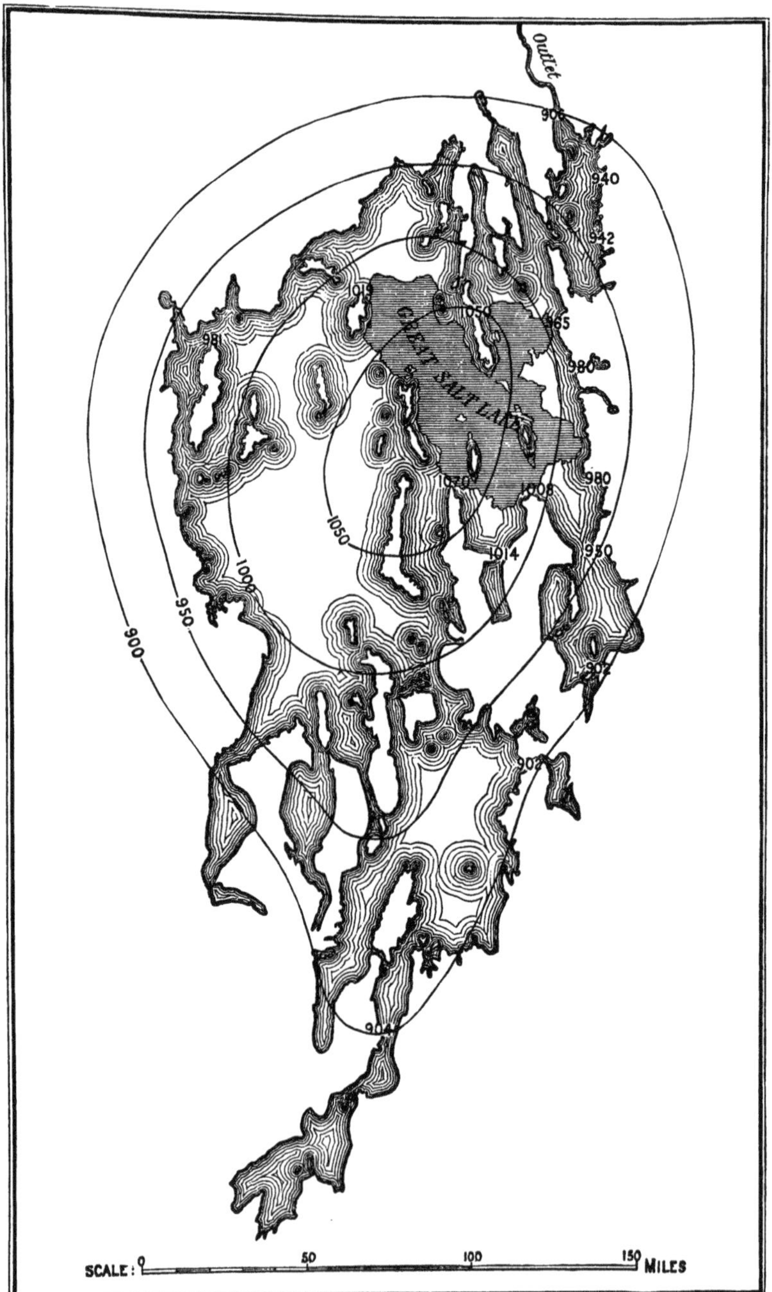

Figure 4

Theoretic curves of post-Bonneville deformation. Plate L from U.S. Geological Survey Monograph 1 (Gilbert 1890 p. 374c). Units are in feet above Great Salt Lake.

(Auerbach 1943). As the result of observing subærially exposed shells, Escalante became the first person to hypothesize in writing that a large lake had previously existed in the area (Ives 1948a). Escalante's information, however, was not available to the people who explored the northeastern Great Basin in the following century. Between autumn 1824 and spring 1825 Etienne Provost, John Weber, or James Bridger discovered Great Salt Lake for European-Americans (Pack 1939 p. 18, Morgan 1953 p. 182-183, Miller 1966 p. 5), but none of these trappers reported evidence of the ancient lake. Captain Bonneville led an expedition to this area for private trading and trapping interests from 1832 to 1835. Members of his party partially explored Great Salt Lake (Irving 1837), however again no mention was made of an ancient lake. A later meeting between Bonneville and Washington Irving resulted in Great Salt Lake being accurately portrayed without outlet and named Lake Bonneville in Irving's (1836) book, *Astoria* (Todd 1961 p. xix). In *Adventures of Captain Bonneville*, Irving (1837) popularized Bonneville's experiences and, therefore, the Great Salt Lake region.

Federally sponsored explorations and surveys made the subsequent early contributions to Bonneville basin studies. During his expedition of 1843-44, Fremont (1845) recognized the endoreic nature of the entire region, named it the Great Basin (p. 175), and, without realizing their possible great antiquity, noted a succession of strandlines above the level of the Great Salt Lake (p. 150 and 154). Finally, while surveying Great Salt Lake in 1849-50, Stansbury (1852) described and correctly interpreted the abundant shoreline and sediment evidence left by the ancient lake. Other geologists and explorers investigated the northeastern Great Basin in the 1850s and 1860s, although publication of some of their written reports was delayed until the 1870s. These investigators identified extensive subærially exposed shoreline features, attributed them to expanded versions of Great Salt Lake that had existed during wetter periods, and suggested that a drier climate had preceded the lacustrine epoch (Beckwith 1855, Engelmann 1876, Simpson 1876, Hague and Emmons 1877, King 1878). During his geological surveys of the early 1870s, Hayden (1871 p. 170-174) assigned the paleolake to Quaternary time and determined that evaporation of that fresh lake left the modern remnant Great Salt Lake. Bradley, a geologist with Hayden's Survey, hypothesized that the ancient lake attained threshold control at four places (Bradley 1873 p. 202).

This was approximately the state of knowledge concerning Lake Bonneville when Gilbert began investigating it as a geologist for the Wheeler Survey in 1872. Before Gilbert's work, knowledge of the lake had not progressed far beyond simple recognition. Previous workers described some lacustrine features in a general manner and advanced a few hypotheses. Most did not attempt to test their hypotheses scientifically, perhaps in part because Lake Bonneville was only one of several topics under investigation during the regional surveys.

Gilbert's era of Lake Bonneville research, 1874-1890

During his work with the Powell Survey from 1875 to 1879, Gilbert (1890 p. 18) intermittently continued the Bonneville research that he had started in 1872 with the Wheeler Survey. When the U.S. Geological Survey was formed in 1879, Gilbert became head of the Great Basin Division. The division's first task was to study Great Basin Pleistocene lakes, and Gilbert assigned himself the study of Lake Bonneville. By continuing Bonneville research, Gilbert (1890 p. 1-2) sought to unravel both the local Pleistocene history and the processes which produced the historical changes. Few modern Quaternary scientists would argue with his rationale, that the recent is a key to the present and past (Gilbert 1890 p. 1-2).

Contributions of Gilbert's Contemporaries

In the period between publication of his first (1874) and last (1890) Bonneville reports, Gilbert dominated Lake Bonneville investigations. Other contributions during this interval generally pale in comparison with Gilbert's work. To Gilbert's (1875) Wheeler Survey report Howell (1875) added only a few details regarding the lake's Escalante Desert arm (Figure 2) and the Provo and Spanish Fork deltas. Packard (1876a) suggested on the basis of hearsay that the Old River Bed (Figure 2), an internal basin threshold, was an external outlet for the ancient lake. He also depicted the deep lake as brackish during overflow (Packard 1876b, p. 681). While Young (1878) described the Red Rock Pass outlet (Figure 2) in detail, Anonymous (1878) and Peale (1878) argued with Gilbert (1878) over priority of its discovery. Peale's (1878, 1879) assertion that the outlet existed 40 mi (65 km) farther north, rather than at the pass itself, provoked Gilbert to scrutinize the divide area carefully. As a result, Gilbert (1880a p. 347) agreed that the point of initial overflow was indeed north of Red Rock Pass, but only by 2 mi (3.2 km). Davis (1883) suggested an orographic rather than a climatic origin for the lake, and Call (1884) described the molluscan fauna of Lakes Bonneville and Lahontan, postulating that their abundance and shell size varied with temperature and salinity changes that accompanied lake oscillations.

Gilbert's Contributions

Gilbert (1875, 1876, 1878, 1880a, 1882, 1883, 1885) published some of his ideas on coastal processes and Lake Bonneville history in separate reports and articles prior to writing his culminating monograph (Gilbert 1890). USGS Monograph 1, however, besides being considered Gilbert's best geologic work, is widely regarded as the most comprehensive and insightful report written on the Pleistocene lake (e.g., Mendenhall 1920, Davis 1927, Pyne 1975, Baker and Pyne 1978, Stokes 1979, Chorley and Beckinsale 1980, Hunt 1980, 1987, Oviatt 1987). The following summary of Gilbert's Lake Bonneville chronology is drawn from that monumental work.

According to Gilbert's chronology (Figure 5), the immediate pre-Bonneville epoch consisted of a long period of aridity. In this epoch alluvial fans and playas dominated the geomorphic system. Change to a more humid climate caused playas in subbasins of the Bonneville basin to expand into lakes. These lakes eventually coalesced, ushering in the first Bonneville epoch of high water. Although Lake Bonneville's water plane oscillated under the closed-basin conditions, the overall trend was one of gradual transgression. The climatically induced lake-level rise may have been augmented by a possible diversion of the

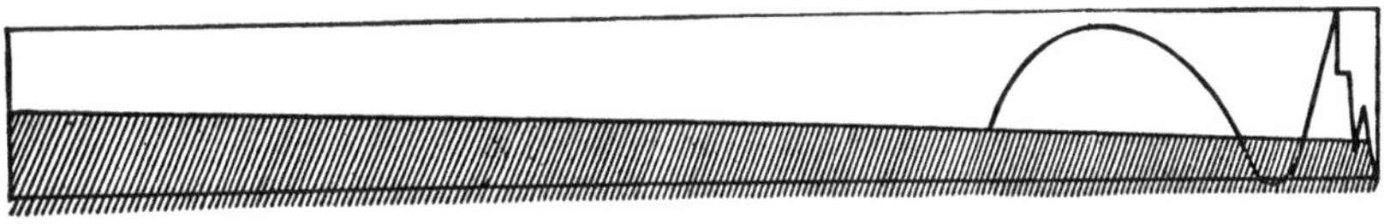

FIG. 34.—Rise and Fall of water in the Bonneville Basin.

Figure 5

Schematic time-altitude diagram of Lake Bonneville by Gilbert (1890 p. 262 Figure 34). The ordinate portrays relative lake level. The abscissa represents time, which is flowing to the right. The upper and lower horizontal lines indicate the Bonneville and Great Salt Lake Shorelines, respectively.

Bear River into the Bonneville basin (Figure 2). The lake rose to within 90 ft (27 m) of the lowest point on the basin perimeter, which was located just north of Red Rock Pass, Idaho (Figure 2). The pelagic yellow clay unit was deposited during this first high-water epoch. An interval of complete or nearly complete desiccation followed. Lake Bonneville then retransgressed. During this second epoch of high water the lake deposited the deep-water white marl unit, which unconformably overlies the yellow clay. In this epoch Lake Bonneville attained the threshold elevation, at which level it formed the Bonneville Shoreline. After a relatively brief period of threshold control, the alluvial material at the threshold failed, resulting in 375 ft (114 m) of rapid downcutting to a bedrock sill. A long period of threshold control at this lower elevation produced the geomorphically distinct Provo Shoreline. Climatically induced regression from the Provo level returned the lake to closed-basin conditions. Gilbert (1890 p. 134 and Figure 34 p. 262) suggested that a slight readvance during the general regression created the Stansbury Shoreline, more than 300 ft (91 m) below the Provo Shoreline (Figure 5). Eventually the lake attained the approximate size of modern Great Salt Lake. Gilbert hypothesized that the two high-water epochs correlated with two Pleistocene glacial maxima, and that both types of phenomena responded to climatic change.

Problems with Gilbert's work

Gilbert's (1890) theory of two Bonneville epochs of high water rests on inconclusive evidence, and therefore represents a major weak point in the chronology. His inference of two high-water epochs separated by a long period of desiccation results from two lines of evidence, one morphostratigraphic and one stratigraphic.

In Monograph 1, Gilbert (1890 p. 199) presented meager morphostratigraphic support for the notion of two deep-water epochs from Preuss (now Wah Wah) Valley, Utah (Figure 2). This evidence consists of an exposure where one set of coastal depositional features is partially overlain by a younger set. The highest morphostratigraphic unit of the older set lies about 90 ft (27 m) below the Bonneville Shoreline. From this single exposure Gilbert (1890) inferred the existence of two separate deep-lake stages. He stated this inference without explaining why he believed the older features represent an earlier lake epoch rather than merely an oscillation on the transgression of the last lake cycle (Gilbert 1890). According to Gilbert's interpretation, the second lake rose 90 ft (27 m) higher than the first.

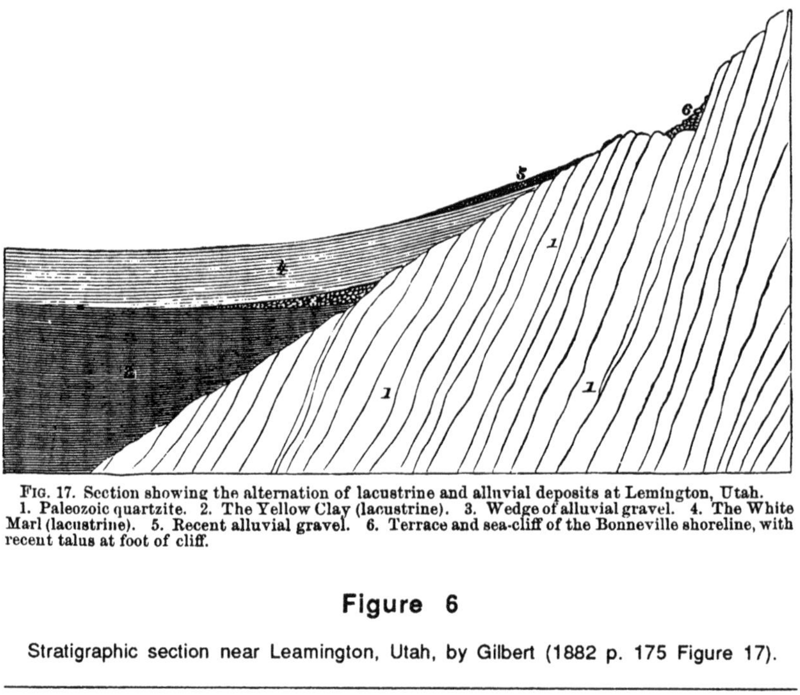

FIG. 17. Section showing the alternation of lacustrine and alluvial deposits at Lemington, Utah. 1. Paleozoic quartzite. 2. The Yellow Clay (lacustrine). 3. Wedge of alluvial gravel. 4. The White Marl (lacustrine). 5. Recent alluvial gravel. 6. Terrace and sea-cliff of the Bonneville shoreline, with recent talus at foot of cliff.

Figure 6

Stratigraphic section near Leamington, Utah, by Gilbert (1882 p. 175 Figure 17).

Stratigraphic evidence for the two epochs of high water comes from three described sections containing the yellow clay and white marl (Gilbert 1890 p. 189-199). Two of these sections occur in the Old River Bed area, the other is located about 40 mi (65 km) to the southeast, near Leamington, Utah (Figures 2 and 6). According to Gilbert's correlation and interpretation of these sections, the lithologically distinct yellow clay and white marl units are separated at the upper River Bed and Leamington sections by an unconformity and subærial deposits. Furthermore, Gilbert suggested that a tufa line about 90 ft (27 m) below the Bonneville Shoreline at Leamington might represent the maximum height of the first deep-water epoch. These relations, coupled with interpreting both the yellow clay and white marl as deep-water deposits, led Gilbert to infer the existence of two lake epochs separated by a period of relative aridity. Weaknesses in Gilbert's stratigraphic evidence include that (1) his interpretation relies on only three stratigraphic sections, (2) the sections are correlated over long distances, and (3)

Gilbert (1890 p. 203-209) admittedly does not adequately explain why the distinctive deep-water deposits of the two lake epochs are so lithologically different.

Gilbert (1890 p. 135) called all shorelines lying between the Bonneville and Provo levels the Intermediate shorelines. He demonstrated (1890 p. 147-157) that the Intermediate shorelines formed sometime before the Bonneville and Provo Shorelines. However, in Monograph 1 Gilbert never explicitly states whether the Intermediate shorelines date from only the first high-water epoch, only the second high-water epoch, or whether they may be of either age. By saying of the Intermediate shorelines that "it was not difficult to see that certain of the shore embankments were referable to an earlier flood than certain others" (Gilbert 1890 p. 152), we might infer that he interpreted some of them as belonging to the first and some to the second high-water epoch of Lake Bonneville.

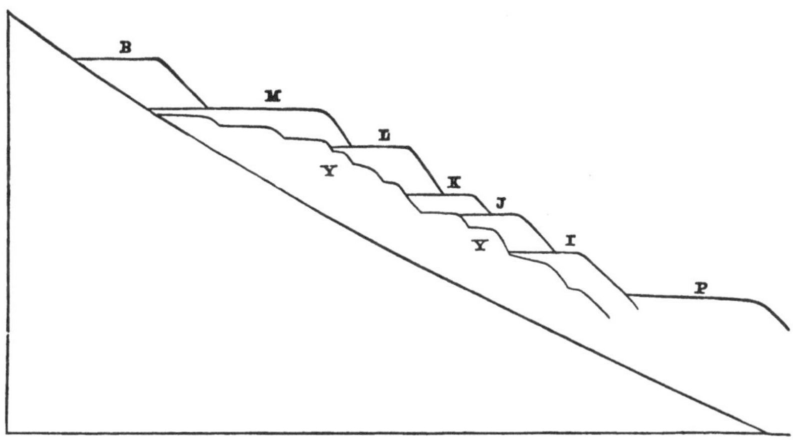

FIG. 19.—Diagram, showing the Overlap and Chronologic Order of the Shore Embankments of the Bonneville Basin.

Figure 7

Morphostratigraphy of the Intermediate shorelines by Gilbert (1882 p. 182 Figure 19).

Gilbert (1882 p. 181-182 and Figure 19) provided more explicit information on the Intermediate shorelines in an earlier report, where he presented a composite cross-sectional profile of the Intermediate embankments (Figure 7). According to Gilbert (1882 p. 182), the Intermediate embankments that appear in the diagram were deposited prior to the Bonneville [B], and Provo [P] embankments, and in the following order: Y, I, J, K, L, and M. Further:

> the series YY records the first rise of the water and is the littoral equivalent of the Yellow Clay. It shows that the water at that time reached an extreme level about ninety feet lower than the Bonneville shore. . . . The overlying banks from B to P record the second rise of the water and are the littoral equivalents of the White Marl (1882 p. 182).

However, "*the series of embankments YY was not discovered in section* . . ." (Gilbert 1882 p. 182) (emphasis added). These statements reveal that Gilbert's Intermediate shorelines consist at least of the transgressive shorelines from the second high-water epoch. In addition, he apparently held open the possibility that morphostratigraphic evidence would be found for shorelines from the first high-water epoch, and if any of these were located between the elevation of the Bonneville and Provo Shorelines, they would also be called Intermediate shorelines. Gilbert's statements in the 1882 report indicate that he favored the scenario of two high-water epochs of Lake Bonneville even before he accepted the morphostratigraphic evidence from Preuss Valley in Monograph 1 (Gilbert 1890). Therefore, in his Lake Bonneville research Gilbert evidently was, as others have suggested, biased by the contemporary bipartite theory of Pleistocene glacial maxima (Hay 1927 p. 139, Antevs 1945 p. 27, Gvosdetsky and Hawkes 1953 p. 3, Jones and Marsell 1955 p. 117). Such bias is rarely found in the work of Gilbert (Davis 1927 p. 134); he typically employed the method of multiple working hypotheses and was conservative in accepting hypotheses (e.g. Gilbert 1886, 1896, 1904). Moreover, Gilbert (1886 p. 285) himself defined scientific observation as endeavoring "to discriminate the phenomena observed from the observer's inference in regard to them, and to record the phenomena pure and simple."

Emphasizing one point of internal weakness in Gilbert's chronology - the inference of two high-water epochs - is not meant to detract from the magnitude of his contribution. Without the aid of modern topographic maps, aerial photographs, significant artificial exposures, or geochronometric techniques, Gilbert mapped the lake's former extent (Figure 8), analyzed its relative history, and applied that history to paleoenvironmental, geophysical, and geomorphic problems. Specifically, his inferences relied on keen geomorphic observations supplemented by stratigraphic evidence; generally, his interpretations were refined by his analytical genius and his method of multiple working hypotheses (Gilbert 1886).

Gilbert's chronology unchallenged, 1891-1944

The first alternative Bonneville scenario was not suggested until 55 years after publication of Monograph 1. During the intervening half century, no one contested the main points of Gilbert's chronology, but some researchers disputed the lake's origin, the synchroneity of high lake levels with Wasatch Front glacial advances, the late Pleistocene age of the lake, and details of the Bonneville Shoreline. In addition, Huntington (1914) and Antevs (1925, 1938) contributed new ideas on pluvial climates.

Several people working between 1891 and 1944 disagreed with some of Gilbert's Lake Bonneville interpretations, but Keyes was the only ungracious dissenter. Keyes (1917, 1918) discounted Gilbert's (1890) notion of a climatic origin for Lake Bonneville and presented instead a poorly supported argument in favor of a diastrophic origin, as first suggested by Davis (1883). Keyes's ideas on the lake were not widely accepted.

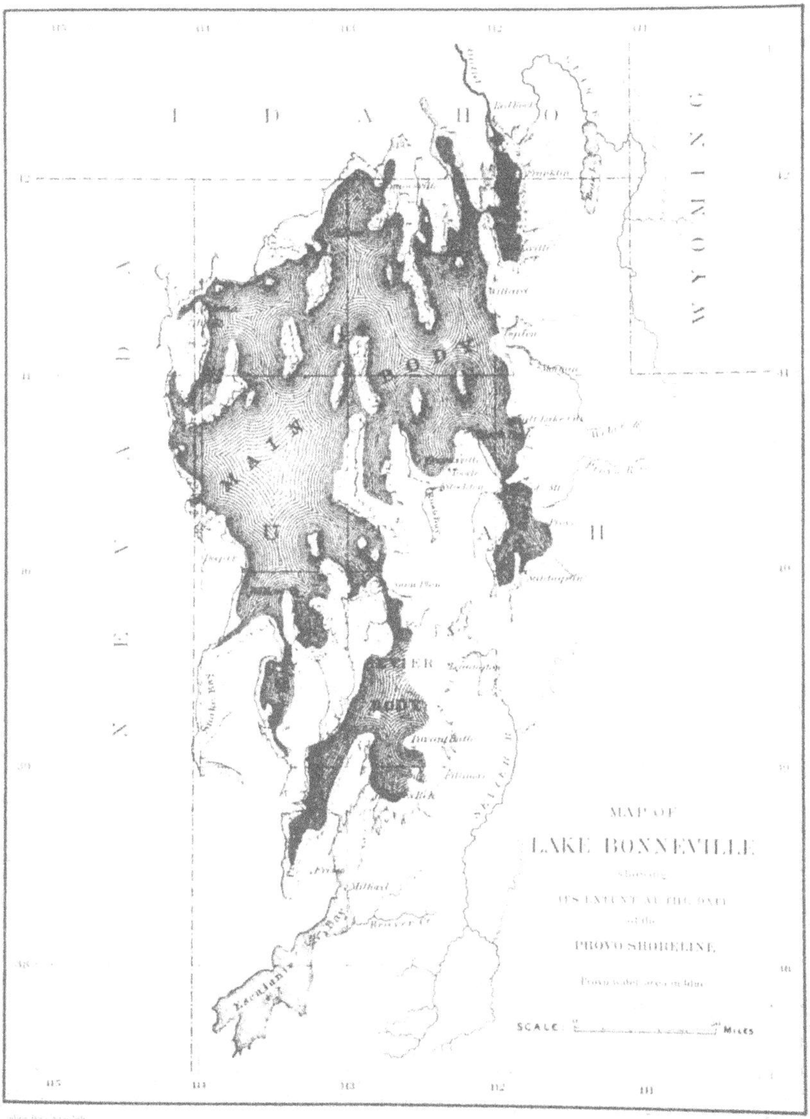

Figure 8

Map from U.S. Geological Survey Monograph 1 showing the extent of Lake Bonneville at the Bonneville and Provo Shorelines (Gilbert 1890 p. 128a Plate XIII).

Along the Wasatch Front where Pleistocene valley glaciers projected out into Jordan Valley (Figure 2), Atwood (1909) and Blackwelder (1931) found evidence in support of Gilbert's hypothesized two periods of ice advance. However, contrary to Gilbert's (1890) correlation of high lake levels with glacial maxima, Atwood (1909 p. 81) inferred that the large, older moraines long preceded the lake epoch. Antevs (1925, 1938) believed that if Lake Bonneville had not become exoreic, it would have peaked slightly after the glacial maximum.

Hay (1927) contested Gilbert's assignment of the second high-water epoch to late Pleistocene time. On the basis of correlative vertebrate fossil evidence, he proposed a Kansan age for the "white marl" lake. Researchers of this period disagreed among themselves as to the probable age of Gilbert's first high-water epoch. Upham (1922) and Hay (1927) suggested a Nebraskan age, whereas Antevs (1925) hypothesized an early Wisconsin age for the "yellow clay" lake. Later, Antevs (1938) correlated late Pleistocene glaciations, and therefore the approximately coeval lake maxima, with the Spitaler-Milankovitch-Köppen temperature chronology. In 1939, Pack (p. 45-46) published his view that the four Pleistocene glacial periods each have a corresponding lake maximum.

A few people in this period disagreed with Gilbert over details of the Bonneville Shoreline. Antevs (1925 p. 74) and Pack (1939 p. 33-34) doubted whether threshold control created the highest shoreline, as Gilbert (1890) had maintained. Antevs and Pack accepted an alternative explanation that attributed the Bonneville Shoreline to a balance between evaporation and precipitation at a subthreshold level. Antevs (1925) reasoned that the alluvial threshold would have been too weak to control the water level for even a brief period without failing. Pack (1939) suggested that the original threshold elevation was up to 75 ft (23 m) above the Bonneville Shoreline. According to Pack, transgression continued after formation of the Bonneville Shoreline, but no significant higher shorelines were formed because no major stillstands occurred during this last transgressive phase. Boutwell (1933 p. 36-37) described the fall from the Bonneville to the Provo Shoreline as a slow descent, thereby apparently discounting Gilbert's (1890) notion of a catastrophic flood. One of the most useful field contributions of this period came in 1944 when Dennis remapped the Bonneville Shoreline in the Escalante Desert. Gilbert (1890) had been unable to adequately explain the anomalously high shoreline that his field assistants had mapped in this arm of the lake. Dennis (1944 p. 124) located the Bonneville Shoreline at its proper elevation, thereby reducing the area of Lake Bonneville's Escalante arm by almost 550 sq mi (1424 sq km).

It was in this time period that researchers first misstated Gilbert's (1882, 1890) definition of the Intermediate shorelines. For example, Hay (1927 p. 138-139) incorrectly paraphrased Gilbert's (1890) ideas concerning the two high-water epochs. According to Hay (1927 p. 139):

> Gilbert concluded . . . that there had been two high-water stages, that which had produced the Intermediate shore lines and laid down the Lower Bonneville yellow clay and a second and higher stage during which the Bonneville shore lines had been carved out and the Upper Bonneville white marls had been deposited.

Hay, therefore, mistakenly inferred from the vague account in Monograph 1 (Gilbert 1890 p. 152), as several subsequent researchers would, that Gilbert defined the Intermediate shorelines as having formed exclusively during the first epoch of high water. In contrast, Boutwell (1933 p. 36) adhered to the more lucid Intermediate shoreline account presented by Gilbert in 1882 (p. 182). Boutwell agreed with Gilbert (1882) that the shorelines of the first high-water epoch remain unidentified. "It is the shore lines of the lake formed in the second humid period that are seen contouring the mountain slopes to-day" (Boutwell 1933 p. 36).

Scarcity of outstanding Lake Bonneville contributions in the interval 1891-1944 is probably attributable to the comprehensiveness of Gilbert's geomorphic work. Although Gilbert studied both Lake Bonneville geomorphology and stratigraphy, his investigations more thoroughly analyzed the landform evidence. Indeed, Hunt (1980 p. 46) ascribed this period of doldrums in Bonneville research to the early 20th century emphasis on landforms rather than deposits. With few exceptions (Russell 1895, Keyes 1917, 1918) ponderous Davisian interpretations of Lake Bonneville are noticeably absent, and this may be due to Gilbert's extensive geomorphic analysis. Davis himself (1883, 1892, 1933) published only three primarily review papers on Lake Bonneville. In these he summarized Gilbert's work and refrained from applying evolutionary terminology. Davis may have respected Lake Bonneville studies as the territory of his primary intellectual rival (Chorley *et al.* 1973 p. 151). More likely, the Davisian model was probably better suited to general accounts of desert or lake basins (Russell 1895) than to a single, already well studied Pleistocene lake.

The slow evolution of new Quaternary methods and technology in this period may also have discouraged studies of Lake Bonneville. It would have been especially difficult to expand upon Gilbert's thorough work without different methods and techniques. Certainly supreme technological innovation occurred for society at large during these years, but not specifically for Quaternary studies. Aerial photographs, accurate large-scale topographic maps, and radiometric dating techniques, for example, were not generally available for the Bonneville basin until after 1950. Moreover, Gilbert probably gained some advantage by traversing most of the basin on horseback rather than by automobile (Hunt 1980 p. 46). On the other hand, we must wonder whether the modest methodological and technical advances discouraged new studies of Lake Bonneville or whether lack of motivation in part suppressed the introduction of new techniques to Lake Bonneville studies.

Gilbert's chronology challenged, 1945-1978

In contrast to the preceding interval, 1945-1978 was marked by rejuvenated interest in Lake Bonneville and the application of new methods and technology to its study. Combining intensive stratigraphic work with Quaternary mapping, soil stratigraphy, and radiometric dating generated fresh viewpoints on Bonneville lacustrine history. As a result, this generation of Bonneville researchers substantially modified Gilbert's history of the lake.

Just before introduction of the new techniques, Antevs (1945 p. 27-28) published the first variant Bonneville chronology on the basis of traditional stratigraphic evidence. Although Antevs agreed with Gilbert concerning the first epoch of high water, reinterpretation of Gilbert's Old River Bed sections led him to infer a different history for the second high-water epoch. In his view, the Bonneville Shoreline formed under closed- rather than open-basin conditions. The lake then rose to the slightly higher threshold elevation. Like Gilbert, Antevs believed that overflow caused threshold failure and incision to a bedrock sill at the Provo level. Unlike Gilbert (1890 p. 134 and Figure 34 p. 262), who depicted an entirely post-Provo Stansbury Shoreline, Antevs maintained that the Stansbury Shoreline formed between two occupations of the Provo Shoreline. According to Antevs, excessive evaporation caused the water level to pause only briefly at the Provo Shoreline before dropping more than 300 ft (91 m) to the level where it formed the Stansbury Shoreline. Afterwards, the lake transgressed back to the threshold and occupied the Provo Shoreline for a second time. Once subsequent researchers became armed with the new techniques, this initial modification of Gilbert's Lake Bonneville chronology had numerous competitors (Table 1).

In probably the most influential Bonneville publication of this era, Hunt *et al.* (1953) introduced detailed Quaternary mapping into Bonneville basin studies. This mapping required the careful examination and logical classification of Lake Bonneville deposits. Hunt *et al.* (1953 p. 4) applied standard principles of bedrock mapping to unconsolidated materials in their northern Utah Valley study area (Figure 2). They divided the Lake Bonneville group (after Gilbert 1875) into three formations distinguished by lake stage. Each formation was further subdivided into members based on dominant lithology. Like many pioneering works, problems with their formational scheme, particularly the lack of type sections, were eventually recognized (e.g. Bright 1963 p. 69, Morrison 1965a p. 14). Nevertheless, Hunt *et al.* (1953) provided an organizational framework for Lake Bonneville deposits that was widely applied. Their work encouraged others to examine closely both the vertical and lateral changes in Bonneville basin sediments, and to classify them into established formations and members.

Hunt *et al.* (1953) accepted Gilbert's (1890) chronology and designated the Bonneville and Provo formations as consisting of sediments deposited during Bonneville and Provo Shoreline time, respectively. Hunt *et al.* (1953 p. 40-41) apparently meant to assign deposits of the first high-water epoch to the Alpine formation (Hunt 1980 p. 50). However, instead of explicitly doing so, they described the Alpine formation as composed of deposits from Intermediate shoreline time (Hunt *et al.* 1953 p. 17). As previously discussed, some Intermediate shorelines could have formed during the first high-water epoch, whereas others definitely formed during the transgressive phase of the second high-water epoch (Gilbert 1882 p. 182). As Hay (1927) had done 26 years earlier, Hunt *et al.* (1953 p. 17) thought that Gilbert (1890) had specified the Intermediate shorelines as belonging solely to the first epoch of high water. As a result, the Alpine formation was ambiguously defined. Because of the tremendous influence of their report and the resulting widespread use of the Alpine formation (Hunt *et al.* 1953) (subsequently spelled with both parts of the binomial capitalized), this error became firmly entrenched in the Bonneville literature.

Another important development in this era was soil stratigraphy. Again, Hunt was instrumental in introducing this innovation into Lake Bonneville studies (Hunt and Sokoloff 1950, Hunt *et al.* 1953). When correctly identified in stratigraphic sections, buried soils (geosols) reveal periods of lake recession and subærial exposure. If datable material is available, geochronometry may provide reliable geosol ages. In the absence of datable material, researchers of this period confidently used degree of soil development to estimate geosol age or duration of exposure.

Table 1

Major Proposed Lake Bonneville Chronologies, 1945-1978

Author(s)	Primary Evidence
Antevs 1945, 1948	Deduction from climate and glacial theory, stratigraphy
Ives 1948b	Geomorphology
Ives 1951	Stratigraphy
Antevs 1952	Deduction from climate and glacial theory, stratigraphy
Bissell 1952	Stratigraphy, soil stratigraphy
Gvosdetsky and Hawkes 1953	Soil stratigraphy, deduction from climate and glacial theory
Antevs 1955	Deduction from climate and glacial theory, stratigraphy, radiometric dating
Marsell and Jones 1955	Stratigraphy
Feth and Rubin 1957	Stratigraphy, radiometric dating
Eardley *et al.* 1957	Stratigraphy, geomorphology, core stratigraphy, soil stratigraphy
Broecker and Orr 1958	Radiometric dating, stratigraphy, geomorphology
Morrison 1961a, 1961b	Stratigraphy, soil stratigraphy
Bissell 1963	Stratigraphy, soil stratigraphy
Bright 1963	Stratigraphy, radiometric dating, geomorphology
Morrison 1964	Stratigraphy, soil stratigraphy
Morrison 1965a	Stratigraphy, soil stratigraphy
Morrison 1965b	Stratigraphy, soil stratigraphy, radiometric dating
Morrison 1965c, 1966	Stratigraphy, soil stratigraphy, radiometric dating
Broecker and Kaufman 1965	Radiometric dating, geomorphology, stratigraphy
Bissell 1968	Deduction from climate and glacial theory, stratigraphy, geomorphology, soil stratigraphy
Morrison 1975	Stratigraphy, soil stratigraphy

Gvosdetsky and Hawkes (1953, 1954, 1956) primarily used fossil soils to identify major recessions of Lake Bonneville. Their interpretation of Lake

Bonneville history consists of five discrete high lake levels (pre-Bonneville, Bonneville, Provo, Stansbury I, and Stansbury II) separated by four periods of soil formation (Gvosdetsky and Hawkes 1953 p. 63) (Figure 9). They correlated

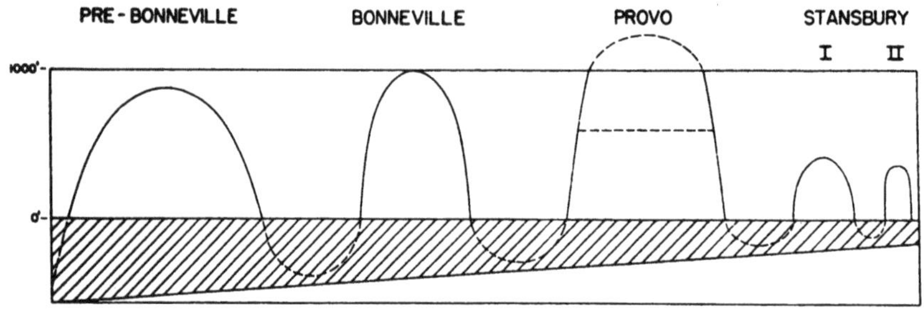

Figure 9

Time-altitude diagram of Lake Bonneville as postulated by Gvosdetsky and Hawkes (1953 p. 62, Figure 15b). Reprinted with permission from the Utah Engineering Experiment Station.

the high lake stages with Nebraskan, Illinoian, Kansan, Wisconsin 1, and Wisconsin 2 glacial maxima, respectively, and the soils with the interglacials and a Wisconsin interstadial. Their admittedly tentative chronology (Gvosdetsky and Hawkes 1953 p. 65) received little acceptance probably because the correlations lack objective age control. Gvosdetsky and Hawkes's application of Russian paleopedologic methods to Lake Bonneville studies, however, demonstrated the potential utility of soil stratigraphy to this field.

Morrison (1961a, 1964, 1965a, 1965b) most strongly advocated the value of soil stratigraphy for distinguishing lake-level fluctuations, although he supplemented it with inferred unconformities, subærial deposits, and radiocarbon dates for constructing iterative models of Lake Bonneville's chronology. In the course of his Bonneville work, Morrison named the Dimple Dell, Promontory, Graniteville, and Midvale Soils, which he believed preceded, separated, and followed the three formations in his version of the Lake Bonneville group (Morrison 1964, 1965b) (Figure 10). Because of his overconfidence in deducing soil age from degree of soil development and his reliance on primarily two field areas, only one of his geosols remains in common usage today.

In Morrison's (1965b) view (Figure 10), major Lake Bonneville oscillations occurred within the Alpine, Bonneville, and Draper Formations as well as between them. Like Antevs (1945), Morrison (1965b) postulated that the lake fell from the Bonneville Shoreline to a low level without a major pause at the Provo Shoreline. Morrison maintained that the Graniteville Soil developed during this recession. He hypothesized that after the low stage, Lake Bonneville rose almost to the elevation of the Provo Shoreline in Draper Formation time.

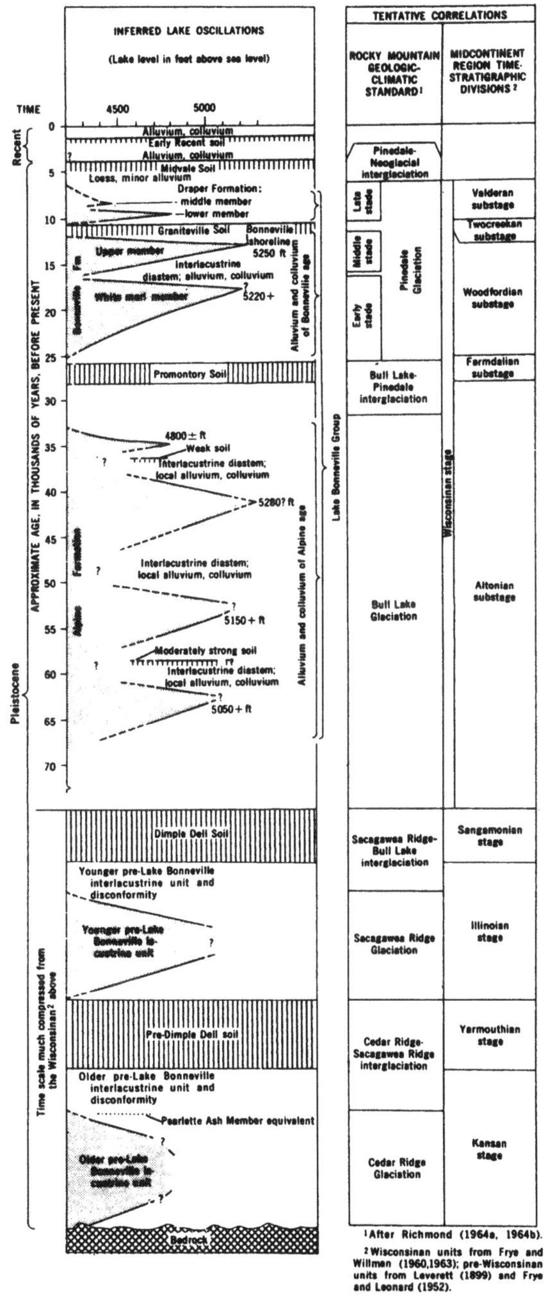

Figure 10

Diagram showing Lake Bonneville oscillations and tentative correlation of Quaternary deposits by Morrison (1965b p. 112, Figure 2).

Despite the numerous competitor Bonneville histories, Morrison's chronologies were the most frequently cited from the mid-1960s to the late 1970s. He employed the latest techniques and was well published. Moreover, R.L. Nielson (pers. commun. 1988) suggested that the popularity of Morrison's Lake Bonneville chronologies may be partially attributable to the design of his chronology diagrams (e.g. Figure 10). Morrison's detailed and seemingly confident portrayal of complex lake fluctuations and glacial correlations present at least a superficially satisfying picture.

Radiometric dating began to impact Lake Bonneville studies by the mid-1950s (Antevs 1955 p. 328). The prospect of fitting an absolute time scale to the history of Lake Bonneville stimulated a burst of lake research. Problems with radiocarbon dating Lake Bonneville materials emerged almost immediately. Eardley *et al.* (1957 p. 1171) and Feth and Rubin (1957 p. 1827) noted that single geomorphic features often yielded inconsistent ages. The scatter of dates suggested to Feth and Rubin (1957 p. 1827) that the Bonneville chronology was extremely complex. Jennings (1957 p. 96) declared that there existed "justifiable doubt as to the validity of radiocarbon dating (e.g. Hunt 1955a), as so far developed." Broecker and Orr (1958) analyzed sources of error in radiocarbon dating and, by employing what they considered to be reliable radiocarbon dates, presented the first Lake Bonneville time-altitude graph (hydrograph) to quantitatively portray thousands of years of lake history on the abscissa. With the flood of often-conflicting radiometric dates, debates ensued over almost every point in Lake Bonneville chronology. Broecker and Kaufman (1965) discussed occurrence and recognition of secondary contamination of carbonate samples, and concluded that tufa dates especially should be used with caution. In spite of their work, conflicting dates still left theorists with the option of choosing as valid only those dates that confirmed their favored chronology.

Problems involved in applying Quaternary mapping, soil stratigraphy, and radiometric dating to Lake Bonneville studies led to the proposal of many divergent chronologies in the period from 1945 to 1978 (Table 1). In 1965, Morrison (1965a, 1965b, 1965c) alone published three histories of Lake Bonneville fluctuations, although this timing was at least partially the result of publication delays (Morrison and Frye 1965 p. 20). Nevertheless, perhaps some investigators of this period proposed chronologies prematurely. On the other hand, by presenting their ideas to the scientific community, there evolved a communal set of multiple working hypotheses. Today's novice Lake Bonneville student may perhaps best consider these multiple chronologies as a kind of "noise" reflecting the confusion and enthusiasm of the period.

An era of conservatism, 1979-1988

Researchers in the latest period of Lake Bonneville chronology studies exhibit a strong conservative tendency that is probably a reaction against the excessive interpretive license displayed in the preceding interval. Lake Bonneville time-altitude diagrams published in this era generally portray only well supported fluctuations (Figure 11), while holding tentative inferences to a minimum. Moderate disagreements over the chronology, however, still exist largely due to

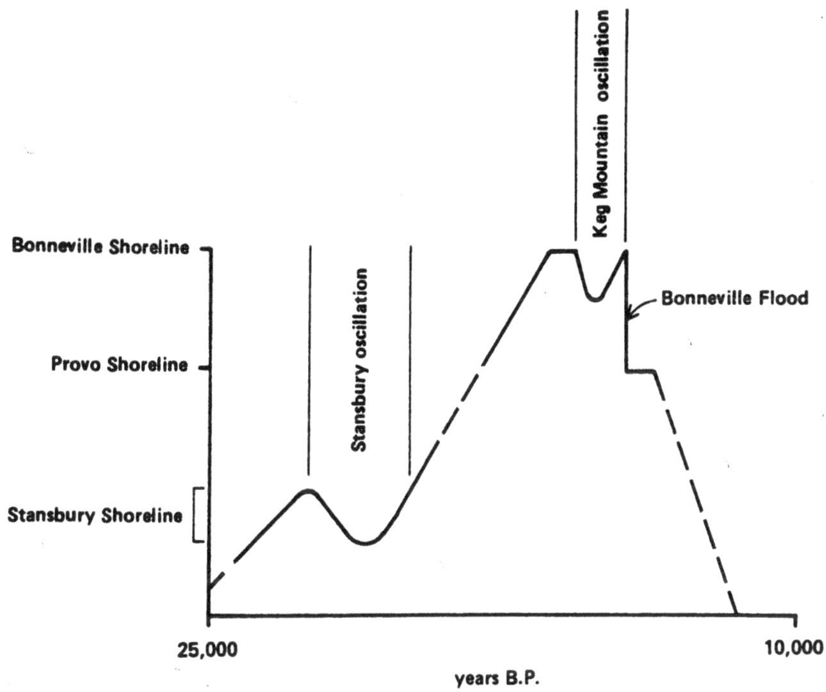

Figure 11

Schematic time-altitude diagram of Lake Bonneville by Oviatt (1984 p. 4, Figure 2). Reproduced with permission of the author.

varying interpretations of traditional stratigraphic and geomorphic evidence (e.g. Scott *et al.* 1983b, Currey and Oviatt 1985), but also because of different inferences derived from Great Salt Lake sediment core versus shoreline morphostratigraphic data (e.g. Spencer *et al.* 1984, Currey and Oviatt 1985). Technical progress has added new methods, including amino-acid racemization and stable-isotope analysis, to the storehouse of Lake Bonneville research tools. Both the neoconservatism and the modern methods have led to a reexamination of and partial return to Gilbert's (1890) Bonneville chronology.

Technical and methodological advances in Lake Bonneville studies made from 1979 to 1988 include (1) rigorous application of stratigraphic methods (e.g. Scott *et al.* 1983b, Currey *et al.* 1983a, Oviatt 1987), (2) cautious use of soil stratigraphy (Scott and Shroba 1980, Shroba 1980), (3) refined understanding of radiometric dating (e.g. Smith and Street-Perrott 1983, Street-Perrott and Harrison 1985), (4) detailed analysis of sediment cores (Spencer *et al.* 1984), (5) aminostratigraphy (Scott *et al.* 1983b, McCoy 1987), (6) use of tephrochronology (Spencer *et al.* 1984, Oviatt and Nash 1989), (7) stable-isotope analysis (McKenzie *et al.* 1982), and (8) thermoluminescence dating techniques

(McCalpin 1986). Equally important, however, has been a renewed emphasis on traditional geomorphology and morphostratigraphy, as originally employed by Gilbert (1882, 1890). For example, Currey (1980a, 1980b, 1980d, 1980e, 1982), Currey *et al.* (1983a), and Currey and Oviatt (1985) used shoreline isostatic-deformation data together with morphostratigraphy to determine the amount and duration of hydro-isostatic events, and thereby to resolve details of lake-level fluctuations during Bonneville and Provo Shoreline time. These researchers provide the most notable recent examples of the continued value of geomorphology to Lake Bonneville chronology studies.

An onslaught of challenges to the methods and conclusions of the preceding period in Bonneville studies began in 1979. In that year, Scott (1979a, 1979b) published the first papers to argue specifically against the validity of the Alpine and Draper Formations (Hunt *et al.* 1953, Morrison 1964, 1965a, 1965b, 1965c). Rather than emphasizing the ambiguous original definition of the Alpine formation (Hunt *et al.* 1953), as discussed above, Scott (1979a, 1980) focused on the emerging awareness that people (e.g. Varnes and Van Horn 1961, Bissell 1963, Morrison 1965a, 1965b, 1965c) had mapped both Bonneville and pre-Bonneville lake deposits as Alpine Formation. Scott and Shroba (1980) highlighted this error by presenting evidence that Morrison's pre- and post-Alpine geosols, the Dimple Dell and Promontory Soils, respectively (Figure 10), were probably equivalent. Scott *et al.* (1983b) and McCoy (1987) used aminostratigraphy to support the contention of a mismapped Alpine Formation. In addition, sediments within the Alpine Formation that Morrison inferred were subærial, and thus evidence for intra-Alpine oscillations, were now interpreted as subaqueous debris flows or minor changes in lacustrine facies (Scott *et al.* 1983b). Scott and other workers (e.g. Scott *et al.* 1982, 1983b, McCoy 1987) suggested that rejecting the Alpine Formation would be the most feasible way of dealing with the resulting confusion.

Morrison (1964, 1965a, 1965b, 1965c) described the Draper Formation as displaying evidence of a lake-level rise from a low stage to 4747 ft (1447 m) within the last 10,000 yrs. This peak, however, had been problematic since it was first proposed because it conflicts with an 11,453 ± 600 yr B.P. radiocarbon date (C-609) on Danger Cave (Figure 2) sheep dung from an elevation of 4311 ft (1314 m) (Jennings 1957 p. 93). Numerous subsequent radiocarbon dates on subærial materials from Danger Cave support the validity of the sheep dung age. Scott (1979a) contributed to the resolution of this problem by noting that the lower member of the Draper Formation at its type locality is nonlacustrine. Several subsequent papers advanced strong additional lines of evidence supporting the nonexistence of the Draper Formation (e.g. Currey 1980c, Krusi and Patterson 1980, Miller *et al.* 1980, Scott 1980, Oviatt 1987).

Rejecting the Alpine and Draper Formations necessitated redefinition of the Bonneville basin's late Quaternary stratigraphic units. Names for the last two deep-lake cycles and their deposits evolved in an awkward fashion. Scott (1980 p. 553) informally referred to the sediments of the most recent deep-water cycle as the Bonneville lake cycle deposits; he termed sediments of the penultimate deep-lake cycle "deposits of the next-to-the-last lake cycle." In 1981, McCoy called the final and penultimate deep-lake cycles the Cottonwood and Little Valley lake cycles, respectively. Later, Scott *et al.* (1982, 1983a) used Jordan Valley and

Little Valley lake cycles for the latest and penultimate Bonneville basin Pleistocene lakes.

Publication of the 1983 North American Stratigraphic Code (North American Commission on Stratigraphic Nomenclature (NACSN)) occurred during this period of major change in Lake Bonneville thought, and provided a formal means by which to redefine Bonneville basin stratigraphic units. Bonneville basin deposits are more feasibly grouped into allostratigraphic than lithostratigraphic units because a logical set of Bonneville basin sediments, for example deposits of a single lake cycle, has time-transgressive (diachronic) boundaries and lithologic heterogeneity. Currey *et al.* (1984b p. 227) first applied the concept of allostratigraphy, as explained in the new code (NACSN, 1983 p. 865-867), to sediments of the Bonneville basin. They defined the Bonneville Alloformation as composed of the sediments of the last deep-lake cycle and the Bonneville Episode as the time represented by those sediments. However, as Hunt *et al.* (1953) had neglected to do when defining the Alpine formation 31 years earlier, Currey *et al.* (1984b) failed to designate a type locality and type area for the Bonneville Alloformation. Finally, McCoy (1987 p. 106-108) formally named the Little Valley Alloformation and denoted type localities and sections for both the Bonneville and Little Valley Alloformations in Little Valley, Utah (Figure 2).

Abandoning the Alpine and Draper Formations also meant renouncing the complex Lake Bonneville histories that employed the discarded formations. In their place, researchers postulated relatively simple complete or partial chronologies, depending on the available evidence (e.g. Currey 1980a, Scott 1980, Scott *et al.* 1983b, Spencer *et al.* 1984, Oviatt 1984, McCoy 1987). As part of their ongoing research, Currey and Oviatt (1987 unpublished diagram) recently formulated the most detailed reconstruction of this period (Figure 12). Lake Bonneville chronologies of the current era resemble Gilbert's (1890) view of the lake's fluctuations (Figure 5) more than they resemble the reconstructions of the preceding period (see Figures 9 and 10).

According to the model presented by Currey and Oviatt (1985, 1987 unpublished) (Figure 12), the Bonneville lacustral cycle began between 32,000 and 25,000 yr B.P. (Mehringer 1977, Spencer *et al.* 1984). During this closed-basin transgressive phase the lake formed various shorelines, including Gilbert's (1890) Stansbury and Intermediate shorelines. Inferences from Great Salt Lake cores (Spencer *et al.* 1984) support the strong coastal stratigraphic evidence of Currey *et al.* (1983a) that the Stansbury Shoreline is transgressive and not regressive as Gilbert portrayed it (1890 p. 134 and Figure 34 p. 262). After a double oscillation around the Stansbury level from about 22,500 to 20,500 yr B.P., Lake Bonneville reached external threshold control at an elevation of 5092 ft (1552 m) near Zenda, Idaho (Figure 2). Initial occupation of the Bonneville Shoreline occurred after 16,400 yr B.P., and perhaps as late as 15,500 to 15,000 yr B.P. (D.R. Currey pers. commun. 1988). Progressive isostatic depression and possible periods of short-lived regression below the Zenda threshold resulted in the formation of three successively higher morphostratigraphic components of the Bonneville Shoreline complex. As portrayed in Figure 12, climatic conditions caused the lake level to drop about 145 ft (44 m) before rising to the Zenda threshold between about 15,900 and 15,000 yr B.P. More recent research, however, indicates that this Keg Mountain oscillation may have occurred as

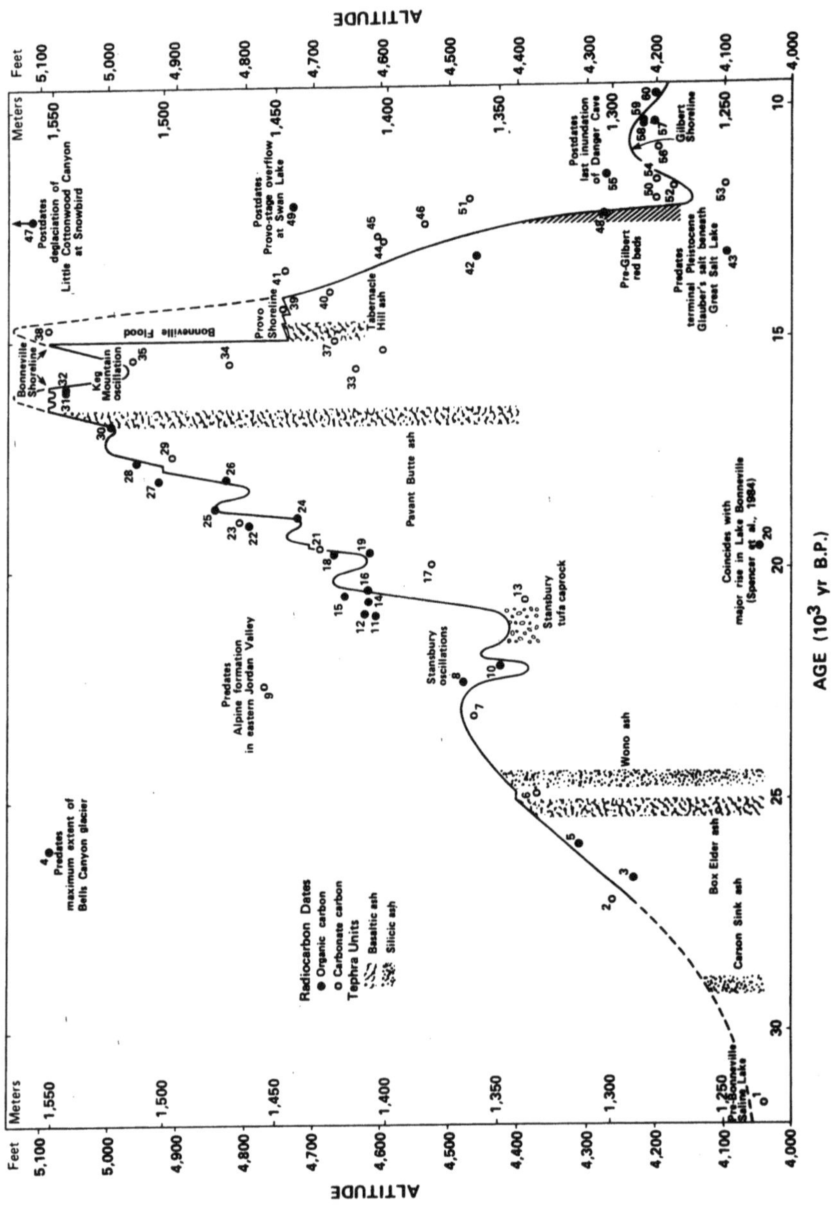

ALTITUDE
Feet
Meters
AGE (10³ yr B.P.)
Radiocarbon Dates
● Organic carbon
o Carbonate carbon
Tephra Units
Basaltic ash
Silicic ash
Bonneville Shoreline
Keg Mountain oscillation
Bonneville Flood
Provo Shoreline
Tabernacle Hill ash
Pavant Butte ash
Stansbury oscillations
Stansbury tufa caprock
Wono ash
Box Elder ash
Carson Sink ash
Pre-Bonneville Sailing Lake
Gilbert Shoreline
Pre-Gilbert red beds
Postdates deglaciation of Little Cottonwood Canyon at Snowbird
Postdates Provo-stage overflow at Swan Lake
Postdates last inundation of Danger Cave
Predates terminal Pleistocene Glauber's salt beneath Great Salt Lake
Predates maximum extent of Bells Canyon glacier
Predates Alpine formation in eastern Jordan Valley
Coincides with major rise in Lake Bonneville (Spencer et al., 1984)

recently as 15,000 to 14,500 yr B.P. (D.R. Currey pers. commun. 1988). Pre-oscillation discharge over the divide had apparently greatly reduced the strength of the threshold by headward erosion (Currey 1980b). When the water plane again reached Zenda the alluvial material failed and the enormous Bonneville Flood (Gilbert 1890, Malde 1968, Jarrett and Malde 1987) ensued. In probably less than a year, the overflow point eroded about 355 ft (108 m) vertically and retreated headward about 2 mi (3.2 km) to near Red Rock Pass. With the lake level controlled by this new elevation, the Provo Shoreline was formed. Approximately 14,000 yr B.P. climatic amelioration caused the lake to return to closed-basin conditions. The lake regressed rapidly to at least 4137 ft (1261 m), which is about 63 ft (19 m) below the modern elevation of Great Salt Lake. Shortly after 11,000 yr B.P., the lake readvanced to an elevation of 4250 ft (1295 m), where it created the Gilbert Shoreline (Eardley *et al.* 1957). Since retreating from the Gilbert Shoreline about 10,000 yr B.P., Great Salt Lake has oscillated around an elevation of 4200 ft (1280 m).

Currey and Oviatt constrain this model with numerous radiocarbon dates and with geomorphic, stratigraphic, and morphostratigraphic information from many sites around the Bonneville basin. Their work exemplifies the trend of the period toward integrating careful geomorphic, stratigraphic, and morphostratigraphic observations with cautious use of the latest chronometric technology. Fundamental changes from Gilbert's (1890) reconstruction consist of (1) including only fluctuations from the last deep-lake cycle, during which both the yellow clay and white marl were deposited (Oviatt 1987), (2) changing the Stansbury Shoreline from post-Provo to pre-Bonneville Shoreline time, (3) displaying a much improved understanding of the transgressive lake-level oscillations, (4) recognizing a short-lived, 145-ft (44m) oscillation during Bonneville Shoreline time, and (5) adding the Gilbert Shoreline.

Conclusions

As in the initial study of many classic Quaternary features, the use of geomorphic observation was almost exclusive in the pre-Gilbert era of Lake Bonneville studies (1776-1873) when simple shoreline recognition and description dominated the work. Later (1874-1890), Gilbert employed especially geomorphic analysis but also stratigraphic and morphostratigraphic techniques. His comparatively restricted use of stratigraphy was quite appropriate and understandable given the early stage of Bonneville basin studies when he entered the field. Gilbert accomplished so much with his keen geomorphic observations that subsequent major Lake Bonneville chronology studies were not undertaken for over 50 years after the publication of Monograph 1.

Figure 12 (<-- see opposite page)

A recent time-altitude diagram of Lake Bonneville by Currey and Oviatt (1987 unpublished diagram). Reproduced with permission of the authors.

Compared to the almost stagnant immediate post-Gilbert era of Lake Bonneville chronology studies (1891-1944), Bonneville researchers of the mid-20th century period (1945-1978) suggested numerous versions of the lake's history, but only after embracing new methods and techniques and greatly increasing the use of stratigraphy. Most of these chronologies differ substantially from Gilbert's interpretation. In general, these researchers widely and overconfidently used newly developed relative and absolute dating techniques in an effort to determine the age of Lake Bonneville deposits. At the same time they largely ignored traditional geomorphic approaches. Confusion arose as single Lake Bonneville features yielded multiple ages (e.g. Eardley *et al.* 1957 p. 117, Feth and Rubin 1957 p. 1827) and as interpretations of Bonneville lacustrine history failed to converge. Detailed basinwide inferences drawn from these applications of stratigraphic, soil stratigraphic, and radiometric-dating techniques failed to indicate a consistent outline of the Lake Bonneville chronology. This lack of consensus revealed more about the methods and techniques and the manner in which they were employed, than about the chronology of Lake Bonneville.

By the late 1970s and early 1980s several researchers had become disillusioned with the way in which Quaternary mapping, stratigraphic, soil stratigraphic, and radiometric techniques had been applied to Lake Bonneville chronology studies (e.g. Scott 1979a, 1979b, McCoy 1981, Scott *et al.* 1982, 1983b). Lack of adequate stratigraphic definitions, mismapping of Quaternary units, overzealous application of soil stratigraphy, interpreting minor facies changes as major unconformities, and conflicting radiocarbon dates were considered to be fundamental weaknesses in those previous studies. As the result of these problems, the models from the preceding period were discarded. Moreover, researchers in the present period of Lake Bonneville studies (1979-1988) reexamined the techniques that the previous generation of workers had so confidently embraced. These techniques are now applied more cautiously to Lake Bonneville chronology studies (e.g. Scott and Shroba 1980, Scott *et al.* 1983b, Oviatt 1987). In addition, recent workers have reestablished the importance of geomorphology in Bonneville basin research (Currey 1980a, 1980b, 1980d, 1980e, 1982, Currey *et al.* 1983a, 1984b, Currey and Oviatt 1985). These trends have helped the chronology studies to reach a closer consensus (e.g. McCoy 1981, Scott *et al.* 1983b, Currey and Oviatt 1985). In this latest period of Bonneville studies, both research trends and chronology diagrams have moved closer to those of Gilbert.

Acknowledgments

The research for this paper was partially supported by the University of Utah Research Committee. I am very grateful to K. J. Tinkler, M. T. Greene, D. R. Currey, C. G. Oviatt, and T. N. Burr for helpful discussions and thoughtful comments regarding an earlier draft of this paper.

References

Anonymous, 1878. Probable ancient outlet of the Great Salt Lake. *American Journal of Science*, 15, 68.

Antevs, E., 1925. On the Pleistocene history of the Great Basin. *Carnegie Institute of Washington Publication* , 352, 51-114.

Antevs, E., 1938. Climatic variations during the last glaciation in North America. *Bulletin of the American Meteorological Society* , 19, 172-176.

Antevs, E., 1945. Correlation of Wisconsin glacial maxima. *American Journal of Science* , 243A, 1-39.

Antevs, E., 1948. Climatic changes and pre-white man. *Bulletin of the University of Utah* , 38, 168-191.

Antevs, E., 1952. Cenozoic climates of the Great Basin. *Geologische Rundschau*, 40, 94-108.

Antevs, E., 1955. Geologic-climatic dating in the West. *American Antiquity*, 20, 317-335.

Atwood, W.W., 1909. Glaciation of the Uinta and Wasatch Mountains. *United States Geological Survey Professional Paper*, 61.

Auerbach, H.S., 1943. Father Escalante's journal with related documents and maps. *Utah Historical Quarterly*, 11, 1-142.

Baker, V.R. and Pyne, S.J., 1978. G.K. Gilbert and modern geomorphology. *American Journal of Science*, 278, 97-123.

Beckwith, E.G., 1855. Explorations for a route for the Pacific Railroad, of the line of the forty-first parallel of north latitude. In *Reports of explorations and surveys to ascertain the most practicable and economic route for a railroad from the Mississippi River to the Pacific Ocean 1853-4*. Congressional Documents, 33rd Congress, 2nd Session, Senate Executive Document No. 78.

Benson, L.V. and Thompson, R.S., 1987. The physical record of lakes in the Great Basin. Pages 241-260 *in* **W.F. Ruddiman and H.E. Wright, Jr.**, eds., *North America and adjacent oceans during the last deglaciation*. Geological Society of America, *The geology of North America*, Vol. K-3.

Bills, B.G. and May, G.M., 1987. Lake Bonneville: constraints on lithospheric thickness and upper mantle viscosity from isostatic warping of Bonneville, Provo, and Gilbert stage shorelines. *Journal of Geophysical Research*, 92 (B11), 11493-11508.

Bissell, H.J., 1952. Stratigraphy of Lake Bonneville and associated Quaternary deposits in Utah Valley, Utah. *Geological Society of America Bulletin*, 63, 1358.

Bissell, H.J., 1963. Lake Bonneville: geology of southern Utah Valley, Utah. *United States Geological Survey Professional Paper* , 257B.

Bissell, H.J., 1968. Bonneville - an ice-age lake. *Brigham Young University Geology Studies*, 15, no. 4.

Blackwelder, E., 1931. Pleistocene glaciation in the Sierra Nevada and Basin Ranges. *Geological Society of America Bulletin*, 42, 865-922.

Boutwell, J.M., 1933. Excursion 1 - Wasatch Front. Pages 32-45 *in* **J.M. Boutwell**, ed., The Salt Lake region, *International Geological Congress, XVI session, Guidebook 17 - Excursion C-1*. Government Printing Office, Washington, D.C.

Bradley, F.H., 1873. Report of Frank H. Bradley, geologist. Pages 189-271 *in* **F.H. Bradley**, *Sixth annual report of the United States geological survey of the territories, embracing portions of Montana, Idaho, Wyoming, and Utah; being a report of progress of the explorations for the year 1872*. Government Printing Office, Washington, D.C.

Bright, R.C., 1963. *Pleistocene Lakes Thatcher and Bonneville, southeastern Idaho.* Unpublished Ph.D. thesis, University of Minnesota.

Bright, R.C., 1966. Pollen and seed stratigraphy of Swan Lake, southeastern Idaho, its relation to regional vegetation history and to Lake Bonneville history. *Tebiwa,Journal of the Idaho State University Museum* , 9, 1-47.

Broecker, W.S. and Kaufman, A., 1965. Radiocarbon chronology of Lake Lahontan and Lake Bonneville II, Great Basin. *Geological Society of America Bulletin*, 76, 537-566.

Broecker, W.S. and Orr, P.C., 1958. Radiocarbon chronology of Lake Lahontan and Lake Bonneville. *Geological Society of America Bulletin* , 69, 1009-1032.

Burr, T.N., In preparation *Hydrographic modelling of open basin threshold-controlled shorelines of Lake Bonneville.* Unpublished M.S. thesis, University of Utah.

Call, R.E., 1884. On the Quaternary and Recent mollusca of the Great Basin, with descriptions of new forms. *United States Geological Survey Bulletin*, 11, 13-67.

Chorley, R.J. and Beckinsale, R.P., 1980. G.K. Gilbert's geomorphology. *Geological Society of America Special Paper*, 183, 129-142.

Chorley, R.J., Beckinsale, R.P., and Dunn, A.J., 1973. *The history of the study of landforms*, Volume 2--*The life and work of William Morris Davis.* Methuen & Co., Ltd., London.

Crittenden, M.D., Jr., 1963a. Effective viscosity of the earth derived from isostatic loading of Pleistocene Lake Bonneville. *Journal of Geophysical Research*, 68, 5517-5530.

Crittenden, M.D., Jr., 1963b. New data on the isostatic deformation of Lake Bonneville. *United States Geological Survey Professional Paper*, 454E.

Crittenden, M.D., Jr., 1967. Viscosity and finite strength of the mantle as determined from water and ice loads. *Geophysical Journal of the Royal Astronomical Society*, 14, 261-279.

Currey, D.R., 1980a. Coastal geomorphology of Great Salt Lake and vicinity, Utah. *Geological and Mineral Survey Bulletin*, 116, 69-82.

Currey, D.R., 1980b. Events associated with the last cycle of Lake Bonneville--Idaho, Nevada, and Utah. *American Quaternary Association Abstracts and Program*, 6, 59-60.

Currey, D.R., 1980c. Radiocarbon dates and their stratigraphic implications from selected localities in Utah and Wyoming. *Encyclia, Journal of the Utah Academy of Sciences, Arts, and Letters*, 57, 110-115.

Currey, D.R., 1980d. Tectonostratigraphic and morphostratigraphic hypotheses concerning threshold-controlled Lake Bonneville shorelines, with emphasis on shorelines and associated deposits in the Long Bench-North Ogden area, Utah. *Geological Society of America Rocky Mountain Section Annual Meeting,Guide to Field Trip 4*.

Currey, D.R., 1980e. Threshold-controlled shorelines of Lake Bonneville. *Great Plains-Rocky Mountain Geographical Journal*, 9, 103-104.

Currey, D.R., 1982. Lake Bonneville: selected features of relevance to neotectonic analysis. *United States Geological Survey Open-File Report*, 82-1070.

Currey, D.R., Atwood, G., and Mabey, D.R., 1984. Major levels of Great Salt Lake and Lake Bonneville. *Utah Geological and Mineral Survey Map*, 73.

Currey, D.R. and Madsen, D.B., 1974. Holocene fluctuations of Great Salt Lake. *American Quaternary Association Abstracts and Program*, 3, 74.

Currey, D.R. and Oviatt, C.G., 1985. Durations, average rates, and probable causes of Lake Bonneville expansions, stillstands, and contractions during the last deep-lake cycle, 32,000 to 10,000 years ago. Pages 9-24 *in* **P.A. Kay and**

H.F. Diaz, eds., *Problems of and prospects for predicting Great Salt Lake levels*. University of Utah Center for Public Affairs and Administration, Salt Lake City.

Currey, D.R., Oviatt, C.G., and Czarnomski, J.E., 1984. Late Quaternary geology of Lake Bonneville and Lake Waring. *Utah Geological Association Publication*, 13, 227-237.

Currey, D.R., Oviatt, C.G., and Plyler, G.B., 1983a. Lake Bonneville stratigraphy, geomorphology, and isostatic deformation in west-central Utah. *Utah Geological and Mineral Survey Special Studies*, 62, 63-82.

Currey, D.R., Oviatt, C.G., and Plyler, G.B., 1983b. Stansbury Shoreline and Bonneville lacustral cycle. *Geological Society of America Abstracts with Programs*,15, 301.

Davis, W.M., 1883. Lake Bonneville. *Science*, 1, 570-571 and 573.

Davis, W.M., 1892. The ancient shore-lines of Lake Bonneville. *Goldthwaite's Geographical Magazine*, 3, 1-5.

Davis, W.M., 1927. Biographical memoir Grove Karl Gilbert, 1843-1918. *Memoirs of the National Academy of Sciences*, volume 21, fifth memoir.

Davis, W.M., 1933. Geomorphology. Pages 6-14 *in* **J.M. Boutwell**, ed., The Salt Lake region, *International Geological Congress, XVI session, Guidebook 17 - Excursion C-1*. Government Printing Office, Washington, D.C.

Dennis, P.E., 1944. Shorelines of the Escalante Bay of Lake Bonneville, Utah *Academy of Science Proceedings*, 19-20, 121-124.

Eardley, A.J., 1938. Sediments of Great Salt Lake, *Utah. American Association of Petroleum Geologists Bulletin*, 22, 1305-1411.

Eardley, A.J., 1964. Lake Bonneville. *Utah Geological and Mineral Survey Bulletin*, 69, 69-78.

Eardley, A.J., Gvosdetsky, V.M, and Marsell, R.E., 1957. Hydrology of Lake Bonneville and sediments and soils of its basin. *Geological Society of America Bulletin*, 68, 1141-1201.

Engelmann, H., 1876. Geological report of country from Fort Leavenworth to the Sierra Nevada. Appendix I *to* **J.H. Simpson,** *Report of explorations across the Great Basin of the Territory of Utah for a direct wagon-route from Camp Floyd to Genoa, in Carson Valley, in 1859*. U.S. Army Engineer Department, Washington, D.C.

Feth, J.H. and Rubin, M., 1957. Radiocarbon dating of wave-formed tufas from the Bonneville basin. *Geological Society of America Bulletin*, 68, 1827.

Flint, R.F., 1974. Three theories in time. *Quaternary Research*, 4, 1-8.

Fremont, J.C., 1845. *Exploring expedition to the Rocky Mountains in the year 1842, and to Oregon and North California in the years 1843-44*. Congressional Documents, 28th Congress, 2nd Session, Senate Document 174.

Gilbert, G.K., 1874. Preliminary geological report. Appendix D *to* **G.M. Wheeler,** *Progress-report upon geographical and geological explorations and surveys west of the one hundredth meridian in 1872*. Government Printing Office, Washington, D.C.

Gilbert, G.K., 1875. Report upon the geology of portions of Nevada, Utah, California, and Arizona, examined in the years 1871 and 1872. Pages 17-187 *in* **G.M. Wheeler,** *Report upon geographical and geological explorations and surveys west of the one hundredth meridian*. Government Printing Office, Washington, D.C.

Gilbert, G.K., 1876. On the outlet of Great Salt Lake. *American Journal of Science*, third series 11, 228-229.

Gilbert, G.K., 1878. The ancient outlet of Great Salt Lake; a letter to the editors. *American Journal of Science*, third series 15, 256-259.

Gilbert, G.K., 1880a. The outlet of Lake Bonneville. *American Journal of Science*, third series 19, 341-349.

Gilbert, G.K., 1880b. The Wasatch a growing mountain. *Bulletin of the Philosophical Society of Washington*, 2, 195.

Gilbert, G.K., 1882. Contributions to the history of Lake Bonneville. Pages 167-200 *in* **J.W. Powell,** *Report of the Director of the United States Geological Survey*. Government Printing Office, Washington, D.C.

Gilbert, G.K., 1885. The topographic features of lake shores. Pages 69-123 *in* **J.W. Powell,** *Fifth annual report of the United States Geological Survey 1883-84*. Government Printing Office, Washington, D.C.

Gilbert, G.K., 1886. The inculcation of scientific method by example, with an illustration drawn from the Quaternary geology of Utah. *American Journal of Science*, third series 31, 284-299.

Gilbert, G.K., 1890. *Lake Bonneville*. United States Geological Survey Monograph 1.

Gilbert, G.K., 1896. The origin of hypotheses, illustrated by the discussion of a topographic problem. *Science*, 3, 1-13.

Gilbert, G.K., 1904. Domes and dome structures of the High Sierra. *Geological Society of America Bulletin*, 15, 29-36.

Gvosdetsky, V. and Hawkes, H.B., 1953. Reappraisal of the history of Lake Bonneville. *Utah Engineering Experiment Station Bulletin*, 60, volume 43.

Gvosdetsky, V. and Hawkes, H.B., 1954. Fossil soils as a key to the past climates of the Bonneville basin. *Geological Society of America Bulletin*, 65, 1394.

Gvosdetsky, V. and Hawkes, H.B., 1956. Middle stream terrace of the Bonneville basin--a key to the history of Lake Stansbury. *Utah Engineering Experiment Station Bulletin*, 81, volume 47.

Hague, A. and Emmons, S.F., 1877. Descriptive geology. Volume II *of* **C. King,** *Report of the geological exploration of the fortieth parallel*. Government Printing Office, Washington, D.C.

Hay, O.P., 1927. *The Pleistocene of the western region of North America and its vertebrated animals*. Carnegie Institute of Washington Publication, 322B.

Hayden, F.V., 1871. *Preliminary report of the United States geological survey of Wyoming, and portions of contiguous territories*. Government Printing Office, Washington, D.C.

Howell, E.E., 1875. Report on the geology of portions of Utah, Nevada, Arizona, and New Mexico, examined in the years 1872 and 1873. Pages 227-301 *in* **G. M. Wheeler,** *Report upon geographical and geological explorations and surveys west of the one hundredth meridian*. Government Printing Office, Washington, D.C.

Hubbs, C.L. and Miller, R.R., 1948. The zoological evidence. *Bulletin, of the University of Utah*, 38, 17-166.

Hunt, C.B., 1955. Radiocarbon dating in the light of stratigraphy and weathering processes. *The Scientific Monthly*, 81, 240-247.

Hunt, C.B., 1980. G.K. Gilbert's Lake Bonneville studies. *Geological Society of America Special Paper*, 183, 45-59.

Hunt, C.B., 1987. Physiography of western Utah. *Utah Geological Association Publication*, 16, 1-29.

Hunt, C.B. and Sokoloff, V.P., 1950. Pre-Wisconsin soil in the Rocky Mountain region, a progress report. *United States Geological Survey Professional Paper*, 221G.

Hunt, C.B., Varnes, H.D., and Thomas, H.E., 1953. Lake Bonneville: geology of northern Utah Valley, Utah. *United States Geological Survey Professional Paper*, 257A.

Huntington, E., 1914. The climatic factor as illustrated in arid America. *Carnegie Institute of Washington Publication*, 192.
Irving, W., 1836. *Astoria*. Carey, Lea, and Blanchard, Philadelphia.
Irving, W., 1837. *Adventures of Captain Bonneville, or scenes beyond the Rocky Mountains of the far west*. A. and W. Galignani and Co., Paris.
Ives, R.L., 1948a. An early report of ancient lakes in the Bonneville basin. *Journal of Geology*, 56, 79-80.
Ives, R.L., 1948b. The outlet of Lake Bonneville. *Scientific Monthly*, 67, 415-426.
Ives, R.L., 1951. Pleistocene valley sediments of the Dugway area, Utah. *Geological Society of America Bulletin*, 62, 781-797.
Jarrett, R.D. and Malde, H.E., 1987. Paleodischarge of the late Pleistocene Bonneville Flood, Snake River, Idaho, computed from new evidence. *Geological Society of America Bulletin*, 99, 127-134.
Jennings, J.D., 1957. Danger Cave. *Society for American Archaeology Memoir* 14.
Jones, D.J. and Marsell, R.E., 1955. Pleistocene sediments of lower Jordan Valley, Utah. *Utah Geological Society Guidebook to the Geology of Utah*, 10, 85-112.
Keyes, C.R., 1917. Orographic origin of ancient Lake Bonneville. *Geological Society of America Bulletin*, 28, 351-374.
Keyes, C.R., 1918. Lacustral record of past climates. *Monthly Weather Review*, 46, 277-280.
King, C., 1878. Systematic geology. Volume I *of* **C.King**, *Report of the geological exploration of the fortieth parallel*. Government Printing Office, Washington, D.C.
Krusi, A.P. and Patterson, R.H., 1980. Problems in Lake Bonneville stratigraphic relationships in the northern Sevier basin revealed by exploratory trenching. *United States Geological Survey Open-File Report*, 80-801, 509-518.
Madsen, D.B., 1980. The human prehistory of the Great Salt Lake region. *Utah Geological and Mineral Survey Bulletin*, 116, 19-31.
Madsen, D.B. and Currey, D.R., 1979. Late Quaternary glacial and vegetation changes, Little Cottonwood Canyon area, Wasatch Mountains, Utah. *Quaternary Research*, 12, 254-270.
Madsen, D.B. and Kay, P.A., 1982. Late Quaternary pollen analysis in the Bonneville basin. *American Quaternary Association Program and Abstracts*, 7, 128.
Malde, H.E., 1968. The catastrophic late Pleistocene Bonneville Flood in the Snake River Plain, Idaho. *United States Geological Survey Professional Paper*, 596.
Marsell, R.E. and Jones, D.J., 1955. Pleistocene history of lower Jordan Valley, Utah. *Utah Geological Society Guidebook to the Geology of Utah*, 10, 113-120.
McCalpin, J., 1986. Thermoluminescence (TL) dating in seismic hazard evaluations: an example from the Bonneville basin, Utah. Pages 156-176 *in Proceedings of the 22nd Symposium on Engineering Geology and Soils Engineering*.
McCoy, W.D., 1981. *Quaternary aminostratigraphy of the Bonneville and Lahontan basins, western U.S., with paleoclimatic implications*. Unpublished Ph.D. thesis, University of Colorado.
McCoy, W.D., 1987. Quaternary aminostratigraphy of the Bonneville basin, western United States. *Geological Society of America Bulletin*, 98, 99-112.
McKenzie, J.A., Eberli, G., Kelts, K., and Pika, J., 1982. Late Pleistocene-Holocene climatic record in carbonate sediments from the Great Salt

Lake, Utah: documentation of lake-level fluctuations using oxygen-isotope stratigraphy. *Geological Society of America Abstracts with Programs*, 14, 562.

Mehringer, P.J., Jr., 1977. Great Basin late Quaternary environments and chronology. *Desert Research Institute Publications in the Social Sciences*, 12, 113-167.

Mendenhall, W.C., 1920. Memorial of Grove Karl Gilbert. *Geological Society of America Bulletin*, 31, 26-64.

Miller, D.E., 1966. Great Salt Lake: an historical sketch. *Utah Geological Society Guidebook to the Geology of Utah*, 20, 3-24.

Miller, R.D., Van Horn, R., Scott, W.E., and Forester, R.M., 1980. Radiocarbon date supports concept of continuous low levels of Lake Bonneville since 11,000 years B.P. *Geological Society of America Abstracts with Programs*, 12, 297-298.

Morgan, D.L., 1953. *Jedediah Smith and the opening of the West*. The Bobbs-Merrill Co., Inc., New York.

Morrison, R.B., 1961a. Correlation of the deposits of Lakes Lahontan and Bonneville and the glacial sequences of the Sierra Nevada and Wasatch Mountains, California, Nevada, and Utah. *United States Geological Survey Professional Paper*, 424D, 122-124.

Morrison, R.B., 1961b. New evidence on the history of Lake Bonneville from an area south of Salt Lake City, Utah. *United States Geological Survey Professional Paper*, 424D, 125-127.

Morrison, R.B., 1964. *Soil stratigraphy: principles, applications to differentiation and correlation of Quaternary deposits and landforms, and applications to soil science*. Unpublished Ph.D. thesis, University of Nevada, Reno.

Morrison, R.B., 1965a. Lake Bonneville: Quaternary stratigraphy of eastern Jordan Valley, south of Salt Lake City, Utah. *United States Geological Survey Professional Paper*, 477.

Morrison, R.B., 1965b. New evidence on Lake Bonneville stratigraphy and history from southern Promontory Point, Utah. *United States Geological Survey Professional Paper*, 525C, 110-119.

Morrison, R.B., 1965c. Quaternary geology of the Great Basin. Pages 265-285 *in* **H.E. Wright, Jr. and D.G. Frey**, eds., *The Quaternary of the United States*. Princeton University Press, Princeton, New Jersey.

Morrison, R.B., 1966. Predecessors of Great Salt Lake. *Utah Geological Society Guidebook to the Geology of Utah*, 20, 77-104.

Morrison, R.B., 1975. Predecessors of Great Salt Lake. *Geological Society of America Abstracts with Programs*, 7, 1206.

Morrison, R.B. and Frye, J.C., 1965. Correlation of the middle and late Quaternary succession of the Lake Lahontan, Lake Bonneville, Rocky Mountain (Wasatch Range), southern Great Plains, and eastern Midwest areas. *Nevada Bureau of Mines Report* 9.

Nakiboglu, S.M. and Lambeck, K., 1982. A study of the earth's response to surface loading with application to Lake Bonneville. *Geophysical Journal of the Royal Astronomical Society*, 70, 577-620.

Nakiboglu, S.M. and Lambeck, K., 1983. A re-evaluation of the isostatic rebound of Lake Bonneville. *Journal of Geophysical Research*, 88, 10439-10447.

North American Commission on Stratigraphic Nomenclature (NACSN), 1983. North American stratigraphic code. *American Association of Petroleum Geologists Bulletin*, 20, 841-875.

Oviatt, C.G., 1984. *Lake Bonneville stratigraphy at the Old River Bed and Leamington, Utah*. Unpublished Ph.D. thesis, University of Utah.

Oviatt, C.G., 1987. Lake Bonneville stratigraphy at the Old River Bed, Utah. *American Journal of Science*, 287, 383-398.

Oviatt, C.G. and Currey, D.R., 1987. Pre-Bonneville Quaternary lakes in the Bonneville basin, Utah. *Utah Geological Association Publication*, 16, 257-263.

Oviatt, C.G., McCoy, W.D., and Reider, R.G., 1987. Evidence for a shallow early or middle Wisconsin-age lake in the Bonneville basin, Utah. *Quaternary Research*, 27, 248-262.

Oviatt, C.G. and Nash, W.P., 1989. Late Pleistocene basaltic ash and volcanic eruptions in the Bonneville basin, Utah. *Geological Society of America Bulletin*, in press.

Pack, F.J., 1939. Lake Bonneville: a popular treatise dealing with the history and physical aspects of Lake Bonneville. *Bulletin of the University of Utah*, 30, 1-112.

Packard, A.S., Jr., 1876a. On the supposed ancient outlet of Great Salt Lake. *Bulletin of the U.S. Geological and Geographical Survey of the Territories*, second series 1, 413-414.

Packard, A.S., Jr., 1876b. The Great Salt Lake in former times. *American Naturalist*, 10, 675-681.

Passey, Q.R., 1981. Upper mantle viscosity derived from the difference in rebound of the Provo and Bonneville shorelines: Lake Bonneville basin, Utah. *Journal of Geophysical Research*, 86, 11701-11708.

Peale, A.C., 1878. The ancient outlet of Great Salt Lake. *American Journal of Science*, third series 15, 439-444.

Peale, A.C., 1879. Report. Pages 506-646 *in* **F.V. Hayden,** *Eleventh annual report of the United States geological and geographical survey of the territories, embracing Idaho and Wyoming, being a report of progress of the exploration for the year 1877*. Government Printing Office, Washington, D.C.

Pyne, S.J., 1975. The mind of Grove Karl Gilbert. Pages 277-298 *in* **W.N. Melhorn and R.C. Flemal,** eds., *Theories of landform development*, Proceedings of the sixth annual Binghamton Geomorphology Symposium. George Allen & Unwin, London.

Rudy, R.C., Jr., 1973. *Holocene fluctuations of Great Salt Lake, with special reference to evidence from the western shore*. Unpublished M.S. thesis, University of Utah.

Russell, I.C., 1895. *Lakes of North America*. Ginn and Co., Boston.

Scott, W.E., 1979a. Evaluation of evidence for a controversial rise of Lake Bonneville about 10,000 yrs ago. *Geological Society of America Abstracts with Programs*, 11, 301-302.

Scott, W.E., 1979b. Stratigraphic problems in the usage of Alpine and Bonneville Formations in the Bonneville Basin, Utah. *Geological Society of America Abstracts with Programs*, 11, 302.

Scott, W.E., 1980. New interpretations of Lake Bonneville stratigraphy and their significance for studies of earthquake-hazard assessment along the Wasatch Front. *United States Geological Survey Open-File Report*, 80-801, 548-576.

Scott, W.E., McCoy, W.D., and Shroba, R.R., 1983. The last two cycles of Lake Bonneville. *Geological Society of America Abstracts with Programs*, 15, 300.

Scott, W.E., McCoy, W.D., Shroba, R.R., and Rubin, M., 1983. Reinterpretation of the exposed record of the last two lake cycles of Lake Bonneville, western United States. *Quaternary Research*, 20, 261-285.

Scott, W.E. and Shroba, R.R., 1980. Stratigraphic significance and variability of soils buried by deposits of the last cycle of Lake Bonneville. *Geological Society of America Abstracts with Programs*, 12, 304.

Scott, W.E., Shroba, R.R., and McCoy, W.D., 1982. Guidebook for the 1982 Friends of the Pleistocene, Rocky Mountain Cell, field trip to Little Valley and Jordan Valley, Utah. *United States Geological Survey Open-File Report*, 82-845.

Shroba, R.R., 1980. Influence of parent material, climate, and time on soils formed in Bonneville-shoreline and younger deposits near Salt Lake City and Ogden, Utah. *Geological Society of America Abstracts with Programs*, 12, 304.

Simpson, J.H., 1876. *Explorations across the Great Basin Territory of Utah for a direct wagon-route from Camp Floyd to Genoa, in Carson Valley, in 1859*. Government Printing Office, Washington, D.C.

Smith, G.I. and Street-Perrott, F.A., 1983. Pluvial lakes of the western United States. Pages 190-212 *in* **S.C. Porter**, ed., *The Late Pleistocene*, Volume 1, *Late Quaternary environments of the United States*. University of Minnesota Press, Minneapolis.

Spencer, R.J., Baedecker, M.J., Eugster, H.P., Forester, R.M., Goldhaber, M.B., Jones, B.F., Kelts, K., McKenzie, J., Madsen, D.B., Rettig, S.L., Rubin, M., and Bowser, C.J., 1984. Great Salt Lake, and precursors, Utah, the last 30,000 years. *Contributions to Mineralogy and Petrology*, 86, 321-334.

Stansbury, H., 1852. *Exploration and survey of the valley of the Great Salt Lake of Utah*. Congressional Documents, 32nd Congress, U.S. Senate Special Session, March 1851, Executive Document 3. Lippincott, Grambo and Co., Philadelphia.

Stokes, W.L., 1979. Paleohydrographic history of the Great Basin region. Pages 345-351 *in* **G.W. Newman and H.D. Goode**, eds., *Basin and Range symposium and Great Basin field conference*. Rocky Mountain Association of Geologists and Utah Geological Association.

Street-Perrott, F.A. and Harrison, S.P., 1985. Lake levels and climate reconstruction. Pages 291-340 *in* **A.D. Hecht**, ed., *Paleoclimate analysis and modeling*. John Wiley & Sons, New York.

Todd, E.W., 1961. Editor's introduction. Pages xvii-liv *in* **W.Irving**, *The adventures of Captain Bonneville*, edited and with an introduction by E.W. Todd. University of Oklahoma Press, Norman.

Upham, W., 1922. Stages of the ice age. *Geological Society of America Bulletin*, 3, 491-514.

Van Horn, R., 1979. The Holocene Ridgeland Formation and associated Decker Soil (new names) near Great Salt Lake, Utah. *United States Geological Survey Bulletin*, 1457C, 1-11.

Varnes, D.J. and Van Horn, R., 1961. A reinterpretation of two of G.K. Gilbert's Lake Bonneville sections, Utah. *United States Geological Survey Professional Paper*, 424C, 98-99.

Walcott, R.I., 1970. Flexural rigidity, thickness, and viscosity of the lithosphere. *Journal of Geophysical Research*, 75, 3941-3954.

Williams, T.R. and Bedinger, M.S., 1984. Selected geologic and hydrologic characteristics of the Basin and Range province, western United States--Pleistocene lakes and marshes. *United States Geological Survey Miscellaneous Investigations*, Map I-1511D.

Young, W., 1878. Appendix D *in* **G.M. Wheeler**, *Appendix NN of the annual report upon the geographical surveys of the territory of the United States west of the one hundredth meridian*. Government Printing Office, Washington, D.C.

12

Different aspects of Polish geomorphology: palæogeographic, dynamic and applied

Leszek Starkel

The history of a particular discipline in a given country may be examined as the development of ideas, prominent individuals and their schools, or of international relations and explorations. One should also consider the size of the country and its geographical situation, for both factors control the direction and scale of the research which is undertaken. However, it may be examined also as the product of economic and political conditions since these are reflected by accelerations and halts in that development.

The latter approach seems most appropriate to the case of geomorphology created by the Polish nation, for the nation was blotted out on the map of Europe between 1795 and 1918, and the two World Wars strongly affected its science. Poland, a country of medium size on the European scale, is located in the heart of Europe but in a single climatic and morphogenetic zone. Poland's diversified relief and palæogeography lie at the junction of the East European shields with the hercynian region of Western Europe (where they are broken into horsts and depressions), and the alpine region of southern Europe (which is tectonically active). Within the territory of Poland the Scandinavian ice sheets reached their southernmost limits as their margins underwent rhythmic changes from the forest zone to the arctic desert. All of this may help to explain the development of some branches of Polish geomorphology as well some very distinctive interests in palæogeography and dynamic processes.

We may distinguish the following periods in the development of Polish geomorphology:

1815-1899	Origins of Polish geomorphology
1899-1918	Creative transformation of foreign concepts
1918-1939	Study of the native landscape
1939-1945	Science in abeyance (World War II)
1945-1956	Development of new research techniques and theories

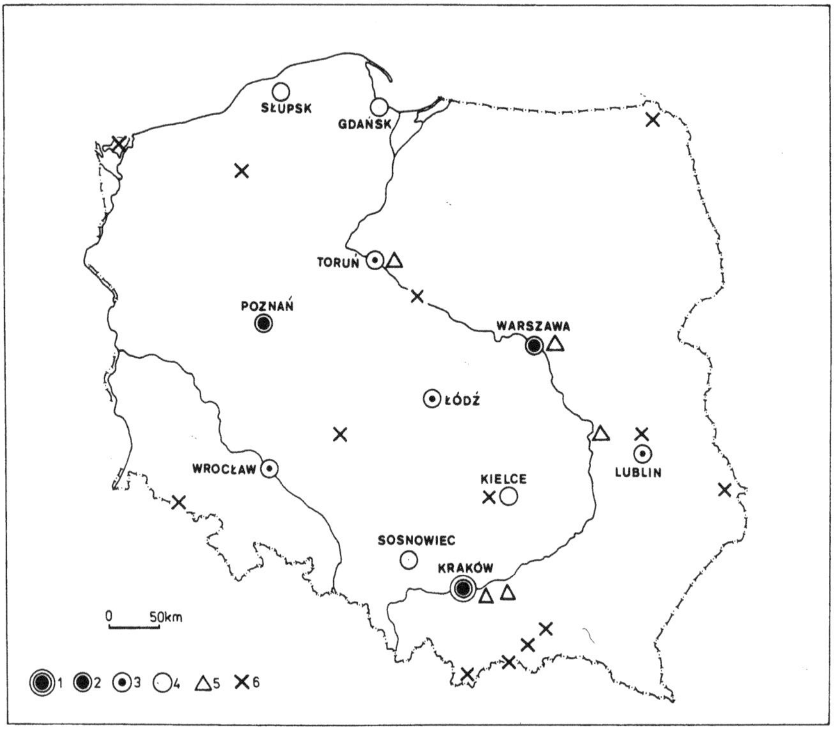

Figure 1

Main geomorphological centres and research stations
1 Oldest University Departments (19th Century) 2 University Departments founded after 1918 3 University Departments founded in 1945 4 New Universities and faculties 5 Polish Academy of Sciences and other research institutes 6 Main field stations: previous and present

---

1956-1968	Rapid progress and international expansion
1968-1977	Weakened development, delayed syntheses, and new searches
1977-	Development of joint studies on palæogeography and on present processes

This paper presents the author's opinions but it is also based on various sources on the history of Polish geomorphology. Especially valuable was the study by L. Czechówna (1969) which presents one century of Polish geomorphology (1840-1939) against the wide backcloth of global science. Critical reviews have also been presented by Halicki (1951), Klimaszewski (1961) and Dylik (1964). The period after World War II has been discussed by Galon (1954), Kostrowicki (1964) and Kozarski (1973). The mechanisms of progress in Polish geomorphology have been considered recently by Kotarba, Kozarski and Starkel (1985). To my co-workers, as well as to colleagues in the Department, I express my gratitude for their comments and discussions. I have also paid attention to the comments of Miller (1961), Tricart (1965), D. St.-Onge (1968), as well as those by Raynal (1975) on the activity of J. Dylik, and by Soviet geographers on the input of Polish scientists in the exploration of Asia (Azatian *et al.* 1969). Figures 1, 2 and 3 summarize the activities of Polish geomorphologists; within Poland (Figure 1), in the world (Figure 2) and through time (Figure 3).

Origins of Polish Geomorphology 1815-1899

Descriptions of Polish landscapes date back to the sixteenth century. The foundations of industry based on mineral resources at the turn of the eighteenth and nineteenth centuries in central Poland (later on the Russian sector) caused a great interest in earth science. The first characterization of the geology and relief of the Polish mountains and uplands was published by the famous geologist S.Staszic (1815).

At a time when Playfair, Lyell, Dan, Humboldt, Agassiz and Torell were laying the foundations of their theories, Poles lived under hard social and cultural conditions. In the part of Poland occupied by the Austrians, W. Pol (1840, 1842), poet and geographer, was the first to describe in detail the northern slopes of the Carpathians and to emphasize the importance of field observations of both surface forms and geomorphic processes. He founded the first chair of Geography at the Jagellonian University in Kraków in 1849 (Figure 1). At the same time the geologist L. Zejszner (1849) was discussing fluvial activity in the Carpathians, and in 1856 he demonstrated the former existence of glaciers in the Tatra Mountains.

By the mid-1860s favourable conditions were created under the Austrian occupation so that by 1866, there was established within the Kraków Society (which later became an Academy of Sciences) a Physiographical Commission, and in 1882 a chair of Geography was founded at Lwów University. A. Alth (1867) wrote a guidebook for field orographic investigations and F. Kreutz (1869) studied the glacial deposits and erosional features of the Tatra Mountains.

Information on American work reached Polish centres selectively, and with great delay. Available German studies were focussed on mountains and glaciations and up until 1899 Polish geomorphic work developed very slowly. What was essentially a descriptive phase culminated with two volumes of regional geography on former Polish territory (Rehman 1895, 1904). However, in 1876 Kawczyński published an essay on the erosion and maturity of a river valley in parallel with G. K. Gilbert! W. Teisseyre (1894) indicated the rôle of tectonic processes in the evolution of cuesta landscapes. Geological mapping supplied new data on the Quaternary such that Łomnicki (1895) presented the concept of relief evolution after the retreat of the ice-sheet, and distinguished four phases: humid with lakes, arid, steppe with dunes, and alluvial.

In the Russian sector of partitioned Poland Siemiradzki (1888) published a study of glacial till and marginal spillways, and the school teacher W. Nałkowski (1887) wrote the first textbook of physical geography. Many Polish naturalists, especially those deported to Siberia, made a substantial contribution to the exploration of Asia: Chodzko (1864) gave the first detailed morphometric characterization of the Caucasus Mountains; Czerski (1881), while leading geological expeditions across Eastern Siberia discovered the regularities of uplift and planation on old morphostructures, and the remnants of former glaciations. In the opinion of Obrutchev (1954), Czerski, together with Kropotkin, were the founders of Russian geomorphology. The first map of the extent of Siberian permafrost, produced by Jaczewski (1889), was based on detailed measurements of air and soil temperature, and the recognition of discontinuous permafrost in upland areas. Bohdanowicz (1895) discovered the traces of glaciation in the arid Kuen-lun Mountains (Figure 2).

The birth and first steps of Polish geomorphology, given the general lack of organization (except in the Austrian sector) and the isolation from western countries, should be evaluated as positive. Field studies were begun, the glacial theory was extended and improved, and new ideas originated on the climatic control of relief.

Period of creative transformation of foreign concepts 1899-1918

Growing knowledge and methodological progress together create the conditions for the simultaneous discovery and development of new theories in different countries. The two leading Polish geomorphologists: E. Romer and L.Sawicki were under the influence of German teachers. In addition the ideas of W. M. Davis were disseminated widely after he published his German language textbook in 1912. Nevertheless, the publication of M. P. Rudzki's book *The Physics of the Earth* (1895) showed that Polish scholars were well-versed in the progress of the earth sciences, and in 1899 Rudzki evaluated the thickness of the Scandinavian ice sheet and re-constructed the magnitude of the glacial rebound.

In the same year that Davis published his *Geographical Cycle* paper (1899) Romer published a theory of climatic geomorphology based on his studies of glaciations, planation surfaces, and river terraces in mountains and uplands. Romer stated that "the main geomorphological zones coincide with the climatic ones, and have been conditioned by them" and that the "climatic factor in the creation of the relief distinctly prevails over the tectonic one." Romer distinguished four morphoclimatic zones: one with perennial rivers (and forest covers), another with seasonal rivers (and æolian activity), the permafrost zone, and the glacial one. He indicated also the stages of relief evolution, and the importance of climatic asymmetry in the highlands. This paper was published in Polish (with a short French summary), and remains unknown abroad. He presented (1911) his opinion on the fluvial genesis of high mountain relief with which he opposed the dominant theory of strong glacial erosion. These opinions resulted from his detailed studies of glacial features in the Eastern Carpathians (1905), and other mountains such as the Sichota-Alin and the Alps.

Davis' theory of the *Geographical Cycle* was actively propagated by Sawicki, who after field studies in the limestone highlands of southern Europe enriched it with the idea of a karst cycle (1909a). He indicated the difference in the stage of maturity between the Mediterranean type, in which the processes die out as the caves become filled, and the middle European type which tends to be rejuvenated. In 1910 Sawicki published the first guidelines for geomorphological field studies (in Polish) and proposed the following sequence of observations: geological substratum, forms, then processes. He also presented (1909b) the first synthesis on the relief evolution of the Polish Carpathians, and referring to de Martonne's opinion, he distinguished two Neogene planation surfaces. Among other papers on the planations should be mentioned the paper of Fleszar (1914) who mapped in detail three surfaces, and that of Smoleński (1912). Great interest was also given to the transversal valleys and canyons (Loziński 1908) who also made the first investigation of gypsum karst (1907). During geological mapping both Łomnicki (1903) and Friedberg (1903) found the valley floors of the Carpathians to be filled with postglacial alluvia rich in black oaks.

In the field of active contemporary landforms, Alpine débris cones were classified by Piwowar (1903). Detailed mapping and measurements of recent landslides in 1913 formed the background of a new synthetic dissertation by Sawicki (1916) on the mechanism of movements and the landslide cycle in the flysch Carpathians. Much further afield the question of young tectonic movements was answered in Antarctica by Arctowski (1900) who proved the existence of isostatic movements at the margin of Antarctica while participating in the *Belgica* exploration (Figure 2).

The new concept of periglacial weathering, facies and climate, presented by Loziński (1909, 1912) at the *International Geological Congress* in 1910, was based on comparative studies of block fields and related phenomena in central Europe. The idea was revived again in 1945.

In the Russian sector several Polish geomorphologists and geologists began studies of the preglacial relief in upland zones (Lewiński 1914, Lencewicz 1916), as well as on glaciogenic deposits (Lewiński and Samsonowicz 1918).

Before the restoration of Poland, Polish geomorphologists were active in the main trends of global research, but their research was unknown abroad. The superiority of climatic geomorphology was reflected in the development of the concept of cyclic evolution (by adaptions to special climatic cases) and by the recognition of the main sequence of relief evolution during the Tertiary and Quaternary. Knowledge of lowland relief was weak compared to that of the highlands, but periglacial and landslide studies created the basis for a dynamic geomorphology. In the opinion of Dylik (1964) the weak transfer of American theories was connected also with the lack of examples reflecting the full cycle of relief evolution in Polish territories, a lack caused primarily by frequent changes of climate and thick blankets of Quaternary deposits. Polish geomorphologists performed many investigations in southern Europe and Asia.

Study of the native landscape 1918-1939

Several new ideas and theories were developed in European geomorphology as answers to, and criticism of, the concept of geographic cycles (A. Penck, W. Penck, H. Baulig and others). Detailed investigations were extended to all morphoclimatic zones of the Earth, but Polish geomorphologists worked in isolation, at least up to 1934. They concentrated their study on the landscape of their own country within its new State boundaries. Alongside the existing Departments at Kraków and Lwów, new university departments were founded at Warsaw, Poznań, Vilnius, and the *State Geological Institute* was created, all of which promoted investigations of the Quaternary (Figure 1). In addition new scientific societies were formed, and journals were founded. Geomorphology remained within the institutional framework of physical geography but of the two former leaders Romer turned to cartography and climatology, and Sawicki died in 1929.

In the light of the requirements of the new State of Poland in 1919 Smoleński pointed to the importance of detailed geomorphological maps (*see* Klimaszewski 1961). The first examples were presented by Pietkiewicz (1928) for the glacial landscape, and by Świderski (1938) for the mountainous areas. The latter maps the highest parts of the East Carpathians by means of genetic groups of forms, and it emphasizes their relations to geological structure. The first general map of the morphogenetic zones of Poland was drawn by Sawicki (1925), and there was a later one by Lencewicz (1937). Very extensive morphometric studies of landscape, developed mainly at Lwów centre, unfortunately lost their connection with genetic geomorphology and detracted from their usefulness.

An important component of geomorphology was the work on the Quaternary forms and deposits of the Polish lowland. In spite of Limanowski's theory of monoglacialism (1922), the more common opinion was that there were three ice sheets. It was based on geomorphic criteria (terminal moraines, freshness of forms, Sawicki 1922) which were later supported by the biostratigraphical divisions introduced by Szafer (1928). Additional regional studies of the lowland landscapes were published by Lencewicz (1927), Pawłowski (1930) and Galon (1934). Quaternary glaciations in the Tatra mountains were summarised by Romer (1929) as well as by Halicki (1930) who introdúced sedimentological and petrographic methods to Quaternary geomorphology. The correlation of the Scandinavian and Tatra glaciations was postulated by Klimaszewski (1937). Tokarski (1938) used detailed granulometric methods to prove that winds depositing the loess cover came from the northeast.

The reconstruction of tectonic movements and climatic changes was the aim of many studies that attempted to explain the rejuvenation of the landscape in various periods (Zierhoffer 1927, Czyżewski 1928, Świderski 1934, Klimaszewski 1934, Kondracki 1938). A continuation of the study of karst is marked by Zaborski's pioneer work on piping (1925) and by Malicki's paper in 1938. Lencewicz (1922) described extensive dune fields and proved that they had been formed by westerly winds after the end of the loess deposition. A new inspiration to more complex studies of the Quaternary using botanical evidence came once again from the botanist W. Szafer, as well as from the geologist E. Rühle.

Little attention was paid to present day processes such as floods and snow-melt events, and only the geologists kept working on the typology of the Carpathian landslides (Teisseyre 1936). With the death of Ludomir Sawicki the number of expeditions abroad decreased although in 1936 E. Rhüle described the Caucasian glaciers. A year later A. Jahn studied the coastal zone of Greenland and in 1938 Halicki, Klimaszewski and Ludwik Sawicki began to survey the glacial foreland in West Spitsbergen (Figure 2).

Polish geomorphology in the interwar period made substantial advances in understanding the genesis of Poland's relief, stages of evolution, and Quaternary stratigraphy. International relations expanded after the International Geographical Union meeting was held in Warsaw in 1934 but what was lacking within the country was research coordination, and no theoretical problems of a continental or global scale were undertaken.

Science in abeyance - WORLD WAR II - 1939-1945

During the War all Polish Universities and research Institutes were closed, and their members of staff, and the libraries, were dispersed. Smoleński (Kraków), Pawłowski (Poznań), Lencewicz (Warsaw), and many others, were killed under tragic circumstances.

Development of new research techniques and new theories 1945-1956

The post-War period saw a new phase with the building of new Institutes, the need to become acquainted with newly acquired territories, and the realisation of old ideas (refer to Fig. 1). Alongside existing and re-opened research centres at Kraków, Warsaw and Poznań, new Universities were founded at Wrocław, Łódź, Toruń and Lublin staffed by geographers from Lwów and Vilnius and Professor Romer promoted young geomorphologists to positions as professors of physical geography. Up until 1956 the political situation was not favourable to the extension of international contacts and the new State looked for cooperation from the earth sciences in the evaluation of natural resources. Under such conditions an excellent team of Polish geomorphologists came into being and started to develop in three different directions:

1) *Relief evolution* - based on detailed geomorphological mapping, and later extended to palæogeographic reconstructions.
2) *Dynamic geomorphology* - based on studies of present-day processes and reconstructions of past processes.
3) *Applied geomorphology* - which explained the relief resources for the various branches of the national economy.

These ambitious aims, established first by the *Polish Geographical Society*, and later on, adopted by the *First Congress of Polish Sciences* in 1951, started to be realised by a group of prominent scholars, and through the ægis of new organisational structures (*cf.* Kotarba *et al.* 1985). In 1953 S. Leszczycki founded within the *Polish Academy of Sciences* a new *Institute of Geography*, with three geomorphological sections: in Kraków, Łódź, and Toruń and thus channeled the initial spontaneous developments (Fig. 1). The lack of opportunities to organise expeditions to other continents gave a unique opportunity for the detailed examination of Poland's relief so that by the 1960s, and bearing in mind the extensive pre-War work, its landscape was one of the most thoroughly studied in the world.

The first group of post-War papers included dissertations on regional or special topics, the result of field work of the pre-War period. Among them there were the monographs on the Pleistocene evolution of the Polish Carpathians (Klimaszewski 1948), the Quaternary of Southern Polesje (Krygowski 1947), the extent of the last ice sheet based on the distribution of tunnel valleys (Majdanowski 1950), and later on, in 1956, Jahn's work on the relief and Quaternary of the Lublin Plateau. Klimaszewski, Galon and finally Pietkiewicz (1947) compiled the general geomorphological map of Poland and gave it a new division into regions.

As well as publishing a detailed geomorphological map of part of the Tatra Mountains (1950) Klimaszewski, at the Kraków centre, started detailed geomorphological mapping at the scale of 1:100,000, and later at 1:25,000, an initiative followed by Galon in Toruń. The mapping extends over the whole country. In 1953 Klimaszewski published a preliminary legend (Klimaszewski 1956, and Galon and Roszkówna 1953) for a general map of the Bydgoszcz region at a scale of 1:300,000. Klimaszewski classified the forms according to morphography, morphometry, genesis and age. The main aim of mapping was to reconstruct the landscape evolution in time and to depict this pattern by a sequence of colours. The map appeared to be an important help in economic planning but the collected field data and the cartographic image posed new problems. What was needed was a new definition for many of the forms which were polygenetic and polychronic - so different from the classical definitions described in the textbooks.

Dynamic geomorphology was growing at the same time. It was promoted on the one hand by agronomists and soil scientists studying the rate of soil erosion at field stations (Reniger 1950, Ziemnicki 1950) and on the other hand by investigations of periglacial processes. In 1955 Gerlach (1966) started to measure slope wash and creep, and in 1956 Maruszczak and Trembaczowski surveyed the effects of heavy rainfall. Geologists continued with studies of landslides.

A. Jahn published a series of papers on various processes: loess (1950), cryogenic phenomena in the present-day and Pleistocene periglacial zone (1951), and the denudation balance on slopes (1954). Meanwhile Dylik (1951) had discovered fossil periglacial structures in the zone of older glaciations, and in 1953 he published a concept of the periglacial cycle as the so-called *tentative leading hypothesis* studying parallel forms, correlative deposits and related processes. A. Dylikowa (1952) introduced structural methods into periglacial work and one of the first examples to follow was a study on the sandur plain by Jewtuchowicz (1955). A crowning cap to this seminal work in periglacial geomorphology was the publication in 1954 of the first issue of *Biuletyn Peryglacjalny.*

The field of applied geomorphology was weaker, and it was based upon the continuation of former trends. Besides studies of soil erosion, geomorphological mapping also delivered basic information on the parameters of form and deposits, and helped in the evaluation of the terrain with respect to various land uses. The first geomorphological maps of land quality were drawn by Starkel (1954). In the course of organising a new discipline called *"urban physiography"* Różycka (1955) adapted the detailed geomorphic map as one of the instruments to be used in evaluating natural resources.

Rapid progress and international expansion 1956-1968

The first post-War international meeting at which Poles met with other world geomorphologists took place at the IGU Congress in Rio de Janeiro in 1956. In the following year J. Dylik organised in Poland the first symposium of the Periglacial Commission and the periodical *Biuletyn Peryglacjalny* was accepted as its official organ. After the INQUA Congress in Spain (1957) the next one

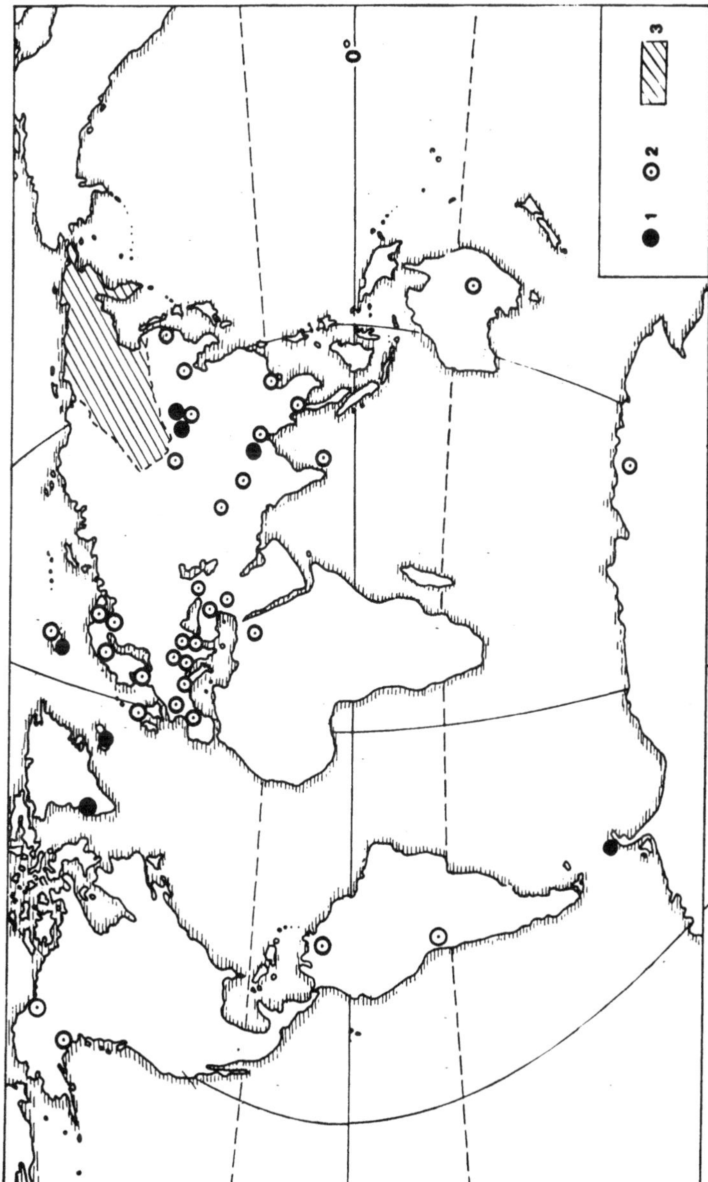

Figure 2 Polish geomorphological expeditions abroad. 1 Areas of frequent Polish expeditions. 2 Areas studied by individual Polish geomorphologists 3 Area of Siberian expeditions by Czerski, Czekanowski and others.

(for 1961) was proposed for Poland. A very high evaluation of various Quaternary investigations served to encourage the Polish team. In 1960 M. Klimaszewski was elected the chairman of the IGU Subcommision on Geomorphological Mapping, and later A. Jahn became chairman of the Commission on Slope Evolution. In 1964 a new journal was founded: *Geographia Polonica* which was published in English, and in 1967 another new journal *Studia Geomorphologica Carpatho-Balcanica* was the organ of the newly organised Carpatho-Balcan Geomorphological Commission. During the International Geophysical Year (1957) a group of Polish geomorphologists went to Spitsbergen, and Polish exploration in that region has continued since that time. In 1964 R. Galon organised a special geomorphological expedition to Iceland (Fig. 2).

During the same period the detailed geomorphological map of Poland expanded to cover about 20% of the country. At a symposium in 1962 many maps were presented along with the first international legend for such maps (Klimaszewski ed. 1963). The legend was regarded as the best available for the presentation of relief evolution (Gilewska 1967) and it inspired a rapid growth of similar mapping in other countries. It was eventually published in the *Encyclopedia of Geomorphology* (article by St. Onge, 1968) and in the same year a legend suitable for detailed world maps was agreed upon by an international team (Klimaszewski 1968).

Geomorphological survey resulted in the development of studies of relief evolution and denudation chronology, for example Klimaszewski's palæogeomorphologic maps of Poland (1958a), and amongst numerous regional syntheses one might mention the studies of cuesta evolution made by Gilewska (1963) and others. Palæogeomorphic maps for various stages of the Quaternary were elaborated by geologists, especially S.Z.Różycki (1965) and his students, as well as by E. Rühle and J.E.Mojski (1968). Similar mapping was also used in studies of periglacial phenomena (Klatka 1962), Holocene transformation (Starkel 1960), types of deglaciation (Bartkowski 1967, Dylik and Galon 1967, and many others), as well as on present-day glaciation (Szupryczyński 1963, and others). B. Rosa (1964) recognised the extent of the Littorina Transgression and the later transformation of the Baltic coastal zone. Careful mapping led to re-evaluation of opinion about the good preservation of the palæogene relief of the uplands (Gilewska 1963) and the character of Neogene relief in the Carpathians (Starkel 1965) and, for example, the rôle of un-rejuvenated valley heads during the course of mountain glaciations was indicated by Klimaszewski (1960b). Moreover the mechanism was recognised by which climatic changes were superimposed on those due to uplift during valley evolution (Starkel 1965, 1969).

Investigations into dynamic geomorphology were divided into two branches. The well-documented structures and textures of periglacial deposits help to document the sequence of changes which occurred during the last cold stage, as well as the roles played by permafrost, congelifluction, wind and wash in the transformation of slopes (Dylik 1958, 1967, Klimaszewski 1958b, Maruszczak 1960), dry valleys (Klatkowa 1965), and loess (Malicki 1961). All of these indicated increasing continentality during the late Vistulian. Krygowski (1964) introduced graniformametric methods in order to distinguish different environments. The use of various methods in parallel led to the recognition that the former periglacial zone was polygenetic (Rotnicki 1966, Klajnert 1966).

On the other dynamic front the measurement of present-day processes was executed in the Arctic as well as in Poland. Czeppe (1961) described the full annual cycle of vertical frost movement and together with Jahn (1961) he evaluated the effect of wash and solifluction on Spitsbergen. Klimaszewski (1960a) published his previous material from Spitsbergen and confirmed the division between frontal and areal deglaciation. The discontinuous belt of permafrost in the Kola Peninsula was described by Olchowik-Kolasińska (1962), and Dzulyński proved by simple laboratory experiments that most cryoturbations may be understood as load structures (Butrym *et al.* 1964).

Measurements of slope processes started to be popular in Poland. The first examples of morphodynamic maps were developed in Łódź (Dorywalski 1958 and others). Gerlach (1966) using his simple "Gerlach's Trough" established the rôle of various types of land use on the rate of slope wash in the Carpathians and parallel studies were carried out on fluvial processes and channel micro-relief (Kaszowski 1973). In 1967 detailed and complex measurements were started at the new field station in Szymbark (Słupik 1973, Gil 1976). Records of winter deflation reaching 20 tonnes/ha. in a transverse depression (Gerlach & Koszarski 1968) helped mould the interpretation that the elongated pans had resulted from periglacial corrasion by wind (Gerlach 1977).

The applied direction bore fruit in dozens of detailed and general maps of economically important areas (e.g. the Silesian Coal Basin). Many of those maps also had special versions for application to agriculture, building, settlement, *etc.* (Klimaszewski 1963, 1968).

The period 1956-1968 brought about a hitherto unknown development in Polish geomorphology. For the first time the Tertiary and Quaternary denudation chronology of different morphostructures was established, together with indications about present rates and trends of change. A synthetic picture of the Quaternary was edited by Dylik and Galon (1967); Rózycki (1967) published an extensive palæogeographical synthesis for central Poland; Gilewska (1964) integrated the history of the karstic plateaux; Krygowski (1961) presented a monograph on the Wielkopolska lowland and Starkel (1960) evaluated the rôle of the Holocene in the relief of the Carpathians and later on in the whole of Europe (1966). In 1962 Malicki published a geomorphological map of the globe: on a morphostructural background he showed the zonal features connected with climate. The first Polish textbook of geomorphology was published in this period by Klimaszewski (1961).

The phase from 1956 to 1968 was characterized by great changes, tempestuous discussions, frequent corrections of methods, the first large scale regional syntheses, and the presence of tens of research workers in the field. In retrospect we see a rather insufficient knowledge of the chronostratigraphy of events and a lack of broad theoretical speculations (Dylik 1964). Nevertheless, to our foreign colleagues the broad front of field work and the fullness of detail was surprising.

Weakened development, delayed syntheses and new searches 1968-1977

There were different causes underlying a weakened development of Polish geomorphology after 1968. Personnel changes began to take place and financial support was reduced and these coincided with the need to use still more precise methods. As a result in that decade many synthetic studies were published and the basis was provided for a new, more interdisciplinary trend in research. A dynamic team was formed in Poznań and began a new journal, *Quæstiones Geographicæ* (ed. S. Kozarski) and at the Silesian University a special periodical was initiated (ed. M. Pulina).

There were difficulties with both the printing and the concepts of geomorphological maps. It was not easy to map polygenetic and polycyclic forms, or primary features which became blurred in form on less resistant rocks, and these difficult questions drew attention to slope deposits. Syntheses were published on the evolution of periglacial slopes in Central Poland (Dylik 1969) and in the Carpathians (Starkel 1969) and the rôle of permafrost was documented even on moraines of the last glaciation (Kozarski 1974). Studies also appeared on karst evolution on chalk (Harasimiuk 1975) and on the variety of structurally controlled relief (Henkiel 1977). On this broad background of synthetic studies mapping continued either for applied purposes, or as the standard basic phase of relief recognition for special studies. All this collective knowledge formed the background for the production of a collective two volume publication, *Geomorphology of Poland* (Klimaszewski and Galon eds. 1972), and in 1974 the general geomorphological map of Poland was finished, with various morpho-structures and chronological categories of forms delineated (ed. Starkel, published 1980).

More attention was paid to the reconstruction of the morphogenetic environments of the Quaternary. In the former periglacial zone various cryogenetic structures were recognised (Gozdzik 1973), and the lithology and stratigraphy of the loess cover was elucidated (Mojski 1969, Maruszczak 1972, Jersak 1973). The latter author showed that during the last glaciation continentality increased eastwards, and a palæogeographic transect for the last cold stage in Europe was presented by Starkel (1977a). Various lithofacial methods were introduced into studies of glacial and glaciofluvial deposits (Kotarba *et al.* 1985) and, for example, Kostrzewski (1975) illustrated the transformation of the weathering mantle in slope and fluvial environments by using detailed sedimentological methods (cf. Krygowski 1964). Studies continued on the present periglacial zone, as well as on glaciers and deglaciation during repeated Spitsbergen expeditions (Baranowski and Jahn, eds. 1975). Galon (1973) published the results of his expedition to Iceland, and Klimek (1972) described the functioning of the sandur plain (Fig. 2). The monograph on the periglacial zone by A. Jahn (English version 1975) may be called the crown of Polish periglacial studies. A regional synthesis on the stratigraphy and palæomorphology of the margins of the Polish uplands was published by Lindner (1977).

Great progress was also made in studies on lateglacial dunes, especially in the Łódź and Poznań centres (Kozarski *et al.* 1969). Rotnicki (1970) showed from the study of a single dune the mechanisms and directions of transformation of its internal structure during the last 13 000 years.

A new inspiration for growing interest in the Holocene came from the meeting of the INQUA Commission in 1972 (Proceedings, 1975). Falkowski (1975) presented his theory on mature and young rivers and the late glacial phase of great palæomeanders was dated in the lowland by Kozarski and Rotnicki (1977), and in the Carpathian foreland by Starkel, who also established the sequence of of rhythmic hydrologic changes (1977b, 1981). The mechanism of environ-mental changes in various climatic zones as a result of the increasing human impact was presented in a monograph published only in Polish (Starkel 1977b). It formed the background for a further development of Holocene studies, including the new IGCP-project 158 on palæohydrology in the temperate zone.

Studies on present-day slope processes were concentrated on mechanics: the measuring of overland flow and infiltration at Szymbark (Słupik 1973, Gil 1976) and the rate of solutional denudation at a zonal scale (Pulina 1974). Kotarba (1976) recognised vertical morphogenetic zones in the Tatra Mountains by measuring solution and physical processes in general, and in the Sudeten Mountains soil creep rates were measured by Jahn and Cielińska (1974). Traditional studies on landslides were enriched by a model of mass movements correlated with changes in rainfall and seepage pressure (Gil and Kotarba 1977).

In the Carpathians repeated transects and the measurement of three types of transport in rivers resulted in monographs on a highland creek (Kaszowski 1973), the rôle of land-use changes (Niemirowski 1974), and the varying rôle of floods in the transport balance (Froehlich 1975). After an extreme rainfall event in the Darjeeling Himalaya Starkel (1972) evaluated its rôle in relief evolution and later (1976) he published a review on the rôle of extreme meteorological events in various world climatic zones. In 1974-1978 physical geography expeditions were carried out to the Khangai Mountains, and the later to Khangai Mountains at their junction with the steppe plateau of Mongolia (led by K. Klimek) at the unique junction of the permafrost, the forest zone and the semi-desert (Fig. 2).

Applied geomorphological studies were reviewed in a textbook by Bartkowski (1977) while the regional synthetic evaluation of form and process is to be found in a paper by Gil and Starkel (1976), in which the need for land use changes in the Carpathians is discussed. Detailed work on the rôle of new water reservoirs was carried out by a team supervised by J. Szupryczyński (ed. 1986).

The decade 1968-1977 was a period in which research technology deepened in the field of historical geomorphology, as well as in present-day processes while at the same time deviations from traditional mapping techniques became evident. There arose an interest in palæogeographic reconstructions, including the Holocene. The state of Polish geomorphology since the 1960s was well presented in new textbooks by Klimaszewski (1978) and Galon (1979). The former, full of data, plays the rôle of a handbook.

Development of joint studies on palæogeography and on present-day processes (1977 to the present)

The last decade is, to a great extent, characterised and determined by Polish geomorphology being open to global problems, palæogeography (INQUA

Commissions, IGCP projects) and contemporaneous changes of the environment (IGU working group, *etc.*) as well as by some internal initiatives such as the formation of national working groups and a central project called 'Evolution of the geographical environment of Poland' (since 1981). Numerous other initiatives have been checked due to technical or financial limitations but in spite of them a new series appeared in 1979, *Quaternary Studies in Poland.* In addition to specialised working groups of the National Committee for Quaternary Research, a Geomorphological Commission of the Polish Geographical Society was founded in 1982.

In contrast, work on preglacial denudation chronology has reduced. Only Gilewska (1987) in a synthetic study emphasized the rôle of tectonic activity on various morphostructures, and Zuchiewicz in his numerous papers (1980) reconstructed the differential uplift in the Carpathians. During the last few years two symposia have been organised: on the Neogene and on early Quaternary correlative sediments in mountain forelands (Dyjor and Jahn 1987, Rutkowski 1987).

Quaternary palæomorphology is discussed in the collective monograph on Poland's Quaternary (ed. Mojski 1984) which was later summarised in an English version (Mojski 1985). Among new concepts should be mentioned a model of glacio-tectonic structures formed during two successive glaciations (Rotnicki 1983). During the international symposium on the Vistulian stage a number of regional syntheses were published (volumes 1 & 3 of *Quaternary Studies in Poland*). The question of whether there were three different ice advances during the Vistulian is still being discussed, primarily on the basis of thermoluminescence dating (TL) (*cf.* Mojski 1985). A detailed analysis of the facies in the marginal zones gave a new picture of deglaciation and palæoclimate during particular recessional stages (Kozarski and Kasprzak 1987). The sequence of climatic changes reflected in loess profiles which include fossil soils, and based on hundreds of TL dates, has been presented by Maruszczak (1980, 1987).

Great progress has also been made in studies on the evolution of river valleys during the last 15 000 years (connected with IGCP Project 158). In a series of volumes concerning the Vistula valley (Starkel ed. 1981, 1982, 1987) the rôle of climatic, glacial, tectonic, eustatic, and anthropogenic factors have been elucidated. In the Warta valley the change from a braided to a meandering river was dated at the Bolling (Antczak 1986). Reconstruction of the fluvial regime are based not only on palæo-channel and sediment parameters (Kozarski 1983, Rotnicki 1983, Kozarski *et al.* 1988) but also on the relationship to measurements of present-day fluvial activity (Rotnicki and Młynarczyk 1987, Rotnicka and Rotnicki 1988). Starkel (1983) presented the typology of rivers and sequences of fluvial change in the temperate zone and demonstrated the coincidence with geomorphological processes in European Mountains (1985). Nowaczyk (1985) showed the synchroneity of dune formation in Middle Europe in the lateglacial. Manikowska (1985) summarised studies of lateglacial soils as palæogeographic indicators and Kowlakowski (1987) extended the hypothesis on the foundations of Holocene soils under periglacial conditions. Numerous studies on lake history finally bore fruit with the discovery at Gościąż Lake of annually laminated sediment representing the last 12 000 years (Ralska-Jasiewiczowa *et al.* 1987).

Work on present-day processes tends towards exact measuring techniques and careful modelling. The rate, and tendency to transformation, of scree slopes by débris flows was determined by Kotarba *et al.* (1987). In the study of rain splash important conclusions were reached on the rôle of surface roughness and gradient (Froehlich and Słupik 1980a, Froehlich 1987). The dominant rôle of cart-roads in sediment transport was shown by Froehlich (1982) who continues studies on sediment delivery by the Cæsium 137 method. The change of rate of chemical denudation with the increase of the catchment area was also determined by Froehlich and Słupik (1980b). Detailed measurements of progressive incision downstream of dams were made by Babiński (1982), while Maruszczak (1984) demonstrated the great contribution to sediment load provided by the loess uplands in the Vistula basin. Baumgart-Kotarba (1987) developed a model for the alluvial plain of the braided Carpathian rivers; a similar study of a meandering river was made by Witt (1979). The rate of movement of coastal dunes was measured by Borówka (1980). Kostrzewski's concept of integrated studies on the circulation of energy and matter in the zone of the last glaciation led to the foundation of a modern field station at Storkowo (Kostrzewski and Zwoliński 1985). It was as late as 1982 before the preparation of a set of maps of present-day processes in Poland began, at a scale of 1:500 000 (the first part is now in press, Bogacki ed.).

The results of the former Mongolian expeditions were published in numerous volumes (Klimek and Starkel 1980 *etc.*) Kowalkowski and Starkel (1984) while studying the vertical zonation discovered a belt of low process activity between the higher cryohumid and the lower cryo-arid zones. The asymmetry of processes was also well documented by complex geo-ecological studies (Starkel and Kowalkowski eds. 1980). Froehlich and Słupik (1978) described a new type of seasonal frost mound in the zone of discontinuous permafrost. An original theory of the formation of granitic rocky floors under conditions of rainfall deficit was presented by Dzułynski and Kotarba (1979). Among other studies on these processes we should mention those on the rate of degradation of nunataks by Pękala (1980) and on the vertical belts of the Hindukush by Kaszowski (1984) (Figure 2).

Special applied studies were continued outside the main branches of geomorphology and is, in part, a result of the recognition, by users of our research, that a knowledge of actual mechanisms, and of long-term tendencies, are equally important for planning and rational land use.

Polish geomorphology is now entering a new phase (Figure 3, next page) of more integrated studies which follows on from a higher level of theory and the use of relatively modern technology. The inspiration comes from detailed field studies, and from a closer cooperation with various disciplines with regard to both present and past changes. The present helps us understand better the mechanism of change and thresholds in the past; evolution in the past helps in turn to discover long-term changes and the phases of various equilibrium states.

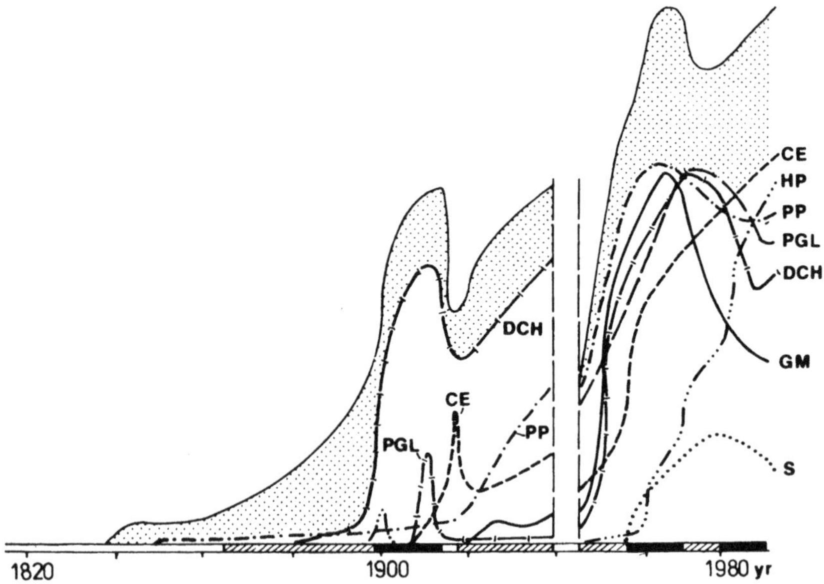

Figure 3

Schematic evolution of Polish geomorphology over time
The steepness of the curve reflects the relative intensity of activity
DCH Relief evolution and denudation chronology **GM** Geomorphological mapping **PGL** Periglacial processes **CE** present-day processes **PP** Pleistocene palæogeography **HP** Holocene palæogeography **S** Syntheses: global, zonal & continental **Stipple**: fluctuations of total activity **Bottom bar:** rate of progress - **dark** = high : **rules** = medium : **white** = slow

Concluding remarks

Polish geomorphology was born in a partitioned country and from the very beginning joined the contemporary climate of science. It supplied creative cyclic theories in the field of karst relief and in the periglacial zone. After 1945 it developed original techniques in geomorphological mapping for the recognition of the history of relief, in the detailed measuring and reconstruction of the periglacial environment, in graniformametric and similar sedimentological analyses, and in studies of present-day slope and fluvial processes.

The parallel reconstruction of the past and measurement of the mechanisms of present processes, as well as examining the inter-relations between slope and fluvial subsystems, are characteristic of the last two decades. The continuation of the palæogeographic approach up to the present helps in the construction of models and the prediction of change while at the same time it is of great value for economic development.

The weaker side of Polish geomorphology reflects the scarceness of global and zonal syntheses as well as theoretical work: all these are needed, together with detailed field investigations, to stimulate progress in the future.

References

For papers marked (in Polish, sum. *or* res. *or* zsf.), the language of the title indicates that an abstract is available in that language with the paper.

A Period 1815-1939

Alth, A., 1867. Instruction for the orographic-geological section of the Physiographic Commission (in Polish). *Spr. Kom. Fizjogr. AU*, Kraków, 1.

Arctowski, H., 1900. Sur l'anciennes extension des glaciers dans le région des terres découvertes par l'expédition Antarctique Belge. *Comptes Rendus de la Académie des Sciences,* (Paris), 131.

Bohdanowicz, K., 1895. Einige Bemerkungen über das System des Kwenlun. *Mitt. Wiener. Geogr. Ges.*, 38.

Chodzko, J., 1864. General view of the orography of the Caucasus (in Russian). *Izwiestia Imperact. Russk. Gieograf. Obsz.*

Czerski, J., 1881. On the remains of former glaciation in Eastern Siberia (in Russian). *Izwiestia. Siber. Otd. Russk. Geograf. Obszcz.*, 11, 4-5.

Czyzewski, J., 1928. Histoire d'une partie de la vallée du Dniestr (in Polish, res.). *Prace geogr. E. Romera,* 10, Lwów, 33-65.

Fleszar, A., 1914, Sur la morphogénese des Karpates situées au Nord de Krosno (in Polish, res.). *Kosmos*,39, Lwów, 99-122.

Friedberg, W., 1903. *Erläuterungen zur geologischen Karte Galiziens* (in Polish, zsf.). Heft XVI, PAU Kraków.

Galon, R., 1934. Die Gestalt und Entwicklung des unteren Weichseltales in Beziehung zum geologischen Aufbau des unteren Weichselgebietes (in Polish, zsf.). *Bad. Geogr. nad Polską - płn-zach.*, 12-13, Poznań.

Halicki, B., 1930. La glaciation quaternaire du versant nord de la Tatra (in Polish and French). *Spraw. Pol. T. Geol.*, tome V, 3/4, 377-600.

Jaczewski, L., 1889. On the perennial frozen grounds in Siberia (in Russian). *Izwiestia Imperat. Russk. Gieograf. Obsz.*, 25, 341-55.

Kawczyński, M., 1876. Comments on general geography and on the landscape of the Lwów region (in Polish). *Kosmos*, 1, 288-98.

Klimaszewski, M., 1934. Über die Morphogenese der Polnischen West-Karpathen (in Polish, zsf.), *Wiad. Geogr.*, 12, Kraków, 30-44.

Klimaszewski, M., 1937. Morphologie und Diluvium des Dunajec Tales von den Pieninen bis zur Mündung (in Polish, zsf.), *Prace Inst. Geogr. UJ,* 18, Kraków, 1-54.

Kondracki, J., 1938. Über die Terassen der unteren Düna, *C. R. Congr. Inter. Geogr.*, Amsterdam, 2.

Kreutz, F., 1869. Tatra Mountains and rift limestones in Galizia (in Polish), *Spr. Kom. Fizjogr. AU,* 3.

Lencewicz, S., 1916. Études sur le Quaternaire du plateau de la Petite Pologne. *Bull. Soc. Neuchâtel de Geogr.*, 25.

Lencewicz, S., 1922. Les dunes continentale de la Pologne (in Polish, zsf.), *Przegl. Geogrf.*, 2., 12-59.

Lencewicz, S., 1927. Glaciation et morphologie du bassin du Vistule moyenne (in Polish, res.), *Prace Państw. Inst. Geol.*, 2, 2, Warszawa, 66-194.

Lencewicz, S., 1937. Map of the geomorphic landscape of Poland (in Polish), *in: Polska* (monograph), Warswaza.

Lewiński,, J., 1914. Die diluvialen Ablagerungen und die präglaziale Oberflächengestaltung des Przemszagebietes, *Prace Tow. Nauk. Warszawa.*, III, Wydz., 7, 1-159.

Lewiński, J. & Samsonowicz, J., 1918, Oberflächengestaltung, Zusammensetzung und Bau des Untergrundes des Diluviums im östlichen Teile des nordeuropäischen Flachlands (in Polish, zsf.), *Prace Tow. Nauk. Wrasz., Wydz.*, 3, 31, 1-179.

Lomnicki, M., 1895-1903. *Erläuterungen zur geologischen Karte Galiziens* (in Polish, zsf.), Heft VII, X, XII, XV PAU, Kraków.

Loziński, W., 1907. Die Karsterscheinungen in Galizisch-Podolien, *Jahrbuch Keis. König. Geol. Reichsanst*, Wien, 57.

Loziński, W., 1908. Die Übertiefung des Täler des palæozoischen Horsts von Podolien, *Föld. Közl.*, 36, Budapest.

Loziński, W., 1909. Über die mechanische Verwitterung des Sandsteinen im gemässigten Klima (in Polish, zsf.), *Rozpr. Wydz. Mat.-Przyr. AU*, Kraków, 9, 1, 1-25.

Loziński, W., 1912. Die periglaziale Facies der mechanischen Verwittenrung. *C. R. XI Congr. Geol. Int. Stockholm*, 1910.

Malicki, A., 1938. The gypsum karst of the Podolian Pokucie (in Polish). *Prace Geogr. E. Romera*, 18, Lwów.

Nałkowski, W., 1887. *Outline of Physical Geography* (in Polish); Warszawa.

Pawłowski, S., 1931. On relief features and geomorphic division of Great Poland (in Polish). *Bad. Geogr. nad Polska NW*, 6-7, Poznań.

Pietkiewicz, S., 1928. Esquisse morphologique de la partie de Suwałki (in Polish, res.). *Przegląd Geogr.*, 8, Warszawa, 168-222.

Piwowar, A., 1903. *Über Maximalböschungen Trockener Schuttkegel und Scutthalden.* Dissert. Univ. Zurich.

Pol, W., 1840. Letters from field trips across country (in Polish), in: Dzieła W. Pola, 5, 137-211, (Lwów 1878).

Pol, W., 1842. *Look at the Northern slope of the Carpathians from the natural point of view* (in Polish), in: Dzieła W. Pola, 3, (Lwów 1877).

Rehman, A., 1895-1904. *Lands of the former Poland and neighbouring Slavian countries described from the physico-geographic point of view* (in Polish). Lwów.

Romer, E., 1899. Influence of climate upon the earth's surface forms (in Polish, sum.). *Kosmos*, 24, 243-271.

Romer, E., 1905. L'époque glaciaire dans le massif du Świdowiec - dans les Karpates Orientales (in Polish, res.). *Rozprawy Wydz. Mat. Przyr. AU.* A, 46.

Romer, E., 1911. Mouvements epeirogénique dans le haut bassin du Rhône et l'évolution du paysage glaciaire. *Bull. Soc. Vandoise. Sc. Nat. Ser.*, 5, 47.

Romer, E., 1929. L'époque glaciaire des Tatras (in Polish, res.). *Prace Geogr.*, 11, Lwów, 1-186.

Rudzki, M.P., 1895. *The Physics of the Earth* (in Polish). Lwów.

Rudzki, M.P., 1899. Deformationen der Erde unter dem Last des Inlandeises. *Bull. Int. Acad. Polon. Sc. et Lettres*, Cracovie.

Sawicki, L., 1909a. Physiographische Studien aus den Westgalizischen Karpathen (in Polish, zsf.). *Arch. Naukowe.*, Lwów, 2,1,5.

Sawicki, L., 1909b. Zur Frage des geographischen Zyklus im Karste. *Mitteil. Geogr. Ges.*, Wien, 52.

Sawicki, L., 1910. Outline of the programme of physiographic investigation (in Polish). *Ziemia*, 1, s.5-7.

Sawicki, L., 1916. Die Szymbarker Erdrutschung und andere westgalizische Rutschungen des Jahres 1913. *Bull. Internat. Ac. Sc. Cracovie*, ser. A., 314-38.
Sawicki, L., 1922. Details on the Middle-Polish terminal moraines (in Polish). *Rozpr. PAU, Wydz. Mat.-Przyr.*, 61, Ser. A, 1-25.
Sawicki, L., 1925. Carta demonstrativa morfogenetica della Polonia. *C. R. Congres Internat. Geogr.*, Caire, 3.
Siemiradzki, J., 1888, Changes of the river courses in the youngest geological period (in Polish). *Wszechświat* 7, 742-3.
Smoleński, J., 1912, On the morphogenesis of the low Beskid Mts (in Polish). *Ks. Pam. 2 Zjazdu Lekarzy i Przyr.* Polskich 1911. Kraków.
Staszic, S., 1815. *On the mineral resources of the Carpathians and other mountains and lowlands of Poland* (in Polish), Warszawa.
Szafer, W., 1928. Entwurf einer Stratigraphie des polnischen Diluviums auf floristischer Grundlage (in Polish, zsf.). *Rocznik Pol. Tow. Geol.*, 5, 21-33, 2 figures.
Świderski, B., 1934. Aperçu sur la morphologie des Karpates du flysch (in Polish, res.). *Przegl. Geograficzny* XIV, z. 1-2.
Świderski, B., 1938. Géomorphologie de la Czarnohora (Karpate orientales Polonaise (in Polish, res.). *Wyd. Kasy. im. Mianowskiego*, Warszawa.
Teisseyre, W., 1894. Palæomorphology of the Podolian Plateau (in Polish). *Spr. Kom. Fizj. AU*, 29.
Teisseyre, H., 1936. Matériaux pour l'étude des éboulements dans quelques de la régions des Karpates et des Subkarpates (in Polish, res.). *Rocznik PT Geol.*, ts XII, 135-92, 19 figures.
Tokarski, J., 1938. Physiographic des podolischen Lösses nach dem Problem seiner Stratigraphie. *Verhandl. der III Int. Quartär-Konf.*, INQUA, Wien.
Zaborski, B., 1925. On the pseudokarstic phenomena in loess deposits (in Polish). *Księga Pam. XII Zjazdu Lekarzy i Przyr.* Polskisch.
Zejszner, L., 1849. *Podhale Basin and Northern slope of the Tatra* (in Polish). Bibl. Warszawska.
Zejszner, L., 1856. Über eine Längsmoräne im Tale des Biały Dunajec bei dem Hochofen in Zakopane in der Tatra. *Sonderberichte Akad.* Wiss., Wien, 21.
Zierhoffer, A., 1927. The northern margin of Podolia in relation to the relief of the Cretaceous surface (in Polish, sum.). *Prace geogr. E. Romera*, 9, Lwów, 61-95.

B references for the period after 1945.

Antczak, B., 1986. Channel pattern conversion and cessation of the Warta river bifurcation in the Warsaw-Berlin pradolina and the southern Poznań gap section during the Late Vistulian (in Polish, sum.). *UAM Poznań, seria geografia*, 35, 1-111.
Babiński, Z., 1982. The influence of the water dam in Wloclawek on fluvial processes of the Vistula River (in Polish). *Prace Geogr. Inst. Geogr. Przestrz. PAN*, 1-2.
Baranoswski, S. & Jahn, A., (eds). 1977. Results of investigations of the Polish Scientific Spitsbergen Expeditions 1970-1874, *Acta Univ. Wratisl.*, Vol I, No 251.
Baranowski, S., Jahn, A., (eds). 1975. Results of investigations of the Polish Scientific Spitsbergen Expeditions 1970-1974. *Acta Univ. Wratisl.*, Vol II, No 387.
Bartkowski, T., 1967. Sur les formes de la zone marginale dans la Plaine de la Grande Pologne (in Polish, res.) *PTPN, Prace Kom. Geogr.-Geol*, VII, 1, Poznań

Bartkowski, T., 1977. *Research methods in Physical Geography* (in Polish sum). PWN, Warszawa, 1-543.
Baumgart-Kotarba, M., 1987. Formation of the coarse gravel bars and alluvial channels, braided Białka river, Carpathians, Poland. *International Geomorphology*, part 1, ed. V. Gardiner, 633-48, New York:Wiley.
Bogacki, M. (ed.), (in press). Present day geomorphic changes of relief in SE Poland (in Polish, sum.). *Prace Geogr. IGiPZ PAN*.
Borówka, K.R., 1980. Present-day transport and sedimentation processes and resulting phenomena on a coastal dune area (in Polish, sum). *PTPN, Prace Kom. Geogr.-Geol.*, XX, Poznań 1-126.
Butrym, J., Cegła, J., Dzułyński, S. & Nakonieczny, S., 1964. New interpretation of "periglacial" structures. *Folia Quaternaria*, 17, Kraków.
Czeppe, Z., 1961. Annual course of frost ground movements at Hornsund (Spitsbergen) 1957-1958. (in Polish sum). *Prace Inst. Geogr. UJ*, 24, Kraków, 1-74.
Dorywalski, M., 1958. An example of morphodynamic map. *Acta Univ Łódź., 8, Łódź*, 67-97
Dyjor, S. & Jahn, A., (eds.), 1987. *Pliocene and Eopleistocene river patterns and their coarse clastic sediments* (in Polish). Oddział PAN Wrocław.
Dylik, J., 1951. Some periglacial structures in Pleistocene deposits of middel Poland. *Bull. Soc. Sci.*, Cl. 3, Vol. III, 2.
Dylik, J., 1953. Du caractère péri-glaciaire de la Pologne centrale. (in Polish res.). *Acta Univ Łódź., 4, Łódź*, 1-109.
Dylik, J., 1958. Notion et méthodes de la géomorphologie dynamique. *Acta Univ Łódź., 8, Łódź*, 23-66.
Dylik, J., 1967. The main elements of Upper Pleistocene palæogeography in Central Poland. *Biuletyn Peryglacjalny*, 18, Łódź, 85-115.
Dylik, J., 1969. Slope development under periglacial conditions in the Łódź region. *Biuletyn Peryglacjalny*, 18, Łódź, 381-410.
Dylik, J. & Galon R. (eds), 1967. *The Quaternary of Poland*. PWN.
Dylikowa, A., 1952. De la méthode structurale dans la morphologie glaciaire (in Polish res.), *Acta Univ Łódź., 8, Łódź*, 23-66.
Dzułyński, S. & Kotarba, A., 1979. Solution pans and their bearing on the development of pediments and tors in granite. *Zeitschrift für Geomorphologie*, NF, 23,172-91.
Falkowski, E., 1975. Variability of channel processes of lowland rivers in Poland and changes of the valley floors during the Holocene. *Biul. Geolog. UW.*, 19, Warszawa, 45-78.
Froehlich, W., 1975. The dynamics of fluvial transport in the Kamienica Nawojowska River (in Polish sum). *Prace Geogr. IG i PZ PAN*, Wrocław, 1-122.
Froehlich, W., 1982. The mechanism of fluvial transport and waste supply into the stream channel in a mountainous flysch catchment (in Polish, sum.), *Prace Geogr. IG i PZ PAN*, 143, Warszawa, 1-144.
Froehlich, W., 1987. Influence of slope gradient and supply area on splash; scope of the problem. *Zeitschrift für Geomorphologie*, NF., Suppl. 60, 105-14.
Froehlich, W. & Słupik, J., 1978. Frost mounds as indicators of water transmission zones in the active layer of the permafrost during the winter season (Khangay Mts., Mongolia). *Proceedings of the Third International Conference of Permafrost*, Edmonton, Canada, I, 188-93.
Froehlich, W. & Słupik, J., 1980a. The pattern of the areal variability of the runoff and dissolved material during the summer drought in flysch drainage basins, *Quæstiones Geographicae*, 6, 11-34.

Froehlich, W. & Słupik, J., 1980b. Importance of splash in erosion process within a small flysch catchment basin. *Studia Geomorph. Carp. Balc.*, XVI, Kraków, 77-112.

Galon, R. & Roszkówna, N., 1953. The general geomorphological map of the Bydgoszcz District. (in Polish sum.). *Przegl. Geogr.* 25, 3, Warszawa, 79-91.

Galon, R., (ed.). 1973. Scientific results of the Polish Geographical Expedition to Vatnajökull (Iceland). *Geogr. Pol.*, 26, 1-311.

Galon, R., 1979. *Landforms - Outline of geomorphology* (in Polish), Warszawa, 1-394.

Gerlach, T., 1966. Développement actual des versant dans le bassin du Haut Grajcarek. *Prace Geogr. IG iPZ PAN,* 52, Wrocław, 1-111.

Gerlach, T., 1977. The rôle of wind in present-day soil formation and fashioning of the Carpathian slopes. *Folia Quaternaria*, 49, 93-113.

Gerlach, T. & Koszarski, L., 1968. The rôle of the wind in the contemporary morphogenesis of the Lower Beskid range (Flysch Carpathians), *Studia Geomorph. Carp.-Balc.* II, Kraków, 85-114.

Gil, E. 1976. Slope wash of flysch slopes in the region of Szymbark. *Dok. Geogr.* 2, Warszawa, 1-65.

Gil, E. & Kotarba, 1977. Model of the side slope evolution in flysch mountains (an example drawn from the Polish Carpathians). *Catena*, 3,4, Giessen, 233-248

Gil, E. & Starkel, L., 1976. Complex physico-geographical investigations and their importance for economic development of the flysch Carpathian area. *Geographia Polonica*, 34, 47-61.

Gilewska, S., 1963. Relief of the Mid-Triassic escarpment in the vicinity of Bedzin (in Polish sum). *Prace Geogr. IG PAN,* 44, Wrocław, 1-135.

Gilewska, S., 1964. Fossil karst in Poland. *Erdkunde*, 18, Bonn, 124-135.

Gilewska, S., 1987. The Tertiary environment of Poland. *Geographia Polonica*, 53, 19-41.

Gozdzik, J., 1973. Origin and stratigraphical position of periglacial structures in Middle Poland (in Polish sum.). *Acta Geogr. Łódź,* 31, Łódź, 1-117.

Harasimiuk, M., 1975. Relief evolution of the Chelm Hills in the Tertiary and Quaternary. *Prace Geograficzne IG PAN*, 115, Warszawa, 108.

Henkiel, A., 1977. Relationship between relief and geological structure of the flysch units and deep basement in the Outer Carpathians. (in Polish sum.), *Rozpr. habil. UMCS,* Lublin, 1-100.

Jahn, A., 1950. Loess: its origin and connection with the climate of the glacial period. (in Polish sum.) *Acta Geol. Poland,* I, 3, 257-310.

Jahn, A., 1951. Cryoturbation phenomena of the present-day and Pleistocene periglacial zone (in Polish sum.). *Acta Geol Poland,* II, 1/2, 159-290.

Jahn, A., 1954. Balance de denudation du versant. (in Polish res.). *Czas. Geogr. IG PAN*, 7, Wrocław, 38-64.

Jahn, A., 1956. Geomorphology and Quaternary history of the Lublin Plateau (in Polish sum.), *Prace Geograficzne IG PAN*, 7, Warszawa, 1-453.

Jahn, A., 1961. Quantitative analysis of some periglacial processes in Spitsbergen, *Zesz. Nauk. Uniw. Wrocl.*, Ser. B, Wrocław, 5, 1-54.

Jahn, A., 1975. *Problems of the Periglacial Zone.* PWN, Warszawa. 1-223.

Jahn, A. & Cielińska, M. 1974. Soil movements on the slopes of the Karkonosze Mountains, (in Polish, sum.), *Acta Univ. Wratisla.*, 236, Prace geogr. se. A, 1, 5-24, Wrocław.

Jersak, J., 1973. Lithology and stratigraphy of the loess of the Southern Polish Uplands (in Polish sum.), *Acta Geogr. Łódź.*, 32, Łódź

Jewtuchowicz, S., 1955. Structure of Outwash Plain (in Polish sum.), *Acta Geogr. Łódź.*, 5, Łódź

Kaszowski, L., 1973, Morphological activity of the mountain stream (with Biały Potok in the Tatra Mts as example). *Prace Geogr. IG UJ, 31*, Kraków, 1-101.

Kaszowski, L., 1984. Vertical differentiation of the dynamics of the mountain relief in the Hindu Kush Munjan Mountains, Afghanistan. *Studia Geomorph. Carp.-Balcanica*, 18, 73-94.

Klajnert, Z., 1966. Origin of Domaniewice Hills and remarks on the mode of waning of the Middle Polish ice sheet. *Acta Geogr. Łódź.*, 23, Łódź, 1-134.

Klatka, T., 1962. Champs de pierre de Łysogóry. Origine et age, (in Polish res.) *Acta Geogr. Łódź.*, 19, Łódź, 1-141.

Klatkowa, H., 1965. Vallons en berceau et vallées sêches aux environs de Łódź, *Acta Geogr. Łódź.*, 12, Łódź, 1-124.

Klimaszewski, M., 1948. Polish Western Carpathians during the Quaternary. (in Polish sum), *Prace Wrocław Tow. Nauk.*, t. IV ser B., 7.

Klimaszewski, M., 1950. Morphology of the head of the White Water valley in the Tatra Mountains (in Polish, sum.), *Ochrona Przyrody*, 19, Kraków, 37-57.

Klimaszewski, M., 1953. The problem of the geomorphological field mapping of Poland (in Polish, sum.), *Przegl. Geogr.*, 25, 3, 16-32.

Klimaszewski, M., 1956. The principles of the geomorphological survey of Poland, *Przegl. Geogr.*, 28 (Supplement), Warszawa.

Klimaszewski, M., 1958a. The geomorphological development of Poland's territory in the pre-Quaternary period (in Polish, sum.), *Przegld. Geogr.*, 30, 1, Warszawa, 3-43.

Klimaszewski, M., 1958b. Modern views on the development of the Karstic relief (in Polish, sum.), *Orzegl. Geogr.*, 30, 3, Warszawa, 421-38.

Klimaszewski, M., 1960a. Geomorphological studies of the Western part of Spitsbergen between Kongsfjord and Eidembutka. *Zesz. Naut. UJ, Prace Geogr.*, S.N. 1, 91-166.

Klimaszewski, M., 1960b. On the influence of pre-glacial relief on the extension and development of glaciation and deglaciation of mountainous regions. *Przegl. Geogr.*, 32, Supp. Warszawa, 41-19.

Klimaszewski, M., 1961. *General geomorphology* (in Polish). PWN, Warszawa

Klimaszewski, M., (ed.), 1963, Problems of geomorphological mapping. *Prace Geogr. IG PAN*, 46.Warszawa, 1-179.

Klimaszewski, M., 1968. Problems of the detailed geomorphological map, in *The unified key to the detailed geomorphological map of the world.* Folia Geogr., Ser. Geogr.-Phys, Vol. 2 (Part 1), 40p., Kraków.

Klimaszewski, M., 1971. The effect of solifluction processes on the development of mountain slopes in the Beskidy (Polish Carpathians), *Folia Quaternaria*, 38, Kraków, 1-18.

Klimaszewski, M., 1978. *Geomorfologia.* PWN, Warszawa.

Klimaszewski, M. & Galon R., 1972. *Geomorphology of Poland* (in Polish), 2 volumes, PWN, Warszawa.

Klimek, K., 1972. Present-day fluvial processes and the relief of the Skeidararsandur plain (Iceland) (in Polish sum.), *Prace Geogr. IG PAN.* 94, Wrocław, 1-139.

Klimek, K. & Starkel, L. (eds.), 1980. Vertical zonality in the Southern Khangai Mountains (Mongolia). *Prace Geograficzne IG iP Z PAN*, 136, Warszawa.

Kostrzewski, A., 1975. Granulometry of the granite weathering materials from the mountainous areas in Europe (in Polish sum.), *Seria Geografia*, 9, Adam Mickiewicz University press, Poznań, 1-132.

Kostrzewski, A. & Zwoliński, Z., 1985. Chemical denudation rates in the upper Parseta catchment, Western Pomerania; research methods and preliminary results, *Quæstiones Geographicae*, Special Issue 1, Poznań.

Kotarba, A., 1976. The rôle of morphogenetic processes in modeling of high-mountain slopes (in Polish, sum.), *Prace Geogr. IG i PZ PAN.*, 120, Wrocław, 1-128.

Kotarba, A., Kaszowski, L. & Krzemień K., 1987. High-mountain denudation system of the Polish Tatra Mountains. *Geographical Studies IG i PZ PAN*, Special Issue 3.

Kowalkowski, A., 1987. Holocene soil evolution of Polish Territory. *Quæstiones Geographicae*, 13, Poznań.

Kowalkowski, A. & Starkel, L., 1984. Altitudinal belts of geomorphic processes in the southern Khangai Mountains (Mongolia). *Studia Geomorph. Carp-Balcanica*, 18, 95-115.

Kozarski, S., 1974. Evidence of Late-Würm perma-frost occurrence in North-West Poland. *Quæstiones Geographicæ*, 1, 65-86.

Kozarski, S., 1983. River channel adjustment to climatic change in west central Poland. In: *Background to Palæohydrology*. K.J.Gregory (ed.), J. Wiley: Chichester, 321-354.

Kozarski, S., Gonera, P. & Antczak, B., 1988. Valley floor development and palæohydrological changes: The Late Vistulian and Holocene history of the Warta River (Poland), in *Lake, Mire and River Environments during the last 15000 years*, eds. Land and Schlüchter, Balkema, 185-203.

Kozarski, S., Nowaczyk, B., Rotnicki, K., Tobolski, K., 1969. The eolian phenomena in west-central Poland with special reference to the chronology of phases of eolian activity, *Geographia Polonica*, 17, 231-248.

Kozarski, S. & Kasprzak, L., 1987. Facies analysis and depositional models of Vistulian ice-marginal features in north-western Poland. *International Geomorphology*, II, ed. V. Gardiner, 693-710, New York: Wiley.

Kozarski, S. & Rotnicki, K., 1977. Valley floors and changes of river channel pattern in the North Polish Plain during the Late-Würm and Holocene, *Quæstiones Geographicæ*, 4, 51-93.

Krygowski, B., 1947. Geological and morphological study on Southern Polesie (in Polish sum.), *PTPN, Prace Kom. Mat.-Przyr*, ser. A, t.5, z.1, Poznań, 1-139.

Krygowski, B., 1961. Physical geography of the Great Polish Lowland, part I, Geomorphology (in Polish sum.), *PTPN, Kom. Fizjogr.*, Poznań, 1-139, 201.

Krygowski, B., 1964. Graniformametria mechaniczna zastosowanie, teoria, *Prace Kom. Geogr.-Geol.*, t. II, z. 4, Poznań, 1-112.

Lindner, L., 1977. Pleistocene Glaciations in the Western Parts of the Holy Cross Mountains (in Polish sum.), *Stud. Geol. Pol.*, Vol. LIII.

Majdanowski, S., 1950. The problem of lake channels in the European plain (in Polish sum.), *Bad Fizjogr. nad Polska Zach.*, 2, z. 1, 35-122

Malicki, A., 1961. Guide-Book of Excursion E: The Lublin Upland, Symposium on Loess. *INQUA VI-th Congress*, Warsaw.

Malicki, A., 1962. *Relief of the Earth* (in Polish). Geografia Powszechna, I, PWN, Warszawa

Manikowska, B., 1985. On the fossil soils, stratigraphy and lithology of the dunes of central Poland (in Polish, sum.). *Acta Geogr. Lodziensia*, 52, 1-137.

Maruszczak, H., 1960. Dépots de couverture de la toudra tachette (in Polish res.), *Ann. UMCS*, Sec. B., 14, 314-350.

Maruszczak, H., 1972. The major genetic and stratigraphic features of loesses in SE Poland (in Polish sum.). Guide to the symposium on *'Lithology and Stratigraphy of loesses in Poland'*, *Wyd. Geolog.*, Warszawa, 89-136.

Maruszczak, H., 1980. Stratigraphy and chronology of the Vistulian loesses in Poland. *Quat. Stud. Pol.*, 2, 57-76.

Maruszczak, H., 1984. Spatial and temporal differentiation of fluvial sediment yield in the Vistula river basin. *Geographia Polonica*, 50, 253-69.

Maruszczak, H., 1987. Loesses in Poland, their stratigraphy and palæogeographical interpretation. *Annales UMCS Lublin*, 1986, 41, 2, sec. B., 15-54.

Maruszczak, H., & Trembaczowski, J., 1956. Geomorphological effects of a cloudburst at Piaski Szlacheckie near Krasnystaw. (in Polish sum.), *Annales UMCS*, Sec B, 11, Lublin, 129-160.

Mojski, J.E., 1969. La stratigraphie des loess de la dernière période glaciaire, *Biuletyn Peryglacjalny*, 20 , 153-177.

Mojski, J.E., (ed.), 1984. Quaternary (in Polish). *Geologia Polski*, t.I. Stratygrafia, część 3b.

Mojski, J.E., 1985. Quaternary, in: *Geology of Poland*, I, Stratigraphy, part 3b, Cainozoic, Wyd. Geolog., Warszawa.

Niemirowski, M., 1974. The dynamics of contemporary river-beds in the mountain streams (in Polish ,sum.), *Prace Geogr. JG UJ*, 34, Kraków, 1-98.

Nowaczyk, B., 1985. The age of dunes, their textural and structural properties against the pattern of Polish atmospheric circulation during the Late Vistulian and Holocene (in Polish, sum.). *Zeszyty Uniw. w. Poznaniu*, seria Geografia, 28.

Olchowik-Kolasińska, J., 1962. Classification génetique des structures de mollisol (in Polish, res.), *Acta Geogr. Univ. Łódź*, 10, Łódź, 1-101.

Pietkiewicz, S., 1947. The morpho-logical divisions of northern and middle Poland (in Polish, sum.). *Czasopismo Geogr.*, 18, Wrocław, 123-69.

Pękala, K., 1980. Relief, present-day geomorphic processes and surficial covers in nunataks in Hornsund, SW Spitsbergen (in Polish, sum.), *Rozpr. habil. UMCS*, Lublin, 1-91.

Pulina, M., 1974. Chemical denudation on the carbonate karst areas (in Polish, sum.). *Prace Inst. Geografii PAN*, 105, Warszawa, 1-159

Ralska-Jasiewiczowa, M., Wicik, B. & Więckowski, K., **1987**. Lake Gościąz- a site of annually laminated sediments covering 12000 years. *Bull. Pol. Acad. Sc. Earth Sciences*, in press.

Reniger, A., 1950., Location, extent and severity of potential soil erosion in Poland. (in Polish, sum.), *Roczn. Nauk Roln.*, 54, Warszawa, 1-99.

Rosa, B., 1963. On the geomorphological evolution of the Polish coast in the light of the former coastal forms (in Polish). *Studia Soc. Sc. Torunensis*, sec. C, 4

Rotnicka, J. & Rotnicki, K., 1988. The problem of the hydrological interpretation of palæochannel pattern, in *Lake, Mire and River Environments during the last 15000 years*, eds. Lang and Schlüchter, Balkema, 205-224.

Rotnicki, K. 1966. The relief of the Ostrzeszów hills as a result of slope development during the Würm (in Polish, sum.), *PTPN, Prace Fom. Geogr.-Geol.*, V 2, Poznań, 260.

Rotnicki, K. 1970. Main problems of inland dunes in Poland based on investigations of the dune at Weglewice (in Polish, sum.), *PTPN, Prace Kom. Geogr.-Geol.*, 11,2, Poznań, 1-146.

Rotnicki, K. 1983. Modelling past discharges of meandering rivers. In: *Background to Palæohydrology*. K.J.Gregory (ed.), J. Wiley: Chichester, 321-354.

Rotnicki, K. & Młynarczyk, K., 1987. Late Vistulian and Holocene channel forms and deposits and their palæohydrological interpretation. *Quæstiones Geographicæ*, 13, Poznań.

Rózycka, W., 1955. Scope and aims of urban hydrology (in Polish, sum.), *Przegl. Geogr.*, 27, 3/4, Warszawa, 501-521.

Rózycki, S.Z., 1965. *Traits principaux de la stratigraphie et de la palæomorphologie de la Pologne pendant le Quaternaire*, Report of the VI th INQUA Congress, Warszawa PWN, vol. 1.

Rózycki, S.Z., 1965. *Traits principaux de la stratigraphie et de la palæomorphologie de la Pologne pendant le Quaternaire*, Report of the VI th INQUA Congress, Warszawa PWN, vol. 1.

Rózycki, S.Z., 1967. The Pleistocene of Middle Poland (in Polish, sum.), Warszawa, 1-251.

Rühle, E. & Mojski, J.E., 1968. *Geological Atlas of Poland at 1:2 000 000*, Inst Geolog., Warszawa.

Rutkowski, J., 1987. Tertiary and Lower Quaternary gravels in the Sandomierz Basin - symposium guidebook (in Polish). Wyd. AGH, Kraków, 1-69.

S łupik, J., 1973. Differentiation of the surface run-off on flysch mountain slopes (in Polish, sum.), *Dok. Geogr.*, 2, Warszawa, 1-118.

Starkel, L., 1954. Importance of geomorphological map for agriculture (in Polish, sum.), *Przegl. Geogr.*, 33, 4, Warszawa, 178-212.

Starkel, L., 1960. The development of the flysch Carpathians during the Holocene (in Polish, sum.), *Prace Geogr. IG PAN*, 22, Wrocław, 1-239.

Starkel, L., 1965. Geomorphological development of the Polish Eastern Carpathians (in Polish, sum.), *Prace Geogr. IG PAN*, 50, Warszawa, 1-157.

Starkel, L., 1966. Post-glacial climate and the moulding of European relief. *Proceedings of Symposium on World Climate from 8000-0 B.C.*, London, 15-33.

Starkel, L., 1969. L'évolution des versants des Carpates a flysch au Quaternaire. *Biuletyn Peryglacjalny*, 18, 349-79.

Starkel, L., 1972. The rôle of catastrophic rainfall in shaping the relief of the Lower Himalaya (Darjeeling Hills), *Geographia Polonica*, 21, Warszawa, 103-147.

Starkel, L., 1976. The rôle of extreme (catastrophic) meteorological events in the contemporaneous evolution of slopes. in: Derbyshire, E. (ed.), *Geomorph-ology and Climate*, 203-246, New York: Wiley.

Starkel , L., 1977a. The palæogeography of Mid- and East-Europe during the Last Cold Stage and West-European comparisons. *Philosophical Transactions of the Royal Society of London*, 241-270.

Starkel, L., 1977b. *The palæogeography of the Holocene* (in Polish, sum.). PWN, Warszawa.

Starkel, L., (ed.), 1980. General Geomorphological map of Poland at 1:500 000 (in Polish, sum.), *Prace Geogr. Inst. Geogr. Przestrz. Zagosp. PAN*.

Starkel, L., (ed.), 1981. The evolution of the Wisłoka valley near Dębica during the Late Glacial and Holocene. *Folia Quaternaria*, 53.

Starkel, L., (ed.), 1982. Evolution of the Vistula river valley during the last 15000 years, part I. *Prace Geogr. IGiPZ PAN*, special issue 1, Warszawa.

Starkel, L., 1983. The reflection of hydrological changes in the fluvial environment of the temperate zone during the last 15000 years, in Gregory, K.J. (ed.), *Background to palæohydrology*, 213-37, New York: Wiley.

Starkel, L., 1985. The reflection of the Holocene climatic variations in slope and fluvial deposits and forms in the European Mountains. *Ecologia Mediterranea*, 11, 1, 91-8.

Starkel, L., (ed.), 1987. Evolution of the Vistula river valley during the last 15000 years, vol II, *Prace Geogr. IGiPZ PAN*, special issue 3, Warszawa.

Starkel, L. & Kowalkowski, A. (eds.), 1980. Environment of the Sant Valley (Southern Khangai Mountains). *Prace Geogr. IGiPZ PAN*, 137.

Szupryczyński, J. , 1963. Relief of the marginal zone of glaciers and types of deglaciation of Southern Spitsbergen Glaciers (in Polish, sum.), *Prace Geogr. IG PAN*, 39,Warszawa, 1-163.

Szupryczyński, J. (ed.), 1986. The Włocławek reservoir - some problems of physical geography (in Polish, sum.). *Dokumentacja Geogr. IGiPZ PAN*, 5, Warszawa.

Witt, A., 1979. Present-day mechanism of flood plain lateral accretion in the middle course of the Warta River, *Quæstiones Geographicæ*, 5, 153-167.

Ziemnicki, S., 1950. Protection of loess soils against erosion, *Badania nad erozja gleb w Polsce*, Warszawa, 155-188.

Zuchiewicz, W., 1980. The tectonic interpretation of longitudinal profiles of the Carpathian rivers. *Rocz. PT Geol.*, 50, 3/4, 311-28.

C Review papers on the history of Polish geomorphology

Azatian, A.A., Bielow, M.I., Gwozdeckyi, N.A., Kamanin, L.G., Murzayew, E.M. & Yuray, R.L., 1969. *History of discovery and exploration of Soviet Asia* (in Russian). Izdat. Mysl. Moskwa.

Czechówna, L., 1969. History of geomorphology in Poland from 1840-1939 with reference to the geomorphology of the world (in Polish, sum.). *PTPN. Prace Kom. Geogr. Geol.*, 9, 4, 1-244.

Dylik, J., 1964. Some remarks on the development of modern geomorphology in Poland (in Polish, sum.). *Czasop. Geogr.* 35, 3-4, Wrocław, 259-77.

Galon, R., 1954. The development of physical geography during the decade of the People's Poland (in Polish, sum.). *Przegląd Geogr.*, 26, 3, 35-52.

Halicki, B., 1951. Le développement de la géomorphologie en Pologne (in Polish, res.), *Wiad. Muzeum Ziemi*, 5, 2, 323-8.

Klimaszewski, M., 1953. The problem of geomorphological field mapping in Poland (in Polish, sum.). *Przegląd Geogr.*, 26, 3, 1-17.

Klimaszewski, M., 1961. *General geomorphology* (in Polish). PWN, Warszawa

Kostrowicki, J., 1964. Polish geography in the last twenty years (in Polish, sum.). *Przegląd Geogr.*, 36, 3, 427-50.

Kotarba, A., Kozarski, S. & Starkel, L., 1985. Geomorphology in Poland - main trends of development in the past three decades, *Quæstiones Geographicæ*, special issue 1, 5-29.

Kozarski, S., 1973. Past achievements of physical geography in Poland and a general perspective on its development. *Przegląd Geogr.*, 45, 3, 459-72.

Miller, J.P., 1961. Physical geography in Poland, *Professional Geographer*, 13, 2.

Obrutchev, W.A., 1954. *History of geological explorations of Siberia* (in *Russian*), Moskwa-Leningrad.

Raynal, R., 1975. Jan Dylik 1905-1973, Le savant, le périglacialiste et l'homme. *Biuletyn Peryglacjalny*, 26, 5-10.

St. Onge., D., 1968. Geomorphic maps, in Fairbridge, R.W. (ed.), *Encyclopedia of Geomorphology*, 389-403. New York: Reinhold.

Tricart, J., 1965. *Principes et méthodes de la géomorphologie*. Paris: Masson.

13

A tribute to John T. Hack by his friends and colleagues

Compiled by W. R. Osterkamp

JOHN TILTON HACK continues to be among the most influential and innovative students of the geomorphology discipline. Because he has played an historic role in the development of twentieth-century geomorphic thought, it seems fitting that we should honor "Johnny" at this Nineteenth Binghamton Geomorphology Symposium dedicated to the history of geomorphology.

This tribute is slanted from the perspective of an admirer who has had the good fortune during many field excursions in recent years to enjoy Johnny's insights into geomorphology, his continued good spirits and sense of humor, and especially his colorful stories about field experiences, colleagues, and his multitude of friends. The tribute is purposely brief on documentation of vitae and awards, which are generally well known, but is weighted towards insights of John Hack -- his intellect, his personality, and his participation in wild stories. As noted throughout, most of these insights have been provided by a number of Johnny's close friends; their recollections were given with genuine enthusiasm, so perhaps thanks are not needed -- but I do appreciate sincerely all the help I've received in preparing this portrait of John Tilton Hack (Frontispiece).

From his birth in 1913, John Hack lived with his family in Chicago until the early years of the Great Depression, when he entered Harvard University. (His father had wanted him to remain in the Chicago area for college, but as has generally been the case through his career, John's determination prevailed.) At Harvard, Johnny took the A.B., M.A., and Ph.D. degrees in 1935, 1938, and 1940. His preference for soft-rock geology, and later geomorphology, was firmly established during his early years at Cambridge. At that time an emphasis on mineral exploration meant that field problems for many students entailed the collection, labelling, and carrying of numerous heavy rock samples back to campus for petrographic studies and classification. Johnny has noted that it took him only one or two times in the field carrying rock samples to decide that there must be a better way to obtain an education. It must have been then that geomorphology began to look appealing, because one cannot carry valleys and ridges to the laboratory. If the rock samples did not help provide direction to

Figure 1

Kirk Bryan, John's close friend and mentor.
The portrait was inscribed by Bryan immediately before John left for Hawaii during World War II. (Photograph provided by Clare Hack.)

Johnny's career, Kirk Bryan did. Bryan had worked for the U.S. Geological Survey in the 1920s and later took the position vacated by William Morris Davis at Harvard. Throughout graduate school, Kirk Bryan was a friend and mentor and principal advisor whom Johnny remembers best for his caring nature (Fig. 1). The two remained close after John's graduate school days, and when in 1942 John and Clare Ferriter married, Kirk Bryan appeared at the newlyweds' door in New York the day after the marriage. As Clare remembers his explaining later, he simply wanted to check Clare out to make sure that she and Johnny were right for each other. Fortunately, the marriage received the Kirk Bryan Award of "Fine!" Of course, less than two decades later, in 1961, John's

Professional Paper, "Studies of Longitudinal Stream Profiles in Virginia and Maryland," won the real Kirk Bryan Award.

John's dissertation research was completed while he was a member of the Awatovi Expedition of Harvard's Peabody Museum during the summers of 1937, 1938, and 1939. The Awatovi Expedition was at an archaeological dig at the eastern edge of the Hopi Indian Reservation in northeastern Arizona. John was one of about 15 staff members and students from Harvard. Others in the group included Watson Smith, an archaeologist who has maintained a warm friendship with Johnny since his early days at Harvard, and, by coincidence, two grandchildren of William Morris Davis.

From the Awatovi work came two of John's earliest papers, published in 1942 by the Peabody Museum. One of these was the result of Johnny's being present when, unexpectedly, prehistoric coal mines were discovered. As the resident geologist, John was drafted into a study of cave geomorphology. As noted by Watson Smith, and with typical modesty by John himself, he demonstrated that the cave was actually a coal mine, and that the Hopis were aware of the thermic value of coal much earlier than were the early inhabitants of Europe. This early cave work led to a sustained interest in caves and archaeology that included studies at a Yuman Indian site near Farson, Wyoming, just before World War II, Russell Cave in Alabama in the early sixties, and Luray Caverns near Luray, Virginia, in the early seventies.

The Awatovi summers also encouraged an interest that Johnny has maintained throughout his career -- that of botany and how it can be useful in geomorphic work. One of his first botanical studies also showed Johnny's ability to use available resources, which were very limited at the Awatovi site. In Watson Smith's words: "There he and I conceived and pursued a joint venture that is little known to science. Behind our tents grew a small pinon tree, and each night before bedtime we met beside it to alleviate the needs of Nature. This was no mere pornographic whim; it was a serious scientific investigation to determine a dendrological formula for the impact of given quantities of uric acid upon tree-growth, expressible in terms of elapsed time. And we found out, for, although I do not remember the exact calculations, I do know that the tree endured for two seasons under stress. Whether this has anything to do with geomorphology, I have no idea."

Following completion of his doctorate in 1940, John accepted a teaching position at Hofstra College (now Hofstra University) in Hempstead, New York. After two years on the faculty, however, John was recruited into the U.S. Geological Survey by another lifelong friend, Wilmot H. (Bill) Bradley, who would serve as Chief Geologist from 1946 to 1960. At that time, Bradley headed the Survey's Military Geology Unit, a terrain- intelligence group headquartered in Washington, D.C., and organized to make landscape analyses, study beaches for amphibious landings, and suggest locations for air strips in both the European and Pacific theaters of the war. The group included earth scientists such as Robert Garrels, Walter White, Frank Swenson, John Cady, James Gilluly, and John Rodgers. A part of the group eventually was sent to Australia to work with the staff of Douglas MacArthur. Its success led to a second group under Philip Shenon being sent to the Pearl Harbor area on Oahu to help with preparations for the invasion of Japan. John Hack was Shenon's assistant in Hawaii, arriving in November 1944.

Members of the terrain-intelligence group in Hawaii were stationed at Schofield Barracks, about 20 kilometers northwest of Pearl Harbor. John Cady, a soil scientist and Johnny's roommate during this period, recalls that they lived in a small house within a cluster of NCO quarters, but took most meals in an underground building of the barracks. The scientific personnel at Schofield Barracks were civilian employees, or "technical representatives" to the military, but were expected to wear Army officer uniforms. Because they were not in the military, however, the uniforms bore no insignia -- no "bars or birds." This created a problem for people like Johnny when chancing on a soldier. Most GIs would see the officer's uniform and start to salute, but the lack of insignia would cause confusion. Usually, though, an awkward salute would be given, returned with one equally awkward. Although Johnny and his colleagues were not officers, they were given assimilated ranks to provide them the rights of officers in case of capture. Most were captains; John was a major.

The work in Hawaii was long -- six days per week -- and generally somber, but was not without pleasant times. Sundays often included relaxation at a beach reserved for officers on the north shore of Oahu, followed by dinner at the Officers' Club. Several, including Johnny, tried their luck with poker occasionally, but "couldn't keep up with the Army people."

In September 1945, Hack and Cady returned to San Francisco, then to Chicago and Washington, D.C., where they resumed their peacetime duties with the Geological Survey; shortly afterward, however, John Cady joined the Soil Survey (now the Soil Conservation Service). Soon Johnny was involved in various administrative duties as an assistant to Bill Bradley, who had just been appointed Chief Geologist of the Survey's Geologic Division. Activities included development of visual displays of the Division's projects and, as Areal Section Chief, the planning for large-scale studies in the Missouri River basin of North Dakota and Montana. During 1946, perhaps as relief to these duties, Johnny increasingly was given, or took, time to pursue his own interests. The first of these studies, begun at the invitation of Cady and others in the Soil Survey who were working on hard-pan soils of the Maryland coastal plain, ultimately resulted in his first Professional Paper, "Geology of the Brandywine Area and Origin of the Upland of Southern Maryland." The companion paper, by C. C. Nikiforoff, concerned the soils of the Maryland coastal plain, the Brandywine area in particular.

Constantine "Niki" Nikiforoff was a soil scientist and Russian refugee of World War I whom John had met through John Cady, and the Soil Survey invitation following their return to the mainland. Niki and his wife were to become very close friends of the Hacks. Nikiforoff had developed principles of soil equilibrium, and these principles, by analogy, were later applied by Hack to Appalachian landscapes; combined with the earlier work of Gilbert, they became the basis of John's concept of dynamic equilibrium. In this post-war period, Hack, with Nikiforoff and Charlie Hunt (Fig. 2), joined in renewed friendships with companions, particularly former graduate students, from Harvard -- Leopold, Denny, Wolman, Goodlett, and others -- to form a group with, as Johnny has implied, a rare rapport. This group, of course, was the source of many classic papers, and much of the work was centered in the Appalachian Ridge and Valley

Figure 2.

"Johnny doesn't think much of Charlie Hunt's story." The scene is a salt pan in Death Valley, 1956. Frank Barnett stands between John and Charlie. (Photograph supplied by Charles Denny.)

Province. As is generally known, Johnny's interests and contributions were mainly concerned with the Great Valley of Virginia, the Shenandoah Valley -- work that began late in 1951 and has never really ended.

Luna Leopold preserved in verse part of the esprit de corps enjoyed by this group. In 1958, he joined Denny and Nikiforoff (a former cabbie in New York City) for a tour led by Johnny of his Appalachian study sites. This was but one of many field trips and meetings for which these four assembled. As Luna suggests in the following poem, the collective caution was not confined to rail crossings.

THE CAUTIOUS ONES

We drive along the road, we stop at every cut,
We're boys who must be showed, we won't be in a rut.

Don't like your fossil soil, nor even buried ones,
From your A2 we recoil, we're skeptic sons-of-guns.

We're cautious as can be, we approach a railroad track,
Any engines do we see?, we must help our Johnny Hack.

First warning is a drawl, real low, of Boston breed,
Watchdog Denny gives his all, "All clear to left, John, proceed."

But Leopoldovitch-Nikiforoff want to stop right on the track,
We observe both near and off, "I tink okay, friend Johnny Hack."

But we take our pen in hand, I tink free-wheeling authors we,
Polygons in yellow sand, Dessawon horizon, A or B.

We find minerals retrograde, find "ascertain development,"
"Ze equilibrium delayed, by ascertain envelopment."

Verry cautious boys are we, Niki, Denny, Johnny Hack,
Leopoldovitch - we cannot see your buried soil or artifact.

The written products of John Hack's research are known to everyone in the discipline, but less well known are his efforts -- some successful, some a bit less than successful -- to devise improved field techniques for geomorphology. Reds Wolman tells of a friendly competition that he had with Johnny some time before Johnny became heavily engaged with his classic studies of the Shenandoah Valley. Borrowing from Rosiwal techniques of petrology, Wolman had just developed his pebble-counting method to measure particle sizes of coarse sediment. John had been experimenting with small- diameter sieves at several sites to obtain similar particle-size data. As Reds tells the story, they "sat on a gravel bar at Seneca Creek below the gauge (at Dawsonville, Maryland), ate our sandwiches, and I told him how I might measure gravel size on the bar, by measuring 100 pebbles at intervals. Johnny said he thought the sieves would work better and faster, so we decided that after lunch we would have a contest. At the word 'go,' I grabbed my ruler and he his sieves and we started measuring. I finished, sat down and calculated the cumulative frequency curve, got up and announced to John median size and a few other measures. He was diligently shoveling, sieving, and weighing with a little spring balance." Faced with this predicament, Johnny protested: "You can't be finished. That's impossible. What did you do?" When he reaches a decision or changes his mind, however, John typically does so with firmness. He added, "I will use that. It's much easier." And a check of USGS Professional Paper 294-B (p. 48) will show that he meant what he said. Understandably, when Johnny has spoken of this episode in recent years, it has not been the contest that he emphasizes. Rather, the contest is just one feature related to a pleasant week that Wolman spent with him in the Shenandoah while Reds was gathering data for his pebble-count paper.

Having been convinced that the Wolman pebble-count method was worthy, John decided to adjust the technique for his own convenience. A modification was the use of a logarithmically-scaled meter stick to measure and classify pebbles into various size categories quickly. He was sufficiently vocal about this scientific advance that it prompted Charlie Hunt, at the instigation of Charlie Denny, to memorialize the idea in verse for a Pick and Hammer show:

Now Johnny Hack's divining rod is really mighty fine,

I wonder why, I wonder why.
It classifies the gravels on a scale of one to nine,
I wonder why, I wonder why.
This quantitative new approach in geomorphology,
Puts it on a firmer basis than it ever used to be.
It tells which stones are biggest just in case you couldn't see,
I wonder why, I wonder why.
(Verse 284 of "Wonderments," Washington, D.C., May 2, 1958)

Although Johnny taught at Hofstra, and much later (1982-85) at George Washington University, it is generally agreed that he always seems more at ease among trees in a forest than among people in a lecture hall. He is at his best in front of people, perhaps, when his subject is one that has excited his curiosity, but he seems most at home when in the field. Three separate anecdotes from three separate sources appear to emphasize this aspect of his personality.

In the spring and summer of 1973, both at a *Geological Society of America* meeting in Knoxville, Tennessee, and in *Science*, a major "breakthrough" in southeastern Pleistocene geology had been reported. The dramatic news was evidence of glaciation -- U-shaped valleys, moraines, and grooved rocks -- on Grandfather Mountain in North Carolina. John and Wayne Newell, who was working closely with Johnny at the time, were piqued with skepticism, and despite a travel freeze, drove to Boone, North Carolina, at John's expense to investigate. By Wayne's account, "--- with maps and field equipment, we climbed up a valley supposed to contain till and striated rocks. The till proved to be colluvium. The striated-grooved rocks all bore the same parallel grooves. Each had a 3/4-inch radius. Some were crusted with rusted iron particles. In the bushes we found rusty lengths of 3/4-inch cable and the clinkers from a coal-fired steam boiler. The grooves had been carved, not by ice, but by a steam-driven log skidder earlier in the century. Local, long-time residents that John knew confirmed that the "glacial valley" had been logged off and grooves from the skidder cables running over the ground and the rocks were common." John and Wayne returned to Washington, and with immense satisfaction gave an informal communication to the *Geological Society of Washington*. The talk included photographs of steel cables resting in the grooves, and a piece of cable was presented as the Grandfather Mountain Glacier. Their efforts won them the coveted "Great Dane Award."

An incident during an interoffice seminar by Johnny on his Shenandoah Valley studies demonstrates both a tendency toward uneasiness in lecture- hall situations and an ability to smile at himself. Dick Hadley was working in Washington, D.C., at the time, and recalls that midway through the presentation John showed a slide of a regression relation, with well- scattered data, involving longitudinal stream profiles. The regression line extended below a boxed explanation in the upper-right corner of the illustration. To John's embarrassment, someone asked the purpose of an extraneous dot in the explanation. Sheepishly, John admitted that it was an "outlier." Everyone, especially John, agreed that he had hit on an ingenious way of ignoring incompatible data while maintaining scientific integrity.

A tell-tale, to-the-point comment indicating his preference for fieldwork, however, was recorded by Reds Wolman at the time he and Johnny were counting pebbles in the Shenandoah Mountains. Seated on the headwater banks of the Tye River, a beautiful stream on a beautiful summer day, John drew a sandwich from his lunch bag and exclaimed, "That's what I like about geology -- picnics!"

One of the better known Hackian stories, related to many of us by John himself, has been termed by Wayne Newell "John at the Turkey Farm," or, for field-trip purposes, "the Turkey Farm stop." The incident, according to Charles Denny, who was present, occurred during a Friends of the Pleistocene trip in 1958 that was led by Johnny and his Little River basin colleague, John Goodlett. The Turkey Farm stop was actually a road cut exposing chert residuum that caps many of the limestone hills in the Shenandoah Valley. Above the road cut was a large sloping field that was fenced. Johnny climbed the bank of the cut to speak to the sixty or more Friends as they assembled. As soon as he began to speak, however, a large white turkey appeared behind the fence and started gobbling. John paused, and so did the turkey. But when he resumed, two more turkeys joined the gobbling. A second pause only allowed hundreds more curious turkeys to appear, all echoing John's enlightened words. The gobbling drowned out everything, and as Charlie Denny writes, "--- the Friends were reduced to a laughing group."

Bob Sigafoos, who was also on the field trip, remembers another "Friends of the Pleistocene" incident. The site was a scree slope crossed by the Appalachian Trail in Shenandoah National Park. John had been citing evidence that the quartzite blocks of the scree slopes showed recent movement. This viewpoint was in conflict with prevailing dogma that the scree was inactive -- a relic of periglacial Pleistocene processes. A strong proponent of this opinion was on the trip, and was loudly expressing his differences with John. The matter was largely settled, however, when a Friend happened to ask a Park Ranger assisting with the trip how often it was necessary to clear the trail. The reply was succinct: "Once a year."

After serving 5 years as Assistant Chief Geologist, Johnny returned to field studies in 1971. From this work came a number of reports, including two Professional Papers and contributions to the Decade of North American Geology series. As of this writing (February 1988), John continues to be active, working on vegetation mapping of the Great Falls of the Potomac area, the geomorphic distribution of white oaks in the Shenandoah Valley (an interest dating from the Brandywine studies), and preparations for a 1989 International Geological Congress field trip through the Shenandoah Valley.

Charlie Denny, one of Johnny's long-time friends and colleagues, notes that he doesn't have many good John Hack stories because "Johnny has always been on the sober side." That is no doubt a fair characterization, but it might also be slightly misleading to those who have not enjoyed much of John's company. For example, Denny also notes that at the retirement party of "Niki" Nikiforoff, the "Moscow vodka" flowed freely, and some of it was in John's direction. Having on various occasions relaxed with John in a motel room after a day of fun in the Great Valley of Virginia, I can assert that sobriety is among his virtues, but abstinence is not. Contrary to popular belief, John and Cliff Hupp (John's only doctoral student at George Washington University) were not volunteering for martinis in the accompanying photograph (Figure 3). Instead, John claimed that folk legend suggests that holding arms in the air will ward off mosquitoes. The photograph was taken in June 1984, near Passage Creek in the Massanutten Mountain area, where exotic measures against insects are certainly warranted.

Figure 3

John "warding off mosquitoes." Cliff Hupp is helping on John's right. (Photograph by W. R.Osterkamp, June 1984.)

A final insight into the Hack personality is provided by Stan Schumm. Not long after John's dynamic equilibrium had first gained popularity at the expense of Davisian thinking, Stan asked Johnny to review a paper for him. The paper reflected, in relatively strong terms, Stan's agreement at that time with John's questioning of Davis' closed-system approach to geomorphology. Rather than endorse Stan's comments on Davis, he warned in the margin: "Don't belittle your predecessors." In my experiences with John, I've often known him to disagree, sometimes strongly and stubbornly, but to no one have I ever known him to be other than cordial and respectful.

14

A perspective on geomorphology in the twentieth century: links to the past and future

John D. Vitek

Introduction

Geomorphology, in the broadest sense, is the study of earth surface processes and resultant forms, be those forms erosional or depositional. Human land-use patterns have generated the need for geomorphic information because of obvious conflicts that arise with surficial processes. As the number of people on the planet continue to increase, greater knowledge of surficial processes is necessary to minimize the negative impact of these processes on human activity. At the beginning of this century simple, descriptive statements about form and process were derived from field observation whereas at present such observations have been supplemented with the addition of numerical data which are used in complex mathematical and statistical analyses to explain the dynamic surface (Morisawa, 1985). Straightforward observation and description has not been abandoned because remote regions, e.g. Tibet, the Andes Mountains, or the surface beneath the oceans, still represent frontiers for detailed geomorphological analyses as is well illustrated by Hewitt's account of Karakoram geomorphology (Chapter 9).

Writing about the history of geomorphology requires the recognition and interpretation of significant events that all individuals involved in the discipline would recognize. But geomorphologists in the United States have different backgrounds because training has been available through geography or geology programs. Although notable scholars such as W.M. Davis and G.K. Gilbert were active in both disciplines at the turn of the century, R.J. Russell was the last individual to have been elected as President of the *Geological Society of America* (1957) and the *Association of American Geographers* (1948). Whereas no individual has since risen to similar positions of leadership in both organizations, cooperative efforts among geomorphologists, regardless of training, are described later in this chapter. Despite different training, every individual is involved in a series of unique daily activities involving questions and answers about his or her specialty. Particular threads can be identified that bind sets of individuals together to form the structure of the discipline. Over the years, the people, thus the questions, the focus, and the answers change in response to numerous factors.

The threads that form the bonds, therefore, take on an importance in the longevity and success of the discipline. The threads represent institutions, which Kohlstedt (1985) identifies with the synonyms of institute, society, academy, association, library, museum, and laboratory. Within geomorphology, various institutions have evolved to promote cooperation and assist in the advancement of the discipline. Recently, specialization has spurred the establishment of regional, national, and international groups to promote this interaction. Disciplinary growth can be related to societal needs and to individual personalities rather than simply the appeal of the natural environment. Current practitioners were drawn to the field, presumably, by scholars, teachers, and researchers whose enthusiasm for the discipline attracted attention. The process of becoming a geomorphologist began, for most, in college because earth science as a subject in high school does not attract many prospective practitioners. For the very small percentage of the population that selected geomorphology, one can probably point to an individual whose personality and intellect provided the attractive stimulation. Once "hooked" the dynamic aspects of the science provide continual re-inforcement that an astute professional choice was made. Because people are the most important thread in the advancement of the discipline, a generalized perspective about people represents the point of departure for this chapter.

People

James Hutton. John Playfair. Louis Agassiz. T.C. Chamberlin. G.K. Gilbert. J.W. Powell. William M. Davis. R.D. Salisbury. Walther Penck. N.M. Fenneman. A.K. Lobeck. D.W. Johnson. Kirk Bryan. Richard J. Russell. This brief list of some of the most significant historic figures who have influenced the development of geomorphology includes the names of individuals that all will recognize (or should recognize). Their contributions to geomorphology have been widely cited because their observations and assessments were recorded in print and are accessible to anyone interested in the discipline. Moreover, these people established or promoted many of the institutions which form the foundations upon which geomorphology has been built, for the lives and work of these early leaders in the establishment of early institutions are almost indistinguishable (Kohlstedt, 1985).

Whereas the printed record extends much further back in time, modern scholars have recognized that our discipline was formalized in the work of Playfair and Hutton. Chorley, Dunn, and Beckinsale (1964), Thornbury (1964), Tinkler (1985), and Gregory (1985) contain ample references to the classic works produced by many of the early scholars. In contrast to the historical development of the discipline few historians of geomorphology have focused on the individuals acknowledged as disciplinary leaders; honorable exceptions are Chorley, Beckinsale, and Dunn (1973) who described the life of W. M. Davis in great detail, and Pyne (1980) and Yochelson (*ed.* 1980) who examined the life of G. K. Gilbert. Because this chapter is not concerned with topical specializations in geomorphology, references to the development of each topical specialty are not presented. A brief section on research into the history of topical specialties can be found elsewhere in this chapter since details of the past are quickly lost if a well-documented record is not preserved. The efforts of James & Martin (1978), who described the roles of geomorphologists in founding the *Association of American Geographers*, and Drake and Jordan (1985), who edited articles that provide historical perspectives about the development of the *Geological Society of America*, provide critical links to the past.

The individuals listed at the start of this section, and many others that could be, developed topical specialties as individuals but collectively worked within institutional frameworks to establish the conceptual basis of the science of geomorphology. When geomorphology finally acquired recognition as a science is, however, debatable (Walker, 1987), although with several thousand practitioners in a multitude of specialized subdisciplines around the world, an answer to the question: "when did it become a science?" is academic. Caution, however, is warranted in assessing how scholars elsewhere in the world view geomorphology because the European tradition dominates the North American perspective. Language barriers represent one of the greatest obstacles to a more complete and exact global history of the discipline. Printed records, if located and translated, may be able to provide details about the status of knowledge of geomorphology in other cultures. "Lost" records or records not readily available to a large number of scholars hinder the writing of an accurate history.

The future of geomorphology is based upon the ability of current practitioners to transfer their enthusiasm to students. The growth of the discipline and establishment of many specializations has permitted students a wide choice in the selection of mentors. A list of mentors would include many of the names of individuals who are involved in research, actively publishing, and hold faculty positions at universities that award graduate degrees. Also on the list would be teachers of geomorphology at two or four-year institutions because they introduce under-graduate students to the opportunities available in the discipline. Geomorphologists in government and private industry have less contact with students. A notable example of this was G. K. Gilbert compared to W. M. Davis, and in recent times R. E. Horton, R. A. Bagnold and J. T. Hack have had either no, or a very limited, teaching career, even though, it may be argued, students should be familiar with the research produced by any scholar. The quality of the next generation depends, in part, on the knowledge transferred, enthusiasm imparted, and internal drive that each student brings to his/her quest for new knowledge. Many of the leaders for the next forty years are already employed and establishing credentials. The task of mentoring students is the ever present challenge.

Institutions

Collective human activity in science generates the *institution*, which by modern definition is an "established organization or foundation, especially one dedicated to public service or culture." It can also mean "a custom, practice, relationship, or behavioral pattern of importance in the life of a community or society." (*The American Heritage Dictionary*, 1982, 2nd ed., p. 666). Support for the development of geomorphology has come from a variety of institutions and behavioral activities. Organizations include universities, professional societies, and informal groups. Many of the individuals that dominated the early years of geomorphology were also linked with the development of institutions such as the *United States Geological Survey*, the *Association of American Geographers*, and various university departments that established reputations for producing excellent science and new scientists. Examples of noteworthy individuals and programs include W. M. Davis (Harvard), T. C. Chamberlin (University of Chicago), D. W. Johnson (Columbia), L. B. Leopold (U.S.G.S and University of California-Berkeley), Eliot Blackwelder (Stanford), and R. J. Russell (Louisiana State University).

The importance of institutions in the history of science is a field being developed by historians. Recently, Kohlstedt (1985) described how institutions shaped the nature of scientific inquiry. The role of learned societies in the twentieth century has changed because many of their functions have been coveted by other private and governmental institutions (Kohlstedt, 1985). Meetings, publications, and forums from which to influence public policy no longer must be associated with formal groups to be viewed as being significant. Nearly all individuals active in the discipline, however, have roots in the university. The role of the university, the library, professional societies, informal groups, symposia, and publications will be assessed relative to the status of geomorphology.

The University

In the year 1900, the university did not serve as an educational tool for the masses to the same degree that it does today. Because of the difficulty of acquiring the number of students and degrees earned in geomorphology, values related to the university have to be viewed as surrogate measures for the trends in geomorphology. Within the United States (Figure 1), only 5,800 students were enrolled that year in graduate degree programs out of a population of 76 millions in the United States. Only 1,965 graduate degrees were granted in 1900 compared to 317,472 in 1985 Although the population of the U.S. was over 238 millions in 1985, the number of graduate degrees granted is a significant proportional increase compared to those granted in 1900 The increase in the number of universities (Figure 2) represents a growth of 340 percent, a small growth compared to the 3,029 percent increase in number of faculty (Figure 1) and 5,246 percent increase in number of students (Figure 3) (Snyder, 1987). A university education is rapidly becoming a required necessity for scientific involvement with the complex human interaction with the natural environment. As the surface of the earth is modified for human use, knowledge of natural atmospheric, geologic, and geomorphic processes are essential for a relative degree of safety. Each one of us contributes specialized skills to the maintenance of the quality of human life. Geomorphology has shared in this growth, as reflected in the number of practitioners, journals, books, and conferences. As a university education acquires greater importance, the discipline of geomorphology should continue to attract its share of practitioners.

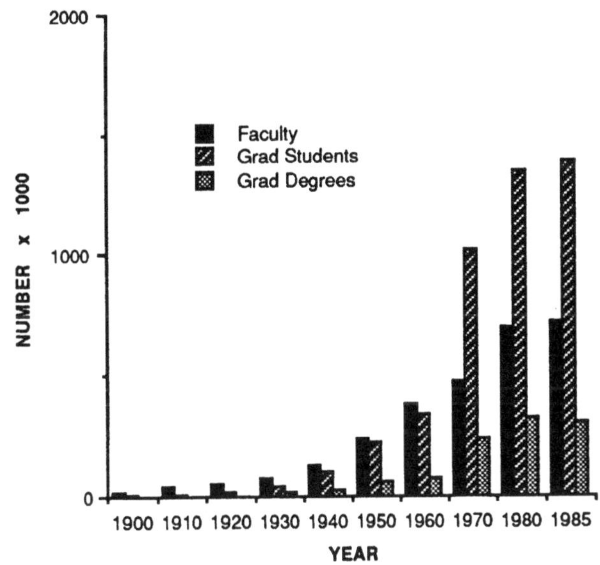

Figure 1

The number of faculty employed, graduate students enrolled, and graduate degrees earned in the United States are shown to illustrate the upwards trends in this century.

Debate about whom should provide geomorphic training within the university should be a moot point. Whereas geomorphology is primarily associated with geography departments at universities throughout the world, in the United States, geomorphology is taught in both geology and geography departments. The emergence of scholars from each program has enhanced the discipline and should be viewed as positive not negative. Briefly consider the role of the atmosphere on geomorphic processes. Landscape development is integrally related to temperature, precipitation, and wind. Earth-sun orbital mechanics influence the receipt of solar energy and thereby introduces climatic change on a cyclic basis. No one would deny the importance of climate to geomorphic processes but unfortunately geomorphologists trained in geology programs are not required and perhaps are not even encouraged to take meteorology or climatology courses. Conversely, geography majors studying geomorphology are not required nor encouraged to take courses in structural geology or mineralogy. Specialization within geomorphology requires merging practitioners with skills from many disciplines to derive answers to the questions that constitute the realm of geomorphology. Techniques from mathematics, statistics, physics, chemistry, climatology, computer science, and other fields are being incorporated into the discipline of geomorphology to advance the research frontier. Butzer (1973) stated that diversity is a source of strength, but only when pluralism is accepted and tolerated. A university facilitates the exposure of a student to a broad range of subjects and viewpoints. A spirit of cooperation does more to advance science than does the negative attitude of assessing whose training is better. Examples of such cooperation are provided in another section in this chapter.

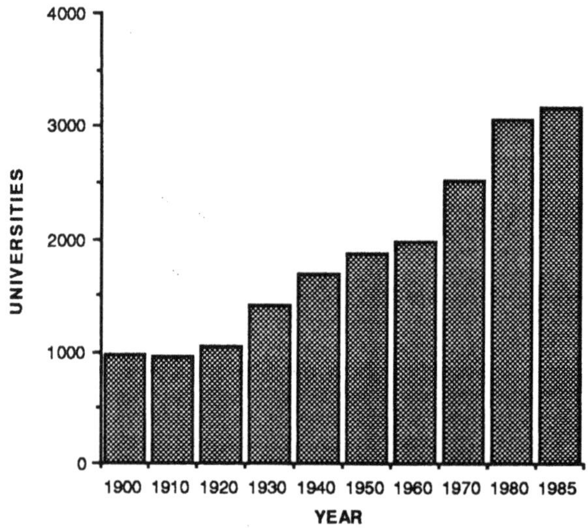

Figure 2

The number of universities in the United States in the twentieth century has risen in response to population increases and the necessity of a college education for many occupations.

Within the University, geomorphology has acquired a general stability in geography departments in contrast to the apparent decline of the subject within geology departments (Costa & Graf, 1984) where dynamic growth in the petroleum industry, hydrogeology, sedimentology, and engineering geology has attracted potential students away from careers in geomorphology. Within geography departments, however, students were attracted to physical geography, especially geomorphology. The movement of students into geomorphology in geography programs represents a change in direction compared to the observations made by Russell (1949) and Bryan (1950). Russell (1949) noted that interest in physical geography in the AAG was down whereas the number of geomorphologists in the GSA was growing. Bryan (1950) observed that physical geographers were a minority of the membership in the AAG. He reported that indifference and hostility could be observed among geographers toward geomorphology. In the 50's, 60's and 70's, a significant number of energetic professors served as mentors and molded students into the research professors active in many universities today. Growth within geography departments was predicted by Graf *et al.*, 1980. Unfortunately, this article in *The Professional*

Geographer is not cited in the standard geologic abstracts and many geomorphologists, trained as geologists, might not be aware of it.

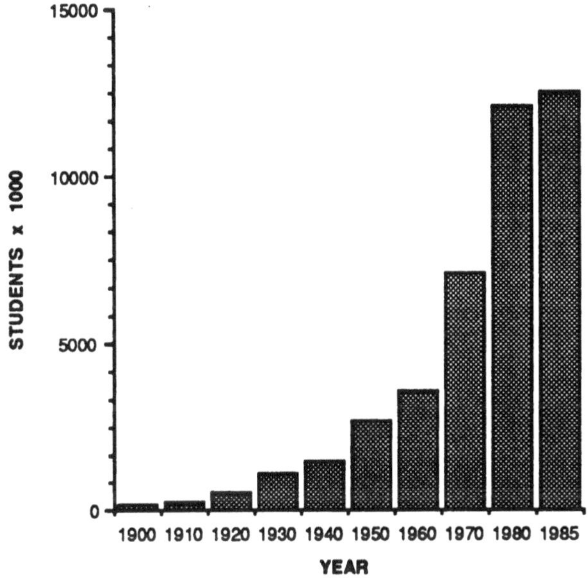

Figure 3

The total number of students in college in the United States has increased rapidly since 1960.

Several recent studies illustrate the spatial distribution of geomorphologists currently practicing; many of whom are affiliated with universities. Graf (1984) examined the status of American field geomorphology by demonstrating the changing distribution of field locations from 1817-1945 as compared to 1946-1980. He concluded that field studies were spatially biased because of the distribution of geomorphologists associated with universities and government agencies. Although the number of practitioners has increased, the distribution is strongly influenced by the academic community and federal agencies. Large areas exist in the USA where few studies have been or are currently being completed related to geomorphology.

Costa and Graf (1984) mapped the distribution of geomorphologists affiliated with the *Association of American Geographers* (AAG), the *Geological Society of America* (GSA), and the total of all geomorphologists. They concluded that a strong regional concentration could be observed, including New York to Virginia, the upper Middle West, and the southwest from Texas to California, and that the sparse population of geomorphologists in other parts of the USA may hinder attendance at regional meetings in those locales. Their research also confirmed that the overlap between members of the GSA and AAG was small (only 4.2 percent belonged to both groups). They urged more cooperation among groups for the benefit of the continued health of the science. Cooperative efforts are described in the section in this chapter on preserving records of the past. With the formation of an international geomorphology organization in 1985, a new focus has emerged that can promote geomorphology and improve its visibility at the global scale (Walker, 1987). It can be the most important thread for linking geomorphologists in the twenty-first century.

The Library

Perhaps the greatest change in the twentieth century involves the amount of knowledge that we have acquired about landforms and related processes. The information explosion in the last several decades signifies a new era in geomorphology. Until every publication has been entered into a computer data base, however, it is virtually impossible to calculate the amount of material that has been published on geomorphology since 1900. Fortunately, scholars have the ability to retrieve and use most published material because of the emphasis that we have placed on libraries as the focal point of information storage.

In response to the information explosion, bibliographic reference services have been developed to help scholars find what has been written on a particular topic. In 1960, Keith Clayton started *Geomorphological Abstracts.* The first issue, which contained 57 abstracts, was completed by volunteers who got a free subscription. Published abstracts were avoided in favor of those completed by the volunteers. A computer generated index was first published in 1966 by which time 3,500 abstracts had been created. During the first year of operation, 233 abstracts were completed in the first half year and 424 in the first full year (1961). These numbers reflect the beginning of the information explosion in geomorphology.

As shown in Table 1,[1] a variety of sources can be used for locating geomorphic literature. Manual searches through volume after volume of indexes or abstracts, however, is about to join the phonograph record as being useable but definitely outdated. With computerized bibliographic databases, a researcher can now enter key words into software programs and produce lists of articles on any topic. The problem of cost for computer searches is rapidly diminishing as a concern. Libraries are beginning to acquire data bases on compact disks, thereby permitting users to perform literature searches on a personal computer. At Oklahoma State, for example, University library patrons can search Agricultural On Line Access (AGRICOLA) free-of-charge. On the compact disk for January 1983 to August 1987 2,680 articles are in the file on the topic of erosion. This total was obtained in seconds, By entering key words, you can determine that 792 specific items exist on soil erosion or 345 items on channels *and* erosion. Only five articles in the database include erosion rates and water. Bibliographic information for each item, including an abstract, can be displayed and printed. Technology has brought a level of efficiency into accessing library materials that was unknown 20 years ago.

---

[1] **Editor's note:** Because of the large number of Tables, with considerable bulk, but essential as a database on institutions, they have all been placed in an Appendix at the end of the paper. Here they can be consulted at leisure without disturbing the flow of the text.

Figure 4

The cover of the first issue of *Geomorphological Abstracts* (with the kind permission of Keith Clayton who founded *Geomorphological Abstracts*)

The library also stores maps and aerial photographs; tools that are invaluable to a geomorphologist. At the turn of the century, most areas were poorly mapped at any scale and surface photography was not a systematic science. Field scientists made their own maps or used very generalized small scale maps. In contrast, 1:24,000 topographic maps exist for many places in the United States (other scales elsewhere!) and are augmented with black and white, color, and color infrared aerial photography at 1:20,000 or larger. Some libraries may have aerial photographs that span 50 years.

This imagery should prove extremely valuable for detecting rates of change in natural and human modified areas. The recent development of digitization with optical scanners may promote change detection derived from historic and modern aerial photographs. Photographs taken by scholars in the late nineteenth century also form the basis for observing surface changes. Interest in the comparison of old and new photographs was great enough to generate the *Repeat Photography Newsletter* (Lambert, 1985).

In summary, the library is perhaps the most useful institution available to geomorphologists. Preserving records of human effort, be that effort publishing, mapping, or photography, is an enormous task that often goes overlooked. As scholars, we often assume that our students will learn to use the library because we have assigned a term paper. Such assumptions should be replaced with positive efforts to insure that every student has the ability to make full use of the knowledge within the library.

Professional societies

At the turn of the century, professional societies were being developed to enhance scholarship. Membership was often a function of being elected based upon criteria established by the society. The inability to travel long distances rapidly and the low membership totals limited the number of individuals who could attend meetings. In contrast, an individual with an interest in a society today can join by paying membership dues. Many organizations distinguish types of membership, i.e., fellow, member, or associate, based on educational background and professional contributions to the discipline. Organizations are able to sponsor international, national, and regional meetings because members have access to rapid transportation systems. Where meetings are held is often related to the willingness of local committees to organize the meeting and the competition among cities for the right to host conferences in facilities designed to promote information exchange. A professional organization grows and remains viable based upon the services it provides to the members. Attending conferences in major cities is viewed as an experience members enjoy.

Geomorphologists have several choices for society affiliations because disciplinary development can be attained through a geology or geography program. Geomorphologists have been affiliated with the *Geological Society of America* and the *Association of American Geographers* during the twentieth century. A list of presidents of these organizations includes the names of prominent geomorphologists (Table 2). Within the AAG, the influence of W.M. Davis in the founding of the society is evident in his three elected terms as president. In the GSA, G.K. Gilbert is the only person to serve two terms as president, and R.J.Russell is the the only person to serve as President of both since the World War II. The growth in all aspects of geology and the dominance of cultural geographers in the AAG may contribute to the inability of geomorphologists to attain the presidency of both organizations

In 1988, the *Geological Society of America* celebrated its 100th anniversary while the *Association of American Geographers* was in its 84th year of operation. In either organization, years passed before geomorphologists were recognized as viable sub-groups. A separate division for geomorphology, created in the GSA in 1955, is currently called the *Quaternary Geology & Geomorphology Division*. The division promotes excellence in research through various awards, including the annual *Kirk Bryan* award for the best research article (Table 3). In 1986 the division began to recognize individuals for career contributions to geomorphology. The first three award winners, Richard P. Goldthwait, Aleksis Dreimanis, and A Lincoln Washburn, have contributed for decades to all aspects of geomorphology, including mentoring students.

Other division activities include awarding scholarships annually for masters and doctoral research. The division promotes the development of special symposia at the annual meeting. Two newsletters per year keep the members informed about events within the division and society of interest to geomorphologists. As the third largest of ten divisions within the GSA, the membership base insures visibility for geomorphologists within the society. Ritter (1988) summarized how the *Geological Society of America* has contributed to the development of geomorphology.

Because people are the driving force in any organization, members recognized by peers to serve in leadership capacities are perceived as individuals who can effectively promote the interests of the members. Table 4 lists the division chairmen who have provided the leadership to enhance the science of geomorphology. The importance of mass and interpersonal communications is recognized by listing the division secretaries (Table 4).

Within the *Association of American Geographers*, a long tradition of individuals active in geomorphology dates back to William M. Davis, G. K. Gilbert, and N. M. Fenneman, founding members of the society in 1904. A volume written for the 75th anniversary of the AAG provided information about Davis but ignored recent aspects of physical geography and geomorphology in the development of the Association (James & Martin, 1978). By the middle of the century Russell (1949) described how physical geography (including geomorphology) had declined within the Association. Although physical geography has become more popular since the early 1950's, it may never again occupy the central position in geography that it did when the society was founded.

Within the association, formal subdivisions or specialty groups, based upon disciplinary interest, were not permitted until 1980. In 1978, however, geomorphologists held a meeting and elected individuals to promote better topical sessions at regional and national meetings. When specialty groups were finally approved, the geomorphologists had firmly established a working group (Table 5). As a minority group in a society dominated by human and cultural geographers (358 members are active in the specialty group compared to 5,718 members in the society), geomorphologists organized themselves in order to present a unified voice in society affairs. An active membership within the *Geomorphology Specialty Group* contributes to the excellent diversity of topical paper sessions at the annual meetings. A newsletter, *Geomorphorum*, keeps members informed about issues and individuals and the *G.K.Gilbert* award, established to recognize outstanding research in the field of geomorphology, has been awarded since 1983 (Table 6).

Other formal societies have been established for geomorphologists. The *British Geomorphological Research Group* (BGRG) was established in 1960 by D.L.Linton (Goudie & Price, 1980). A newsletter, *Geophemera*, was established in 1973 to serve as a link between members, and a journal, *Earth Surface Processes*, was started in 1976 to provide a specialist journal in the subject to a worldwide audience. The journal name was changed to *Earth Surface Processes and Landforms* in 1981, a change which reflects a resurgence of interests in landforms after the process-oriented focus of the sixties and seventies. This group has some distinct geographic advantages over societies in North America because the smaller distances between members in the United Kingdom promotes attendance at meetings. Another group, the *Quaternary Field Studies Group*, was founded in Britain at Birmingham in 1964 and held its first field meeting in Cheshire and North Shropshire. The name was changed to the *Quaternary Research Association* at the Canterbury meeting in 1968, and it currently has 1,000 members (Table 7). Issue number 43 of the *Quaternary Newsletter* carries a number of small articles relating to the foundation of the QFSG, including one that recalls that the initial idea emerged over lunch in the garden of a public house during a 1962 field meeting of the Yorkshire Geological Society, being held at Binbrook, Lincolnshire!

Active participation of geomorphologists in other societies enhances the perception that the discipline is viable (Table 8). Groups such as the *American Association for the Advancement of Science* (AAAS) and the *American Geophysical Union* (AGU) have broad scientific interests, thereby providing other forums in which geomorphologists can assemble. The *American Quaternary Association* (AMQUA) and *Canadian Quaternary Association* (CANQUA), concerned with Quaternary landscapes and processes, are noted for hosting excellent meetings. Scholars from many earth science specialties are active in AMQUA and CANQUA because a broad knowledge base is essential to derive answers to the questions posed about the surface. The quest for gaining broad recognition that geomorphology is a viable discipline now has a focal point. An organization with the potential to attract global attention to geomorphology began as an outgrowth of an international conference in 1985. Members of the BGRG hosted the first international conference on geomorphology at the University of Manchester and developed the mechanisms necessary to form a world-wide organization (Walker & Orme, 1986). The internal structure of this organization is rapidly taking form as the 1989 date of the second conference at Frankfurt/Main approaches. One goal, the development of an international directory of geomorphologists, is progressing as approximately 2,000 entries have been submitted to Dr. Jesse Walker at Louisiana State University (Anon, 1987).

Informal groups

Before formal societies, individuals met to discuss research objectives and findings. Although professional societies were the outgrowth of informal meetings, informal groups have flourished in the twentieth century. Perhaps the oldest informal group promoting various aspects of geomorphology is the *Friends of the Pleistocene*. Informal trips have been led to many field sites for the purpose of disseminating information and generating discussion on a particular subject. In the field, scholars can debate while being able to observe all characteristics of the environment that may be involved in deriving a solution to a particular problem. From the initial group that began in New England in 1934 (Table 9) (Goldthwait, 1988a), six groups are now functioning in the U.S. No dues, minimal registration costs to attend meetings, and the opportunity to acquire knowledge from individuals who have performed the field research create an excellent setting for professionals and students. Although guidebooks have been developed for some of the trips, formal publications are not available from every trip that has been organized. In Goldthwait (1988b), scholars will have a resource that describes the history of the *Friends of the Pleistocene*.

In Ontario the *Association of Ontario Geomorphologists* has had something of a twilight existence. It was once a formal group with a nominal subscription but rare meetings. In the last few years, however, it has held successful one-day meetings in April with a focus on graduate student presentations, and on practical demonstrations of field equipment or laboratory facilities. The latest meeting, at the Scarborough campus of the University of Toronto in April 1988, was distinguished by a demonstration of Rork Bryan's flume, and only one of the formal lectures even mentioned Ontario landscapes: so diversified are the interests of geomorphologists in Ontario.

In 1982, the *American Geomorphological Field Group* hosted its first meeting in Pinedale, Wyoming. With leadership provided by Luna B. Leopold and support from the *U.S. Geological Survey*, this organization began an effort to fill a perceived need for field conferences in the western United States. Presentations are held at research sites to provide participants with an opportunity to observe and question the individuals who did the research. As an informal group, conferences are hosted by volunteers (Table 10). No time interval has been established between conferences nor location although research in the west will probably be the focal point of future meetings. Based upon the attendance, these sessions have been successful. Whereas these meetings are similar to those offered by the *Friends of the Pleistocene*, focus is on current geomorphic processes rather than those associated with a previous climatic regime.

Symposia

The advancement of science depends upon the ability to generate and share new knowledge. Whereas national organizations provide forums at which information exchange can take place, small subdisciplines, such as geomorphology, within major organizations, are often "lost in the crowd" at large conventions. Growth in the number of geomorphologists and a skewed spatial distribution within the

United States and Canada led to the development of special symposia not associated with professional societies.

In 1969, geomorphologists at Guelph University in Guelph, Ontario, Canada established a geomorphology symposium with an emphasis on research methods (Table 11). Six conferences, each organized on a different topic, provided participants with opportunities to interact with colleagues. Papers presented at each conference were published as paperback books by GeoBooks in an effort to disseminate the information widely. Although the series has lapsed in recent years, the Binghamton meeting on Aeolian processes, held at Guelph in 1986, may be seen as a spiritual revival of the series.

The *Binghamton Geomorphology Symposium*, established as an annual meeting in 1970 by Donald R. Coates and Marie Morisawa, has now continued for almost two decades. The idea has survived because of the definition of specific topics for each meeting and the involvement of many prominent geomorphologists (Table 12). During the first decade, participants journeyed to Binghamton, NY for a fall weekend to learn about a specific theme. Participants, speakers, and audiences have come from all over the world. The goal of session organizers has been to attract the best researchers and to have them present in-depth, stimulating lectures with sufficient time for discussion. Success can be attributed to the efforts of Don Coates and Marie Morisawa and the personal relationships that have developed among participants at a meeting devoted solely to some aspect of geomorphology.

In 1978, over 400 participants at the session on *Thresholds in Geomorphology* discussed the future of the symposium. It was announced that after the tenth session in 1979, the symposium, rather than continue as a yearly event in Binghamton, would "go on the road". Overwhelming sentiment was expressed to keep the symposium alive by moving the session each year to a different site as selected by a symposium organizer. During the discussion, John T. Hack stated that this symposium was the first chance that geomorphologists, whether geologists or geographers, had had to interact annually with regard to geomorphic research. Speakers were selected on the basis of what each could contribute from his or her research rather than upon departmental affiliation. After a decade, therefore, the symposium became a road show. A steering committee, consisting of previous organizers, was established to screen proposals and to determine who would host future symposia. A valuable, and permanent, part of the the meeting is the volume of papers delivered at the Symposium which has been published each year by Allen Unwin, now Unwin Hyman. However, that association concludes with this volume.

The first *York University Quaternary Symposium,* organized by William C. Mahaney, was held in 1974. This roughly biennial gathering of Quaternary geomorphologists, geographers, and geologists emphasizes formal paper presentations and informal field trips with purposes similar to the *Friends of the Pleistocene.* After organizing and hosting the first six symposia (Table 13), Bill Mahaney passed the torch to René Barendregt of the University of Lethbridge. The last "*York*" and first *CANQUA*-sponsored meeting was held at the new location in 1985.

Publications

Scholarship involves numerous processes that culminate in the generation of new knowledge. Oral expressions of knowledge can be mis-interpreted and eventually lost, whereas in printed form they can be widely disseminated, interpreted for accuracy, and establish a foundation for future research. In printed form, the information can be viewed as permanent. Recording observations and assessments was as important at the beginning of the twentieth century as it is now. Fledgling societies provided scholars with the opportunity to disseminate information through presentations at annual meetings and in publications produced by the Societies. The same avenues for information dissemination exist today, but many more Societies, journals, and meetings are available. Table 8 illustrates a number of the professional organizations available to geomorphologists. Professional societies, however, are just one agency

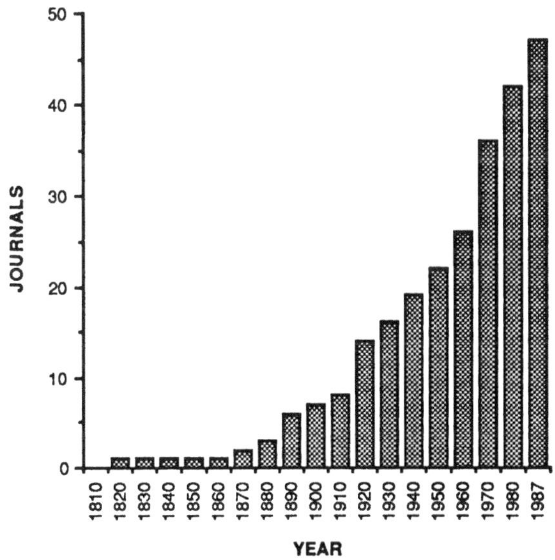

Figure 5

Since the appearance of the American Journal of Science in 1818, the number of new journals in which geomorphologists can publish has increased significantly since 1920. Since 1960, 22 journals, directly related to geomorphology, have been started.

producing publications. Independent publishers have joined with scholars to publish journals not affiliated with societies. The purpose of such publications, besides information dissemination, is to make a profit for the publisher (*e.g.*, Elsevier: *Geomorphology*; John Wiley: *Earth Surface Processes and Landforms*; and Winston: *Physical Geography*).

Figure 5 displays the trend in the growth of journals utilized by geomorphologists. Forty-six percent of the 52 journals included in Figure 5 came into existence after 1960. Only two journals related to geomorphology ceased publication in this century: the *Journal of Geomorphology* (1938-1942) and *Zeitschrift für Geomorphologie* (1925-1943), The latter was re-organized and re-started in 1957. In addition to journals closely affiliated with geomorphology, many other journals are published which contain articles written by geomorphologists. Usually the nature of the research helps a scholar determine the appropriate journal to which an article should be submitted. Journal reputation and the elapsed time from submission to publication are also important considerations in the publication process.

The publishing process is changing to satisfy the demand for published material by using new computer technology. Word processing software permits authors to generate and revise material rapidly. Graphics software coupled with color printers can be used in a manner which enhances the written word. This computer technology is the basis of desk-top publishing. The technology permits a book or monograph to be developed for market quicker than using other printing methods. For example, *Rock Glaciers*, edited by Giardino *et al.* (1987) was delivered to the publisher, Allen & Unwin, as camera ready pages produced with a software program called *Pagemaker*. The book was availabie on the market six months after submission of the manuscript. An increase in the number of scholars generating new knowledge plus the technological revolutions in printing explains the number of journals and books on geomorphic topics that are now available to scholars.

Computer technology is also the basis for rapid, low cost communications termed electronic mail: a technique which permits scholars to strengthen their interactions with each other. Currently, the amount of material being transmitted via this medium is small but should increase as more scholars join the network. Information on how to use BITNET was provided to geomorphologists within the AAG in an article written by David Mark (1986).

Because the process of publishing is dynamic, perhaps all material in the future will be preserved in electronic form. Hardcopy deteriorates and requires space for storage. The utility of hardcopy, however, is that information can be directly transmitted to the reader, provided of course that the reader has the ability to interpret the language of the publication. All information on computer disks is of absolutely no value if devices do not exist which can convert the electronic data into legible characters. Whereas computer disks will play an increasing role in information storage and retrieval, the printed page should retain its value for several generations.

Preserving the records of the past

The directions taken by geomorphology in the twentieth century are preserved from a variety of perspectives in publications ranging from short articles to lengthy books. The majority of these works, including the most recent articles by Marston (1989) and Ritter (1988), examine the discipline on the basis of processes, forms, and methods of analyses. Each individual acquires a reputation as being a particular type of geomorphologist, i.e., a fluvial geomorphologist, a periglacial geomorphologist, a coastal geomorphologist, or any one of the numerous specialties currently being practiced. Ritter (1988) emphasized that interdisciplinary solutions to geomorphic problems will require greater cooperation among geomorphologists in the future. Many activities already confirm that cooperation has begun thereby strengthening the discipline of geomorphology (Walker, 1987).

One form of cooperation between members of the GSA and AAG involves the annual assessment of the status of geomorphology published in *Geotimes*. In recent years the chairs of the geomorphology sections from these societies have co-authored the reviews and for example Baker and Dixon (1988) described the major events of 1987. Awards, grants, meetings highlights, and news items can be found in this yearly report and in recent years reports were provided by Marston and Baker (1987), Giardino and Ashley (1986), Easterbrook and Toy (1984), Andrews and Graf (1983), Morisawa (1982), and Schumm (1981). If *Geotimes* continues this annual review issue, all interested parties will have access to the major events related to geomorphology.

Preserving records of events and information about people has been accepted as a valuable service. A key paradigm in geomorphology states that "*the present is the key to the past*" Conversely, recognition of the threads from the past that provide meaning for the future is essential in the development of the discipline. Published works about the history of geomorphology reflect different view points. As noted earlier Russell (1949) and Bryan (1950) described the decline of geomorphology among geographers. Freeman (1961) assessed how physical geography evolved and emphasized those individuals who made major contributions. Chorley, Dunn, and Beckinsale (1964) provided scholars with a landmark contribution that emphasized the development of geomorphology in the 19th century. Mather (1967) wrote individual chapters on the most noteworthy scholars from 1900-1950. Interestingly, a majority of the scholars who shaped the early directions of geology and geography in this century were geomorphologists and, as I noted earlier, of these Davis (Chorley, Beckinsale and Dunn 1973) and Gilbert (Pyne, 1980, Yochelson (1980) have received detailed analysis because of the lasting impact each had on the development of the discipline. Dury (1972) observed that the techniques, concepts, and aims of geomorphology in the 1960's changed as a result of quantitative techniques. He predicted growth in the future based upon environmental work, the use of remote sensing techniques, and theoretical abstraction in the form of model building. Sarjeant (1980, re-issued in 1987 with two supplementary volumes) developed a comprehensive set of references to the works of the most important scholars who have shaped geology. Gregory (1985) and Hart (1986) included material of an historic nature although the focus of each book had a broader context. The historical works on geomorphology must continue to be assembled under a variety of subheadings, such as biographical, bibliographic, reference books, and collected essays, to assist future scholars with the interpretation of the past.

In honor of the 100th year of *Geological Society of American*, numerous works were commissioned on the history of the society. Within an entire volume devoted to the History of Geology (primarily in America), and with a focus on people and ideas, Morisawa (1985) has reviewed the quantitative revolution in geomorphology, a revolution that led to the formation of a rigorous science. Another GSA volume, devoted to a regional perspective of North America (Graf 1987), documents many of the people who have developed the surficial knowledge of North America. In contrast, the role of geographers in geomorphology has been assessed by Marston (1989) for a book commissioned by the Association of American Geographers. Independent of any society support, Tinkler (1985) documented the development of geomorphology and demonstrated relationships to events in human history, for the need to have precise knowledge of how the surface changes are driven by human utilization of the surface of the earth. Finally, reflections developed by Goldthwait (1988b) on the history of the *Friends of the Pleistocene* are invaluable because the document preserves material from an informal group that has had a wide influence on geomorphic thought. Personal comments by Wahrhaftig (1987) in the Foreword to *Rock Glaciers* provide all scholars with a perspective not available in his widely cited classic paper on rock glaciers (Wahrhaftig and Cox 1959). History is the totality of the events that have shaped the discipline, not simply those which are recorded because of association with formal organizations.

One final note on preserving records of the past should focus on the preservation of the private papers of individuals who have contributed to the rich history of geomorphology. At present, no central depository exists to which geomorphologists could make such a contribution. For example, R. J. Russell's papers are held in the library of the *American Philosophical Society* in Philadelphia. But if historians of the subject wish to use Hoover Mackin's papers or those written by J. P. Miller, how extensive an effort would be necessary to locate and use them? Perhaps professional societies should assess the value of collecting personal papers, thereby enhancing the ability of those interested to interpet accurately the past. Such a depository should accept papers from all individuals to insure than any historical analysis of the data is not biased by too few observations, or by undue selection.

Conclusions

The future of geomorphology is linked to the past through numerous institutions, including societies, universities, and publications, which are the threads that bind the discipline together. The most important link, however, is the mentor that each person had. That individual instilled values and nurtured the development of thought processes in such a manner that each could define problems and seek solutions. The societies, universities, and publications link all geomorphologists to promote the analysis and acceptance (or rejection) of new knowledge. In the past century, a population explosion, a technological revolution, and a quantitative revolution have changed the nature of geomorphology. Giant strides have been made toward accepting differences among practitioners based on the quality of an individual's science rather than his or her training (geology *versus* geography). The complex environment requires interdisciplinary efforts to find solutions to problems that impact human utilization of the surface. The newest thread, the international organization for geomorphologists, may provide a common link to all individuals identified as geomorphologists. The basis for a bright future, however, is a credit to every person who has contributed to the growth of the discipline.

Acknowledgements

I would like to express my appreciation for the help that I received in the development of this paper. Keith Tinkler provided a direction and focus for the paper. S.G.Kohlstedt provided a perspective as an historian; her comments were helpful in organizing the paper. J.R.(Rick) Giardino provided valuable comments on a draft of this paper and created the figures on his user friendly "Mac". David Butler also provided a critical review and made suggestions for material to include. Don Coates, Peter G. Johnson, and Richard A. Marston provided valuable suggestions on an earlier draft. Richard P. Goldthwait was gracious to share material about the Friends of the Pleistocene before his article was published. Allan F. Schneider provided material about the Midwest cell of the Friends of the Pleistocene and Scott Burns shared the list of meetings of the South Central group of the *Friends of the Pleistocene*. Peter Worsley provided Table 7, and other information about the history of the *Quaternary Research Association* incorporated into the text. Finally, the continuous support and encouragement of Wayne A. Pettyjohn (Head, School of Geology) and Norman N. Durham (Dean, Graduate College) is greatly appreciated.

Appendix

TABLE 1: Bibliographic Reference Sources

NAME		DATE STARTED
Bibliography and Index of Geology		1933-44
Bibliography and Index of Geology, Exclusive of No. America		1945-68
Bibliography and Index of Geology (AGI)		1968+
Current Contents (Physics, Chemistry & Earth Sciences)		1961
GeoAbstracts		
Vol. A	Geomorphology	1966-1971
Vol. A	Landforms and The Quaternary	1972
Vol. B	Climatology & Hydrology	1972
Vol. E	Sedimentology.	1972
Vol. G	Remote Sensing & Cartography	1974
Environmental Periodicals		1972
Publications of the U.S. Geological Survey		1879 - 1961
Publications of the U.S. Geological Survey		1962+
Science Citation Index		1955

TABLE 2: Geomorphologists in Leadership Positions

SOCIETY PRESIDENTS

Geol. Soc. of America		Assoc. of American Geographers
1892 & 1909	G.K. Gilbert.	1908
1894	T.C. Chamberlin	
1911	W.M. Davis	1904, 1905, 1909
	R.S. Tarr	1911
	R.D. Salisbury	1912
1935	N.M. Fenneman	1918
1940	E. Blackwelder	
1942	D.W. Johnson	1928
	F.E. Matthes	1933
	W.W. Atwood	1934
1957	R.J. Russell	1948
	J.R. Mackay	1969
1972	L.B. Leopold	
	M.G. Marcus	1987
1984	M.G. Wolman	

TABLE 3: Geological Society of America: Quaternary Geology & Geomorphology Division Awards

KIRK BRYAN AWARD

DATE	WINNERS	DATE	WINNERS
1958	Luna B. Leopold &	1974	Robert V. Ruhe
	Thomas J. Maddock, Jr.	1975	James B. Benedict
1959	Jack L. Hough	1976	Geoffrey S. Boulton
1960	John F. Nye	1977	Michael A. Church
1961	John T. Hack	1978	Richard L. Hay
1962	Anders Rapp	1979	Stanley A. Schumm
1963	Arthur H. Lachenbruch	1980	James A. Clark, William
1964	Robert P. Sharp		E. Farrell, & W. Richard Peltier
1965	Gerald M. Richmond	1981	J. Ross Mackay
1966	Charles S. Denny	1982	Kenneth L. Pierce
1967	Clyde F. Wahrhaftig	1983	Leland H. Gile, John W.Hawley
1968	David M. Hopkins		& Robert B.Grossman
1969	Ronald L. Shreve	1984	Steven M. Coleman
1970	Harold E. Malde	1985	No award given.
1971	A. Lincoln Washburn	1986	Ronald I. Dorn &
1972	Dwight R. Crandell		Theodore M. Oberlander
1973	John T. Andrews	1987	Richard B. Waitt
		1988	Peter W. Birkeland

DISTINGUISHED CAREER AWARD:

1986	Richard P. Goldthwait
1987	Aleksis Dreimanis
1988	A. Lincon Washburn

TABLE 4: Geological Society of America: Quaternary Geology & Geomorphology Division

Division Chairmen:

1955	E. Blackwelder	1966:	R.F. Black	1977	:W.C. Bradley
1956	J.H. Mackin	1967:	H.E. Wright, Jr.	1978:	A.L.Bloom
1957:	J.H. Mackin	1968:	W.D. Thornbury	1979:	W.B. Bull
1958:	G.W. White	1969:	A.D. Howard	1980:	S.A. Schumm
1959:	G.W. White	1970:	D.M. Hopkins	1981:	M. Morisawa
1960:	C.S. Denny	1971:	R.P. Goldthwait	1982:	T.L. Pewe
1961: J	.T. Hack	1972:	R.F. Flint	1983:	J.T. Andrews
1962: J	.G. Fyles	1973:	H.E. Malde	1984:	D.J. Easterbrook
1963:	L.B. Leopold	1974:	D.R. Crandall	1985:	D.F. Eschman
1964:	A.L. Washburn	1975:	R.L. Nichols	1986:	G.M. Ashley
1965:	G.M. Richmond	1976:	L.H. Lattman	1987:	V.R. Baker
				1988:	J.C. Knox
				1989:	D.F. Ritter

Division Secretaries:

1955-56:	R.F. Goldthwait	1967-1970:	S.E. White
1957-60:	H.E. Wright, Jr.	1971-1980:	D.J. Easterbrook
1961-66: L	.H. Nobles	1981-1988:	R.F. Madole
		1988-	J.E. Costa

TABLE 5: Association of American Geographers: Geomorphology Specialty Group: A History

Elected Representatives:
Regional representatives for the promotion and formation of a specialty group:
1978: Richard H. Kesel, Colin Thorn, & John D. Vitek
1979: Percy Daugherty, Colin Thorn, & John D. Vitek

AAG recognized the formation of specialty groups in 1980:

	Chair	Secretary
1980:	John D. Vitek	William L. Graf
1981:	William L. Graf	Stanley W. Trimble
1982:	Stanley W. Trimble	Terrence J. Toy
1983:	Terrence J. Toy	Athol D. Abrahams
1984:	Athol D. Abrahams	John R. Giardino
1985:	John R. Giardino	Richard A. Marston
1986:	Richard A. Marston	John C. Dixon
1987:	John C. Dixon	James S. Gardner
1988:	James S. Gardner	David R. Butler

TABLE 6: Association of American Geographers: Geomorphology Specialty Group Award

G.K.GILBERT AWARD

DATE	WINNERS
1983	J. Ross Mackay
1984	William L. Graf
1985	Athol D. Abrahams
1986	Karl W. Butzer
1987	Derek C. Ford
1988	Ronald I. Dorn

TABLE 7: Quaternary Research Association

Founded in 1964 as the *Quaternary Field Study Group.* Name changed to the *Quaternary Research Association* in 1968.

Past /Present Officers

President

1969-71	R.G. West, F.R.S.
1971-73	L.F. Penny
1973-75	F.W. Shotton, F.R.S
1975-77	J.J. Wymer
1977-79	J.D. Peacock
1979-81	D.Q. Bowen
1981-83	R.W. Hey
1983-85	G.R. Coope
1985-88	J.N. Hutchinson
1988-	J. Rose

Vice-President

1970-72	E.A. Francis
1972-74	D.D. Bartley
1974-76	W.G. Jardine
1976-78	G.R. Coope
1978-80	E.A. Francis
1980-82	C. Turner
1982-84	F.M. Synge
1984-87	J.A. Catt
1987-	P. Worsley

Treasurer

1968-69	E.A. Francis
1969-73	D.Q. Bowen
1973-76	J.N. Hutchinson
1976-80	P. Worsley
1980-84	H. Davies
1984-88	T.D. Douglas
1988-	C.A. Whiteman

Committee Members
Quaternary Field Study Group

1964-66	K.M. Clayton
1964-66	R.W. Hey
1964-67	G.A. Kellaway
1964-67	G.F Mitchell, F.R.S.
1964-68	F.W. Shotton, F.R.S.
1964-68	J.B. Sissons
1966-68	E.A. Francis
1966-69	G.R. Coope
1967-70	D.S. Peake
1967-70	F.M. Synge

Table 7 (continued)

Secretary/Treasurer

1964-68	L.F. Penny
1968	E.A. Francis

Secretary, Quaternary Research Association

1968-70	E.A. Francis
1970-74	W.G. Jardine
1974-78	J.A. Catt
1978-82	J. Rose
1982-86	P.L. Gibbard
1986-	D.H. Keen

Newsletter Editor

1981-85	D.H. Keen
1985-	D.T. Holyoak

Quaternary Research Association

1968-69	R.J. Price
1968-71	J.N. Hutchinson
1968-71	E. Watson
1969-72	E.H. Brown
1969-72	F.W. Shotton, F.R.S.
1970-71	D.D. Bartley
1970-73	G.F. Mitchell, F.R.S.
1971-73	P.A. Mellars
1971-73	W. Tutin, F.R. S.
1971-74	J. A. Catt
1972-74	R.H. Johnson
1972-74	J.D. Peacock
1973-74	A.J. Sutcliffe
1973-76	J.F.D. Bridger
1973-76	G.S. Boulton
1974-77	J. Turner

(This Table was supplied by Peter Worsley, University of Nottingham)

TABLE 8: Major Societies and Informal Groups

American Association for the Advancement of Science	1848
American Geological Institute	1948
American Geomorphological Field Group	1983
American Geophysical Union	1919
American Quaternary Association	1969
Association of American Geographers	1904
Binghamton Geomorphology Symposium	1970
British Geomorphological Research Group	1960
Canadian Quaternary Association	1981
Friends of the Pleistocene	1934
Geological Society of America	1888
International Geomorphological Organization	1985
National Geographic Society	1888
National Speological Society	1941
Quaternary Research Association (was Quaternary Field Study Group from 1964-8)	1968
York/CANQUA Symposium	1974

TABLE 9: Friends of the Pleistocene: Trip Organisers and Locations

Eastern Cell

DATE	LEADERS	AREA
1934	George White & J.Walter Goldthwait	Durham to Hanover, NH
1935	Dick Flint	New Haven to Hartford, CT
1936	Kirk Bryan	South RI to Cape Cod, MA
1937	J.W. Goldthwait, Dick Goldthwait & Dick Lougee	Hanover to Jefferson, NH
1938	Charlie Denny & Hugh Raup	Black Rock Forest, NY
1939	Paul MacClintock Meredith Johnson	Drifts, N. NJ
1940	Kirtley Mather & Dick Goldthwait	West Cape Cod, MA
1941	John Rich	Catskill Mts., NY
	1942.1943.1944.1945...War	**Years. No meetings**
1946	Lou Currier & Kirk Bryan	Lowell-Westford area, MA
1947	Earl Apfel	E. Finger Lakes, NY
1948	D.F. Putnam, Archie Watt, & Roy Deane	Toronto to Georgian Bay, ONT
1949	Paul MacClintock & John Lucke	Pensauken problem, NJ
1950	O.D. Von Engeln	Central Finger Lakes, NY
1951	John Hack & Paul MacClintock	Chesapeake soils & stratigraphy, MD
1952	Dick Goldthwait	Tills, central Ohio
1953	Lou Currier & Joe Hartshorn	Outwash sequences, Ayer quad., MA
1954	Charlie Denny & Walter Lyford	Wellsboro-Elmira-Towanda, PA-NY
1955	Paul MacClintock	Champlain lake and sea, NY
1956	Nelson Gadd	St. Lawrence lowland, QUE
1957	Paul MacClintock & John Harris	St. Lawrence seaway, NY
1958	John Hack & John Goodlett	Appalachians, Shenandoah, VA
1959	Aleksis Dreimanis & Bob Packer	Lake Erie till bluffs, ONT
1960	Ernie Muller	Cattaraugus Co., W NY
1961	Art Bloom	Marine clay & ice margins, SW ME
1962	Cliff Kaye & Phil Schafer	Charlestown moraine & vicinity, RI
1963	Hubert Lee	Lower St. Lawrence, QUE
1964	Cliff Kaye	Marthas Vineyard, MA
1965	Joe Upson	Northern Long Island, NY
1966	Nick Coch & Bob Oaks	Scarps & stratigraphy, SE VA
1967	Hal Borns	Marine & moraines, E ME
1968	Carl Koteff, Bob Oldale & Joe Hartshorn	E. Cape Cod. MA
1969	Nelson Gadd & Barrie McDonald.	Sherbrooke area, QUE
1970	Dick Goldthwait & George Bailey	Mt. Washington region, NH
1971	Gordon Connally	Upper Hudson, Albany NY

Table 9 (continued)

1972	Art Bloom & Jock McAndrews	Central Finger Lakes, NY
1973	Don Coates & Cuchlaine King	Susquehanna & Oswego Val., NY-PA
1974	Bill Dean & Peter Duckworth	Oak Ridges-Crawford Lake, ONT
1975	George Crowl, Gordon Connally & Bill Sevon	Lower Delaware Valley, PA
1976	Bob Jordan' & John Talley	Coastal Plain, DE
1977	Bob Newton	Ossipee quad., NH
1978	Denis Marchand & Ed Ciolkosz	Central Susquehanna Valley, PA
1979	Jesse Craft	NE Adirondack Mts., NY
1980	Bob LaFleur & Parker Calkin	Upper Cattaraugus, Hamburg, NY
1981	Carl Koteff & Byron Stone	Nashua Valley, MA
1982	Pierre LaSalle et al.	Drummondville, QUE
1983	Woody Thompson & Geoff Smith	Ice margins, central ME
1984	Peter Clark & J.S. Street	St. Lawrence lowland, Massena-Malone, NY
1985	Ed Evenson et al.	Great Valley, NJ-PA
1986	Tom Lowell & Steve Kite	Northernmost ME
1987	Carl Koteff, Janet Stone, Fred Larson & Joe Hartshorn	Lake Hitchcock, Conn. Valley, CT-MA

(source: Goldthwait, R.P., 1988)

Southeast Cell

1987	S. Kite, R. Linton, A. Miller, B. Jacobsen,G.M. Clark, M. McCoy, J. Harper, et al.	Potomac Highlands, Canaan Valley Pk, VW
1988	W. Newell, J. Farnsworth, J. Wykoff, J. Owens, & C. Smith	New Jersey coastal plain

(source: Kite, S., 1988)

Midwest Cell

1950	S. Judson	Eastern Wisconsin
1951	H.E. Wright & R.V. Ruhe	Southeastern Minnesota
1952	P.R.Shaffer & H.W. Scholtes	Western Illinois & eastern Iowa
1953	F.T. Thwaites	Northeastern Wisconsin
1954	H.E. Wright & A.F. Schneider	Central Minnesota
1955	R.V. Ruhe	Southwestern Iowa
1956	J.H. Zumberge & W.N. Melhorn	NW Lower Michigan
1957	W.D. Thornbury & W.J. Wayne	South-central Indiana
1958	W. Laird et al.	E. North Dakota
1959	R.F. Black	Western Wisconsin

Table 9 (continued)

1960	A.F. Agnew et al.	E. South Dakota
1961	C. Gravenor et al.	Eastern Alberta
1962	R.P. Goldthwait	Western Ohio
1963	J.C. Frye & H.B. Willman	Western Illinois
1964	H.E. Wright & E.J. Cushing	Eastern Minnesota
1965	R.V. Ruhe et al.	Northeastern Iowa
1966	E.C. Reed et al.	Eastern Nebraska
1967	L. Clayton & T.F. Freers	SC North Dakota
1969	W. Kupsch	Cyprus Hills, Saskatchewan & Alberta
1971	C.K. Bayne et al.	Kansas-Missouri border
1972	W.H. Johnson, et al.	East-central Illinois
1973	E.B. Evenson, et al.	Lake Michigan basin
1975	W.H. Allen et al.	Western Missouri
1976	C.K. Bayne et al.	Meade County, Kansas
1978	R.V. Ruhe & C.G. Olsen.	Southwestern Indiana
1979	L.R. Follmer et al.	Central Illinois
1980	G.R. Hallberg et al.	Yarmouth, Iowa
1981	W.A. Burgis & D.F. Eschman	NE Lower Michigan
1982	J.C. Knox et al.	Driftless Area, Wisconsin
1983	H.K. Bleuer et al.	Wabash Valley, Indiana
1984	R.W. Baker	Western Wisconsin
1985	R.C. Berg et al.	North-central Illinois
1986	W.C. Johnson et al.	Northeastern Kansas
1987	S.M. Totten & J.P. Szabo	North-central Ohio
1988	G.J. Larson & G.W. Monaghan	Southwestern Michigan

(source: Schneider, A.F., 1988)

South Central Cell

1983	V. Holliday	Central Llano Estacado, Texas
1984	T.C. Gustavson	Rolling Plains, Texas Panhandle
1985	B.J. Miller, G.C. Lewis, J.J. Alford, W.J. Day, & S.F. Burns	Loesses in Louisiana & Vicksburg, MS
1986	J. Mossa & W. Autin	Florida Parishes, Southeastern Louisiana
1987	C.R. Ferring, A.J. Crone, S.A. Hall, K.V. Luza, & R.F. Madole	Southwestern Oklahoma
1988	R. Birdseye & S. Aronow	SW Louisiana and SE Texas

(source: Burns, S.F., 1988)

Rocky Mountain Cell

1952	G.M. Richmond	Rocky Mountain National Park, Colorado
1953	G.M. Richmond	Twin Lakes area, Colorado
1954	S.H. Knight	Medicine Bow-Laramie area, Wyoming
1955-57	none	
1958	J.D. Love & J. Montagne	Jackson Hole area, Wyoming
1959	G.M. Richmond	Pinedale and Lander areas, Wyoming

Table 9 (continued)

1960	H.D. Goode & R.B. Morrison	Little Cottonwood Creek area, Salt Lake Country, Utah
1961	J.S. Williams, R.C. Bright, W.J. Carr, D.E. Trimble, A.D. Willard, & V. Parker	Bear Lake Valley, Utah-Idaho to American Falls, Idaho
1962	H.E. Malde, H.A. Powers, D.W. Taylor	Twin Falls to Glenns Ferry, Idaho
1963	J.M. Good, J.D. Love, & J. Montagne	Madison River Valley and Yellowstone RiverValley from Hayden Valley to Pine Creek, Montana
1964	none	
1965	R.B. Morrison	Upper Gila River region, New Mexico-Arizona
1966	J.W. Hawley & L.H. Gile	Rio Grande region (Las Cruces), southern NM
1967	H.E. Malde	Western Snake River Plain, Idaho
1968	C.V. Haynes, L. Aggen-broad, P. J. Mehringer, P.S. Martin, E.C. Lindsay, W.W. Wasley, & E.T. Hemmings	San Pedro Valley and Murry Springs, Arizona
1969	R. Van Horn & E.C. Weakly.	Jordan Valley, Utah
1970	T.L. Pewe & R.G. Updike	San Francisco Peaks (Flagstaff area), AZ
1971	M.F. Sheridan	Bishop-Mono Lakes area, California
1972	G.R. Scott, R. Taylor, & R.C. Epis	Canon City-Westcliffe area (Wet Mountain Valley), Colorado
1973	P.W. Birkeland, R.R. Shroba, J.C. Yount	Mt. Sopris-Thomas Lakes area, Colorado
1974	J. Montagne, C. Montagne, K. Pierce & L. Davis	West Yellowstone-Gallatin River-Three Forks area, Montana
1975	C.C. Reeves, Jr., Goolsby, C. Johnson, E. Johnson & J. Hawley.	Southern High Plains, Texas
1976	C. Beaty, R. Barendregt, J. Dormaar, S. Harris, & A. Stalker	Plains of southern Alberta
1977	B.M. Gilbert	Natural Trap Cave, Lovell, Wyoming
1978	L.D. Aggenbroad	Hoy Spring Mammoth site, Wyoming
1979	G.C. Frison	Agate Basin archeological site, Wyoming
1980	none	
1981	A.R. Nelson, L.A. Piety, R.R. Shroba	Roaring Fork Valley and Twin Lakes and Chalk Creek areas, central Colorado
1982	W.E. Scott, M.N. Machette, R.R. Shroba, & W.D. McCoy	Little Valley, Jordan Valley, and Beaver Basin, Utah
1983	B. Atwater & R. Waitt	Glacial Lake Columbia basin, Washington
1983	S. Coleman, K. Pierce, & M. Fosberg	McCall area, Idaho
1984	M. Reheis, D. Ritter, & R. Palmquist	Northern Bighorn Basin, Wyoming-Montana
1985	M. Gillam & R. Blair	Animas Valley, Colorado-New Mexico

Table 9 (continued)

1986	W. Locke, G.A. Meyer, W. Hamilton, J. Montagne, D.B. Nash, S.F. Personius, K.L. Pierce, & G.M. Richmond	Yellowstone National Park, WY-MT-ID
1987	D. Dethier, C. Harrington, Hawley, P. Karas, K. Kelson, D. Love, C. Menges, S. Wells, J. Wesling, & R.R. Shroba	Northern Rio Grande Rift, New Mexico
1988	L. Anderson, L. Piety, D. Nations, J. Faulds, J. Sturm, & C. Wellendorf	Tonto Basin, central Arizona

(source: Fullerton, D., 1988)

Pacific Coast Cell

1966	D. Easterbrook	Puget and Fraser Lowlands, Washington
1967	G. Smith, E. Leopold, & R. Davis	Searles Lake, California
1968	none	
1969	S. Porter	East central Cascade range, Yakima, and Icicle Creek Valleys, Washington
1970	D. Janda	Cape Blanco, Oregon
1971	B. Curry	Mammoth Lakes area, Sierra Nevada, CA
1972	D. Adams, D. Burke, J. Cummings, & 8 others	Searles Lake, California
1973	D.R. Crandall, D. Mullineaux, & J. Hyde	Mount St. Helens, Washington
1974	none	
1975	J. Bada, G. Carter, D. Nettleton, & G. Borst	San Diego coastal area, California
1976	S. Berryman, W. Duffield, G. Roguemore, & W. Page	China Lake site, California
1977	none	
1978	G. Smith, R. Smith, & R. Hooke	Searles, Panamint, and Death Valleys, CA
1979	R.M. Burke, P.W. Birke-. land, & J.C. Yount	Central Sierra Nevada, California
1980	R. Waitt, Jr.	Columbia River Valley, Washington
1981	E.A. Keller, T.K. Rock-. well, G.R. Dembroff, A.M. Sarna-Wojcicki, K.R. Lajoie, R.F. Yerkes, M.N. Clark, & D.L. Johnson.	Western Transverse Ranges, California
1982	D.R. Harden, D.C. Marron, A. MacDonald	Humboldt Basin, north coastal California
1983	(see Rocky Mountain Cell for joint trips held this year)	
1984	S. Stine, S. Wood, K. Sieh, & D. Miller	Yosemite National Park, Mono Lake, and Mono-Inyo volcanic chain, eastern California
1985	G.R. Hale, J.B. Ritter, & J.C. Dohrenwend	Eastern Mohave Desert, California
1986	R.H. Brady, III, P.R Butler, & B.W. Troxel.	Southern Death Valley, California

Table 9 (conclusion)

1987	J.O. Davis & R.M. Negrini	Pyramid Lake area, Black Rock and Smoke Creek Deserts, Nevada
1988	W.E. Scott, C.A. Gardner, A.M. Sarna-Wojcicki, E.M.Taylor, & B.E. Hill	Mount Bachelor-Three Sisters area, central Oregon Cascades and Bend, Oregon

(source: W.E. Scott, 1988)

TABLE 10: American Geomorphological Field Group; Conferences

Date	Organizer	Location
1982.	L.B. Leopold	Pinedale, Wyoming & vicinity
1983	S.G. Wells	Chaco Canyon country, NM
1984	none	
1985	H. Kellsey, M. Savina & T. Lisle.	Redwood Country, CA
1986	T.C. Pierson	Mount St. Helens, Washington
1987	none	
1988	none	

TABLE 11: The Guelph Geomorphology Symposium: Organisers and Topics

DATE	ORGANIZERS	TOPIC
1969	E. Yatsu & A. Falconer	Research Methods in Geomorphology
1971	E. Yatsu, F.A. Dahms, A. Falconer, A.J. Ward, & J.S. Wolfe	Research Methods in Pleistocene Geomorphology
1973	Barry D. Fahey & Russell D. Thompson	Research in Polar and Alpine Geomorphology
1975	E. Yatsu, A.J. Ward, & F. Adams	Mass Wasting
1977	R. Davidson-Arnott & W.G. Nickling	Research in Fluvial Systems
1980	R. Davidson-Arnott, W.G. Nickling, & B.D. Fahey	Research in Glacial, Glaciofluvial, and Glaciolacustrine Systems

(Volumes published by GeoBooks: Horwich)

TABLE 12: The Binghamton Geomorphology Symposium: Organisers and topics

DATE	ORGANIZERS	TOPIC
1970	Donald R. Coates	Environmental Geomorphology
1971	Marie Morisawa	Quantitative Geomorphology
1972	Donald R. Coates	Coastal Geomorphology
1973	Marie Morisawa	Fluvial Geomorphology
1974	Donald R. Coates	Glacial Geomorphology
1975	Wilton N. Melhorn & Ronald C. Flemal	Theories of Landform Development
1976	Donald R. Coates	Geomorphology and Engineering
1977	Donald O. Doehring	Geomorphology in Arid Regions
1978	Raymond L. Frederking, John D. Vitek, & Donald R. Coates	Thresholds in Geomorphology
1979	Dallas D. Rhodes & Garnet P. Williams	Adjustments of the Fluvial System
1980	Richard G. Craig & Jesse Craft	Applied Geomorphology
1981	Colin E. Thorn	Space and Time in Geomorphology
1982	Robert G. LaFleur	Groundwater as a Geomorphic Agent
1983	Michael J. Woldenberg	Models in Geomorphology
1984	Marie Morisawa & John T. Hack	Tectonic Geomorphology
1985	Athol D. Abrahams	Hillslope Processes
1986	William G. Nickling	Eolian Geomorphology
1987	Larry Mayer & David Nash	Catastrophic Flooding
1988	Keith J. Tinkler	History of Geomorphology
1989	T.W Gardner, W.Sevon, N. Potter Jr.,D.D.Braun & G.H.Thompson	Geomorphic Evolution of the Applachians

TABLE 13: York/CANQUA Quaternary Symposium

Date	Organisers	Topic
1974	William C. Mahaney	Quaternary Environments
1975	William C. Mahaney	Quaternary Stratigraphy
1976	William C. Mahaney	Quaternary Soils
1979	William C. Mahaney	Quaternary Climatic Change
1981	William C. Mahaney	Quaternary Dating Methods
1983	William C. Mahaney	Correlation of Quaternary Chronologies
1985	Rene' W. Barendregt	Paleoenvironmental Reconstruction of the Late Wisconsin Deglaciation and the Holocene

References

Abrahams, A.D., 1984. Landform symposium meets. *Geotimes*, 29(1), 14-15.

Andrews, J.T. & Graf, W.L., 1983. Quaternary and geomorphology. *Geotimes*, 28(2), 35.

Anon., 1987. Summary of newsletters 2 and 3 of the working committee for international collaboration in geomorphology. *Zeitschrift für Geomorphologie*, 31(4), 501-502.

Baker, V.R. & Dixon, J.C., 1988. Geomorphology and Quaternary geology. *Geotimes*, 33 (2), 25-27.

Bryan, K., 1950. The place of geomorphology in the geographic sciences. *Annals, Association of American Geographers*, 40, 196-208.

Burns, S.F., 1988. Personal communication.

Butzer, K.W., 1973. Pluralism in geomorphology. *Proceedings, Association of American Geographers*, 5, 39-43.

Chorley, R.J., Beckinsale, R.P., & Dunn, A.J., 1973. *The History of the Study of Landforms or The Development of Geomorphology*. Vol. Two: *The Life and Work of William Morris Davis*. London: Methuen , 874p.

Chorley, R.J., Dunn, A.J., & Beckinsale, R.P., 1964. *The History of the Study of Landforms*. Volume One: *Geomorphology Before Davis*. London: Methuen & Co. Ltd., 678 p.

Costa, J.E. & Graf, W.L., 1984. The geography of geomorphologists in the United States. *The Professional Geographer*, 36(1), 82-89.

Davis, L.R., 1987. Geomorphologists discuss wind erosion. *Geotimes*, 32 (4), 13-14

Dury, G.H., 1972. Some current trends in geomorphology. *Earth-Science Reviews*, 2, 45-72.

Drake, E.T. & Jordan, W.M. (eds), 1985. *Geologists and Ideas: A History of North American Geology*. Geological Society of America, Centennial Special Volume 1, 525p.

Easterbrook, D.J. & Toy, T.J., 1984. Quaternary geology, morphology thrive. *Geotimes*, 29(10), 13-14.

Freeman, T.W., 1961. *A Hundred Years of Geography*. Chicago: Aldine, 335 p.

Fullerton, D., 1988. Personal communication to R.P. Goldthwait.

Giardino, J.R. & Ashley, G.M., 1986. Geomorphology and Quaternary geology. *Geotimes*, 31(2), 22-24.

Giardino, J.R., Shroder, J.F., Jr., & Vitek, J.D., eds., 1987. *Rock Glaciers*. Boston: Allen & Unwin, 355 p.

Goldthwait, R.P., 1988a. Personal communication

Goldthwait, R.P., 1988b. *Recollections of fifty field reunions of the Friends of the Pleistocene, 1934-1988*. Ohio State University/Byrd Polar Research Center, forthcoming.

Goudie, A. & Price, R.J., 1980. The BGRG 1960-1981. *Area*, 12 (4), 241-244.

Graf, W.L., 1984. The geography of American field geomorphology. *The Professional Geographer*, 36(1), 78-82.

Graf, W.L., 1985. Morphotectonics topic of 15th annual meet. *Geotimes*, 30 (3), 10-11.

Graf, W.L. (ed.), 1987. *Geomorphic Systems of North America*. Geological Society of America, Centennial Special Volume 2, 661 p.

Graf, W.L., Trimble, S.W., Toy, T.J., & Costa, J.E., 1980. Geographic geomorphology in the eighties. *Professional Geographer*, 32(3), 279-284.

Gregory, K.J., 1985. *The Nature of Physical Geography*. Baltimore: Edward Arnold, 262 p.

Hart, M.G., 1986. *Geomorphology: Pure and Applied*. London: George Allen & Unwin, 228 p.

James, P.E. & Martin, G.J., 1978. *The Association of American Geographers: The First Seventy-Five Years*. Washington, Association of American Geographers, 279 p.

Kite, S., 1988. Personal communication to R.P. Goldthwait.

Kohlstedt, S.G., 1985. Institutional history. *OSIRIS*, 2nd series, 1, 17-36.

Lambert, W. (ed.), 1985. *Repeat photography newsletter*. West Texas State University, 2(2), 38 pages.

Mark, D., 1986. Electronic mail for geomorphologists. *Geomorphorum*, 5 , 7-8.

Marston, R.A., 1989. American geographers in geomorphology. in **G.L. Gaile and C.J. Willmott (eds.)**, *Geography in America*, Columbus: C.E. Merrill, 800p.

Marston, R.A. & Baker, V.R., 1987. Geomorphology & Quaternary geology. *Geotimes*, 32(2), 26-27.

Mather, K.F., 1967. *Source Book in Geology*. Cambridge: Harvard University Press, 435 p.

Morisawa, M., 1982. Geomorphology. *Geotimes*. 27(2), 36-37.

Morisawa, M., 1985. Development of quantitative geomorphology; in **Drake, E.T. and Jordan, W.M. (eds.)**, *Geologists and Ideas: A History of North American Geology*, Geological Society of America, Centennial Special Volume 1, 79-107.

Pitty, A.F., 1982. *The Nature of Geomorphology*. London: Methuen, 161 p.

Pyne, S., 1980. *Grove Karl Gilbert: A Great Engine of Research*. Austin: University of Texas Press, 312 p.

Ritter, Dale F., 1988. Landscape analysis and the search for geomorphic unity. *Geological Society of America, Bulletin*, 100 (2), 160-171.

Russell, R.J., 1949. Geographical geomorphology. *Annals, Association of American Geographers*, 39, 1-11.

Sarjeant, W.A.S., 1980. *Geologists and the History of Geology: An International Bibliography from the Origins to 1978*. New York: Arno Press, 5 volumes; 1987, 2 Supp. volumes, Montalca (Florida): R. Krieger.

Schneider, A.F., 1988. Personal communication. February 16, 1988.

Schumm, S.A., 1981. Geomorphology. *Geotimes*, 26(2), 32-33.

Scott, W.E., 1988. Personal communication to R.P.Goldthwait.

Snyder, T.D., 1987. *Digest of Educational Statistics*. OERI, U.S. Department of Education, Center for Education Statistics, 364 p.

The American Heritage Dictionary, 1982, 666 p.

Thornbury, W.D., 1964. *Regional Geomorphology of the United States*. New York: J. Wiley & Sons, Inc., 609 p.

Tinkler, K.J., 1985. *A Short History of Geomorphology*. Totowa: Barnes & Nobles, 317 p.

Wahrhaftig, C., 1987. Foreword: in **Giardino, J.R., Shroder, J.F., Jr., & Vitek, J.D. (eds.)**, *Rock Glaciers*, Boston, Allen & Unwin, pp. vii-xii.

Wahrhaftig, C. & Cox, A., 1959. Rock glaciers in the Alaskan Range. *Geological Society of America, Bulletin*, 70, 383-436.

Walker, H.J., 1987. Potentials for international collaboration in geomorphological research; in **Gardiner, V. (ed)**, *International Geomorphology 1986*, Part I, London. J. Wiley & Sons, 11-24.

Walker, H.J. & Orme, A., 1986. International geomorphology in the 1980's. *Zeitschrift für Geomorphologie*. 30 (4), 503-511.

Yochelson, E.L.(ed), 1980. *The scientific ideas of G.K.Gilbert*. Geological Society of America, Special Paper 183.

15

Afterword

Mott T. Greene

I have been asked by the convenor, Keith Tinkler, to provide a brief afterword to this volume, based on my reading of the papers herein collected, and my participation in the symposium as a commentator on each paper after it was delivered. I would like to say a few words about each of the papers in the order in which it was given, and to suggest what seemed to me interesting questions, novel results, and possible avenues for future research.

The first paper, by Gordon Herries-Davies, had the spirit of a keynote address, reminding working scientists that some of our more cheerful and sanguine characterizations of scientific work in the earth sciences - its "progress" in "discovering the actuality of the past" must be set aside when we undertake to write a critical history of this scientific work. For the "excitement of discovery" we must substitute the sobering reflection that some of the most exciting discoveries of the last 300 years in geomorphology are now a "wreckage of rejected concepts." This, Davies reminds us, is not because our predecessors were deranged or incompetent or their concepts categorically inferior to our own. It is because the normal course of science is not the substitution of an ultimate truths for misguided fallacies, but rather the framing of concepts (human inventions) after "painstaking examination of the appropriate field-phenomena." Under this constraint, one can see that if geomorphology attempts to explain "landforms in terms of sets of natural processes operating within a variety of palæo-environments" then the work of the history of geomorphology is in the first instance "discovering the different types of palæo-environments which various scholars have invoked as the basis for their explanation of given suites of landforms." When we know that, we then need to explore why certain interpretations prevailed at the times they prevailed - what Herries-Davies calls the "contextual approach." The resulting product tells us not just "who was right and who was wrong" in the past, judged by current conceptions, but how and why the answers have changed, and more importantly, how and why the questions have changed. Herries-Davies clearly believes (if he does not explicitly say) that such reflection affords an extremely valuable adjunct to the framing of concepts in the course of scientific work, and that working scientists would do well to practice it.

The second paper, by François Ellenberger, discusses the results of some very sophisticated and detailed work in geomorphology, conducted on actualistic principles, by Boulanger, Montlosier, and Gautier in 18th century France. These results are the more striking because they appeared long before the "origins" of modern geology and geomorphology at the turn of the nineteenth century. Ellenberger's work suggests a number of important conclusions about the way we periodize earth science and award responsibility for its origins. The first of these is that many of the ideas we tend to think of as inaugurating modern geomorphology were actually quite widely known and circulated in the 18th century. The next is that not everyone who accepted these ideas wrote about them, and not everyone who wrote about them published them - Boulanger's ideas were circulated in manuscript. These points lead us to the conclusion that our location of the origins of modern geology at the beginning of the nineteenth century had to do not so much with the invention of ideas, but the appearance of a "geological literature" in which these same ideas came to be published in a different form (the article to a learned journal specific to a given science) and a new problem-context. That is, we must not confuse the origin of scientific ideas with the beginning of a certain pattern of publication which gave these ideas general currency. Moreover, we should be aware that we judge ideas not on their own merits alone but by the company they keep - and that the company an idea keeps (the theory to which a datum is applied, for instance) determines its fate to an extent that makes it very difficult to speak of the correctness of an idea apart from its context.

Dennis Dean's paper was a defense of the priority of James Hutton in the foundation of modern geomorphology, a position challenged repeatedly in recent years by most other historians of the earth sciences who study the 18th and 19th centuries. Hutton's inspiration to do earth science was genuinely brought about by what are today distinctly geomorphic problems: soil transport and removal and fluvial denudation. But his strategy was to move away from landform analysis to underlying causes - a Newtonian predilection for primary over secondary qualities, and a distinct prejudice against the current state of affairs as epiphenomenal. Yet the closing pages of Dean's paper contains a most interesting suggestion - that is, of the influence on geology and geomorphology in the later 19th century of the geographical works of Karl Ritter and Alexander von Humboldt, whom William Morris Davis saw as important to the foundations of geomorphology. "Where," asks Dean, "is the historian who can show us in outline the important Werner-Humboldt tradition?" which Dean sees culminating in the work of Darwin. This directs our attention, as did Ellenberger's paper, to the impact of the pattern of publication. Throughout the latter 18th and all of the 19th century expedition reports, group and individual, were an important source of geological, geographical, zoological, botanical and anthropological information, and even of political intelligence - one thinks of Leopold von Buch's travels in Scandinavia, Humboldt's Mexico, Powell's exploration of the Colorado, Darwin's voyage on the Beagle, Richtofen's travels in China. In the later 20th century this genre is now denominated "travel literature," with a connotation that it is at best popular science or science popularization. This should remind us, perhaps, that *National Geographic Magazine* was once, like *Petermann's Mitteilungen*, not a coffee table book but rather a serious scientific journal. This is to say that just as we must be aware that ideas can gain a wider currency when a new publication pattern is adopted, so must we remember that they can also lose currency when an older pattern of publication of scientific results is abandoned, irrespective of their merit.

Keith Tinkler's paper on 18th century writings on rivers, lakes and the idea of the "terraqueous globe" addresses the question of the emergence of geology as a science in the 18th century given the independent existence on the one hand of theories of the earth with cosmological import, and on the other, detailed (if scattered) field records of natural history from at home (Great Britain) and abroad , with the context for these latter contributions often set implicitly by biblical time, "the Deluge" or some other consideration of natural theology. Tinkler finds a unifying thread providing coherence and even impetus for these scattered field reports in a generalized allegiance to Newton's 'Rules of Reasoning in Philosophy' from *Principia Mathematica.* Of these rules, the most important for geology was the notion of the permissible generalization of a process or a chronology from a local instance to the world as a whole, under the rubric of "Nature's conformability to herself." It is interesting to note that this allegiance seems in Britain to have cut across such well known divides as that between Neptunism and Plutonism. His eventual conclusion is that the ideal of universality promoted by Newtonian successes in the "experimental philosophy" - with its conflation of experiment and observation - only gradually and reluctantly gave way in geology to the less secure idea of uniformity. With this shift came a move in demonstration from "crucial instances" to the inductive demonstration of modal occurrence. This would indicate that as happened in the (failed) 18th century attempt to develop the study of living beings along Newtonian lines, study of the earth also, in the 18th century, went through a Newtonian phase, which it later abandoned. Geology found itself faced with different sorts of questions: what is the means whereby we establish the scope of a generalization based on a field instance? When and how does a single field report become decisive? Other than laboratory modelling, is there much more to an "experimental" test of a geological hypothesis than multiplication of field instances? The solution of these and other questions, Tinkler suggests, is a much more important and complicated story than we have heretofore imagined.

David Alexander's paper takes up the question of Newtonian science within 18th century investigations of the earth, and of the importance of 18th century "travel literature" and its geological descriptions not only for 19th century theorists, but for 20th century geomorphologists looking at recurrent episodes of seismically induced slope instability in Italy. The Calabrian Earthquakes of the Spring of 1783 were extensively investigated by representatives of many European scientific societies, ("the sheer scale of ground failure provoked in Calabria in 1783 makes this event stand out in European scientific history") and their results published in "earthquake anthologies." The Calabrian region, in the aftermath of the great quakes, appeared to offer a crucial locale for "testing" the relative adequacy of the two candidate mid-18th century theories of the cause of seismic disturbances: vulcanist and electrist. Of these the electrist is less well known. Its fortunes in the 18th century make an outstanding example of the alacrity with which, in any period of modern science, the physical principles surrounding a newly investigated phenomenon (in this case static electric discharge) are invoked to explain a wide variety of phenomena heretofore unexplained - in this case the origin of earthquakes.

Brian Bunting's paper on early 19th century concepts of soil shows what a curious hodge-podge might be created by the confluence of political economy (a Malthusian interest in soil improvement and land reclamation), 18th century chemistry (vestiges of the Stahlian theory of earths and principles in soil amendment recipes) and agricultural geology. Among the many questions suggested (if not explicitly raised) by this paper are: were soils seen as vital entities - existing at the border of the organic and inorganic? Or were they precisely an intermediary between organic and inorganic at a time (1830s) when the originators of cell theory were abolishing, for the time being, the organic/inorganic distinction? Were soils to be treated as species, like minerals and plants? If so, what were the principles of classification - Appearance? Composition? Texture? These questions might be usefully investigated.

David Stoddart's paper on the study of Australian (and open ocean atoll) reef geomorphology in the nineteenth century took conference participants out of Europe and to a problem of great theoretical importance for 19th century geology - the question of former sea levels and the possible rise and fall of the sea bottom and the land, as measured by the emergence and subsidence of coral reefs. But Stoddart has an equally interesting story to tell, of the determination of the range and character of reef studies by the character and extent of government support. Studies of the Great Barrier Reef waned quickly after completion of the initial (and Imperial) hydrographic work and did not re-emerge until Australia developed an indigenous scientific elite. Even then, with Australian geologists turned inward toward geological problems with economic consequences, it was foreign geologists who did most of the reef work.

This is notable beyond its intrinsic interest because historians of geology tend to assume that the foundation of an official geological survey must benefit the science. In this case it is clear that the benefits were selective, were not dictated by theoretical concerns, and directed attention away from important problems. Given this state of affairs, one might argue that the shift from expensive, imperial expedition science to local control put a real crimp in studies of coral reefs.

Stoddart has also demonstrated in his highlighting of the early coral reef work of the Frenchman François Peron, published in 1816, (which was treated in a 1984 work by C. Wallace - see Stoddart's bibliography) that knowing the history of British Imperial geology, even in British Imperial dominions, is not the full story of the geology practiced there; Peron's coral reef work, for all its extent and quality, is almost unknown in the English-speaking world.

Kenneth Hewitt's essay on Karakoram geomorphology is an extremely interesting and thoughtful review of the history of work on one of the earth's most remote and dramatic landscapes. Several aspects of his treatment deserve mention. The first is the way in which the military-cartographic-imperial character of the British work in the region, in context of the strictly controlled (for political reasons) access to the region had an enormous dampening effect on our knowledge of the region, and our ability to apply results of researches in the region to global problems. Second, how the difficulty of access from a purely logistic standpoint made most researches "reconnaissances" even well into the twentieth century. Third, that exceptions to this rule are those continental European workers - the von Schlagintweits, von Richtofen, Sven Hedin, Giotto Dainelli, Philips Visser and Jeanette Visser-Hooft who worked in the region over periods of many years and produced multi-volume, synthetic treatments but little read in the English speaking world. Hewitt's excellent summaries of their work add to the force of his conclusion that here is good science about an interesting region which is almost entirely unknown because of "powerful political and nationalist factors influencing what was sought, accepted and disseminated." But still more is at work - for here is also a tradition of geomorphology with problems that turn on ideas and the history, as Hewitt points out, of physical geography - while the history of geomorphology has definitely leaned toward the history and ideas of geology. This is an important issue, soundly directed: geomorphology exists between physical geography and geology and yet is not precisely either. Hewitt does well to question the tendency to treat it as mostly geology, and he makes an argument which defends the regional and geographically based geomorphology which has gone out of fashion since the Second World War. One would like to see him extend this argument even further, and speak to the greater intellectual possibilities of comprehensive regional geographies.

Robert Beckinsale's paper offered an opportunity to review those theories of global scope which have governed the earth sciences, including geomorphology, in the last century. Several compelling generalizations emerged: that all these theories attempt in one or another way to account for the problem of the earth's internal heat and how it escapes; that any general theory of the earth's crust is also a theory of the movement of sea levels (eustatic theories); that theories of global scope are numerous, vary widely in their proposed mechanisms, and have short lives. Particularly important for geomorphology have been theoretical discussions of vertical instability of landmasses - not just orogeny but epeirogeny or (as Beckinsale said) "that lovely word: cymatogeny" - broad upflexures and swellings of great lateral extent. Beckinsale also demonstrated that there has been a move throughout the last century from theories emphasizing cycles to theories which are "directional." Finally, he suggested that there was much to be learned by a study of the major syntheses, including those, like the work of Walthar Penck in the 1920s, which were never widely read. This later point seemed to reinforce Hewitt's contention that there is in geomorphology a large and useful literature which is today all but unread because it does not reflect current conceptions of what geomorphology "ought" to be doing, or was published in a compendious form. Beckinsale also alerts us to a question of great interest in the presentation of global theories: the tendency to use map projections, such as the Mercator, in discussions of continental drift and plate tectonics - for which they are entirely unsuitable - the visual cogency of the argument for continental drift being intuitively clear only on a globe or an equal area projection such as the Mollweide. A careful reading of Beckinsale's paper makes quite clear the importance of cartographic representation in the fate of a given theory.

Dorothy Sack's paper on the history of studies of Lake Bonneville was comprehensively researched and beautifully presented. In tracing the history of studies of (glacial) Lake Bonneville shorelines, from the first discovery by Europeans of the old lake bed, she developed a coordinated set of problems in a way that shows how the history of geomorphology can be at the same time good history of science, and good science.

The canonical interpretation of Lake Bonneville was that of G.K. Gilbert, whose work between 1875 and 1890 set a chronology that stood until the Second World War. Thereafter, Lake Bonneville, as a "classic locale" became both a subject of renewed interest in its own right, and a testing ground for new techniques of shoreline chronology. Sack's work on these reinterpretations shows in a quite strikingly detailed way that trying a new technique on an old problem may not tell you much of interest about the old problem but can tell you a lot about your technique and lead to its further refinement - a polite way of saying that several attempts to unseat the Gilbert chronology unseated the new technique instead. Sack also shows that in challenging Gilbert's orthodox interpretation, workers invented new nomenclatures to describe the shorelines - a way of cementing their work into the Lake Bonneville picture - and also made headway in advancing their interpretations by new and striking methods of graphic representation (which Sack correctly points out do not advance the truth of the interpretation a whit, even if they advance the stock of the proposers). When the dust settled around the new techniques, new and competing nomenclatures and graphic representations, according to Sack's account, there was an uneasy sense that Gilbert's interpretation had not been done in by any of these approaches, and that a decisive encounter would be one which employed traditional methods of field geomorphic analysis - characterized by professionals as a turn away from sweeping generalization and a trend toward more rigor. As a first effort by a young geomorphologist, it is an outstanding piece of historical research.

The next speaker was Leszek Starkel, from the University of Kraków. Starkel's paper raised an important issue in the history of earth sciences and of the sciences generally, which is that of the character and scope of investigations in a given subject area - in this case geomorphology - under different economic and political conditions. Starkel's survey of Polish geomorphology raised into high relief the issue of national science, and of international co-operation in a restricted setting - in a way also highlighted by Stoddart's paper on the history of investigations into the great barrier reef, by Hewitt's on the Karakoram and to a certain extent by Professor Ellenberger. In each of these cases the development of a national survey - often taken as a benchmark for professional maturation - served to limit the theoretical generality of work produced, and even the choice of topics. In periods of geomorphic and geologic nationalism the mental boundaries seem to shrink to the political as in Starkel's example of Dylik's 1964 explanation of the weak reception of Davisian cycle theory - the lack of Polish territorial examples of all phases of a Davisian cycle. This is also a comment about the ubiquity of the demands in the history of geology for the opportunity to see it - whatever it is - first hand, and recalls Derek Ager's remark in *The Nature of the Stratigraphical Record* that the geographic distribution of certain classes of brachiopods in a celebrated series of 19th century memoirs bears a suspicious resemblance to the political borders of the old Austro-Hungarian Empire. Starkel commented that lack of opportunity after WWII for international expeditions gave a unique opportunity for intensive study of the Polish landscape and made it one of the best studied. Poland's international cooperations have been as diverse as any country's in the world - in different historical periods it co-operated scientifically with neighbors to the north (Spitsbergen work), to the East, and to the south (Austria after the 1860s) and today works globally. These periods had different characters and different kinds of productivity - one could point to the theoretical prominence of Poland in certain areas of earth science after the turn of the 20th century exemplified by the work of Romer and Rudzki and Sawicki as a consequence of their careers in a partitioned nation - in their case living in the Austrian-controlled portion. I found the paper fresh and enlightening in this problem area, one often discussed by sociologists of science but rarely actually chronicled by historians.

The final paper of the conference was an appreciation of twentieth century geomorphology by Jack Vitek. Vitek's written paper contains a great deal of useful tabular information on the social and institutional structure of geomorphology in the United States in the twentieth century, and illustrates in a striking way the role of individuals in adding cohesion to institutions in a discipline notorious for its diverse allegiances.

Not delivered at the conference, but included in the volume, is Cunningham's paper on James Forbes's measurements of the movement of glacier ice in the Alps in 1842. Cunningham wishes to demonstrate that it was Forbes who made the first accurate measurements of the movement of an Alpine glacier - and not Agassiz, Hugi, or any of the other claimants to this particular scientific achievement. It is clear from Cunningham's reconstruction of the field-work that this is the case, and we can look forward to a more extensive treatment of the episode in Cunningham's forthcoming biography of Forbes.

I have argued elsewhere ["History of Geology" *Osiris* (n.s.) Volume 1, no. 1, 1985, pp. 97-116] that history of the geosciences will probably be written, for some time yet, mostly by earth scientists themselves. This was certainly the case in the papers delivered at this symposium. As an exploratory effort, it was certainly a success in eliciting thematic concerns which deserve attention not only from practitioners of geomorphology, but from all those interested in the history of earth sciences. Several of the papers made substantial contributions to our knowledge of this branch of earth science history. The convenor and editor of the symposium and volume, Keith Tinkler, himself the author of an excellent introductory history of geomorphology, has gone to great effort to insure that contributions to the volume are not mere "occasional pieces" but works of serious, original, scholarship. In this he has largely succeeded, and he is to be congratulated on the standard he has set here.

Index

Note: *The Afterword has not been indexed: read it! Life dates are supplied where feasible.*

For Product Safety Concerns and Information please contact our EU representative GPSR@taylorandfrancis.com
Taylor & Francis Verlag GmbH, Kaufingerstraße 24, 80331 München, Germany

www.ingramcontent.com/pod-product-compliance
Ingram Content Group UK Ltd.
Pitfield, Milton Keynes, MK11 3LW, UK
UKHW022215190926
475702UK00003B/363
9780367464493